Clinical Microbiology
Diagnosis, Treatment and
Prophylaxis of Infections

THE MICROBIOLOGY OF SKIN, SOFT TISSUE, BONE AND JOINT INFECTIONS

Clinical Microbiology Diagnosis, Treatment and Prophylaxis of Infections

THE MICROBIOLOGY OF SKIN, SOFT TISSUE, BONE AND JOINT INFECTIONS

Edited by

KATERYNA KON

Associate Professor, Department of Microbiology, Virology and Immunology,
Kharkiv National Medical University, Kharkiv, Ukraine

MAHENDRA RAI

Professor and Head, Department of Biotechnology,
Amravati University, Maharashtra, India

ELSEVIER

ACADEMIC PRESS

An imprint of Elsevier

Academic Press is an imprint of Elsevier
125 London Wall, London EC2Y 5AS, United Kingdom
525 B Street, Suite 1800, San Diego, CA 92101-4495, United States
50 Hampshire Street, 5th Floor, Cambridge, MA 02139, United States
The Boulevard, Langford Lane, Kidlington, Oxford OX5 1GB, United Kingdom

Library of Congress Cataloging-in-Publication Data
A catalog record for this book is available from the Library of Congress

British Library Cataloguing-in-Publication Data
A catalogue record for this book is available from the British Library

ISBN: 978-0-12-811079-9
ISSN: 2451-9006

For information on all Academic Press publications
visit our website at https://www.elsevier.com/books-and-journals

Publisher: Sara Tenney
Acquisition Editor: Linda Versteeg-buschman
Editorial Project Manager: Timothy Bennett
Senior Production Project Manager: Priya Kumaraguruparan
Cover Designer: Matthew Limbert

CONTENTS

CONTRIBUTORS

N. Anasane
SGB Amravati University, Amravati, India

L. Ansaloni
Papa Giovanni XXIII Hospital, Bergamo, Italy

J. Bueno
Research Center of Bioprospecting and Biotechnology for Biodiversity Foundation (BIOLABB), Colombia

F. Catena
Parma University Hospital, Parma, Italy

S.V. Chanda
Saurashtra University, Rajkot, Gujarat, India

F. Coccolini
Papa Giovanni XXIII Hospital, Bergamo, Italy

F. Demirci
Faculty of Health Sciences; Faculty of Pharmacy, Anadolu University, Eskişehir, Turkey

M. Dolenc-Voljč
University Medical Centre Ljubljana; Faculty of Medicine, University of Ljubljana, Ljubljana, Slovenia

J. Esteban
IIS-Fundación Jiménez Díaz, UAM, Madrid, Spain

P. Fazii
Spirito Santo Hospital, Pescara, Italy

C. Fusella
Spirito Santo Hospital, Pescara, Italy

I. Gupta
SGB Amravati University, Amravati, India

A. Hamedi
Medicinal Plants Processing Research Center; School of Pharmacy, Shiraz University of Medical Sciences, Shiraz, Iran

C.L. Harris
Western University, London, ON, Canada

K. Husnu Can Baser
Faculty of Pharmacy, Near East University, Nicosia, N. Cyprus

A.P. Ingle
SGB Amravati University, Amravati, India

V. Javdekar
MES Abasaheb Garware College, Pune, India

M.J. Kaneria
Saurashtra University, Rajkot, Gujarat, India

N. Kavanagh
Royal College of Surgeons in Ireland and Trinity College Dublin, Dublin, Ireland

S.W. Kerrigan
Royal College of Surgeons in Ireland and Trinity College Dublin, Dublin, Ireland

M. Kulkarni
Savitribai Phule Pune University, Pune, India

K. Laforet
Western University, London, ON, Canada

M.F. Mahomoodally
University of Mauritius, Réduit, Mauritius

M. Malone
High Risk Foot Service, Liverpool Hospital, South West Sydney LHD, Sydney, Australia and Liverpool Diabetes Collaborative Research Unit, Ingham Institute

R. Marrollo
Spirito Santo Hospital, Pescara, Italy

S.O. Martinez-Chapa
School of Engineering and Science, Tecnologico de Monterrey, Campus Monterrey, Mexico

R. Nigro
Spirito Santo Hospital, Pescara, Italy

F.J. O'Brien
Royal College of Surgeons in Ireland and Trinity College Dublin, Dublin, Ireland

R. Pandit
SGB Amravati University, Amravati, India

P. Paralikar
SGB Amravati University, Amravati, India

R. Parra-Saldívar
School of Engineering and Science, Tecnologico de Monterrey, Campus Monterrey, Mexico

A. Parsaeimehr
School of Engineering and Science, Tecnologico de Monterrey, Campus Monterrey, Mexico

A. Pasdaran
School of Pharmacy, Research and Development Center of Plants and Medicinal Chemistry, Guilan University of Medical Sciences, Rasht; Medicinal Plants Processing Research Center, Shiraz University of Medical Sciences, Shiraz, Iran

C. Pérez-Jorge
IIS-Fundación Jiménez Díaz, UAM, Madrid, Spain

A.A. Prasetyo
Universitas Sebelas Maret, Surakarta, Indonesia

L. Prieto-Borja
IIS-Fundación Jiménez Díaz, UAM, Madrid, Spain

M. Rai
SGB Amravati University, Amravati, India

K.D. Rakholiya
Main Dry Farming Research Station, Junagadh Agricultural University, Targhadia, Gujarat, India

R. Sariyatun
Universitas Sebelas Maret, Surakarta, Indonesia

M. Sartelli
Macerata Hospital, Macerata, Italy

V. Savini
Spirito Santo Hospital, Pescara, Italy

C. Scharf
Endomed Médicos Associados; Hospital Universitário Evangélico de Curitiba, Curitiba, Brazil

A.M. Settembrini
Vascular Surgery, ASST Santi Paolo e Carlo, San Carlo Borromeo Hospital, Milan, Italy

P.G. Settembrini
Professor of Vascular Surgery, University of Milan, Milan, Italy

R.G. Sibbald
University of Toronto, Toronto, ON, Canada

R. Somayaji
University of Calgary, Calgary, AB, Canada

H. Sowani
Savitribai Phule Pune University, Pune, India

S. Suroowan
University of Mauritius, Réduit, Mauritius

P. Yagupsky
Ben-Gurion University of the Negev, Beer-Sheva, Israel

S. Zinjarde
Savitribai Phule Pune University, Pune, India

SECTION I

Skin and Soft Tissue Infections: Diagnosis and Treatment

CHAPTER 1

Soft Tissue Infections: Consideration Regarding Different International Guidelines

M. Sartelli*, F. Coccolini†, F. Catena‡, L. Ansaloni†
*Macerata Hospital, Macerata, Italy
†Papa Giovanni XXIII Hospital, Bergamo, Italy
‡Parma University Hospital, Parma, Italy

1. INTRODUCTION

In 2014 the Infectious Diseases Society of America (IDSA) published new guidelines for the diagnosis and management of soft tissue infections.[1] These new IDSA guidelines replaced those previously published in 2005.[2]

The 2014 World Society of Emergency Surgery (WSES) guidelines represent an additional contribution to the debate and were written by consensus of a group of worldwide specialists, mainly surgeons with interest in surgical infections.[3] The WSES guidelines, written by surgeons for surgeons, have a more practical approach than the IDSA guidelines.

Both sets of guidelines focus on the dramatic emergence of resistance to many of the antimicrobial agents commonly used to treat Skin and soft tissue infections (SSTIs) in the past. Particularly, infections due to *S. aureus* are an important health problem worldwide,[4] and treatment of these infections has become more difficult in the last decade due to the emergence and rapid spread of methicillin-resistant *S. aureus* (MRSA).

MRSA has usually been acquired during exposure to hospitals and other healthcare facilities (healthcare-associated MRSA). However, by the late 1990s, MRSA infections acquired from the community were recognized as a distinct clinical entity affecting young and healthy individuals without healthcare risk factors and with relatively preserved antimicrobial susceptibility patterns.[5]

2. CLASSIFICATION

The practice guidelines of IDSA[1] classify SSTIs into four groups: purulent, nonpurulent, infections associated with bites and animal contact, and surgical site infections.

The Microbiology of Skin, Soft Tissue, Bone and Joint Infections
ISSN 2451-9006, http://dx.doi.org/10.1016/B978-0-12-811079-9.00001-X

In addition, because of an increasing number of immunocompromised hosts worldwide, this set of guidelines addresses the wide array of SSTIs that occur in this population.

The WSES expert panel classified soft tissue infections using the following classification system[3]:

Surgical Site Infections

Incisional

- Superficial
- Deep

Nonnecrotizing SSTIs

- Superficial infections (impetigo, erysipelas, cellulitis)
- Simple abscess, boils and carbuncles
- Complex abscesses

Necrotizing SSTIs (NSTIs)

- Necrotizing cellulitis
- Necrotizing fasciitis
- Fournier gangrene
- Necrotizing myositis

The first group includes surgical site infections (SSIs). These infections represent a separate chapter among the soft tissue infections.[3] They are postoperative infections and because of their multifaceted aspects they are framed into a separate group.

Soft tissue nonsurgical site infections are divided into nonnecrotizing and necrotizing soft tissue infections.

Nonnecrotizing soft tissue infections typically involve one or both of the superficial layers of the skin (epidermis and dermis) and the subcutaneous tissue, such as complex abscesses, but may occasionally involve deeper structures. Generally, these infections are treated by antibiotics alone (nonpurulent superficial infections such as impetigo, erysipelas and cellulitis) or surgical drainage (purulent superficial infections such as simple abscesses, boils and carbuncles). In some cases, complex abscesses need both surgical drainage and antibiotics.

Conversely, necrotizing soft tissue infections (NSTIs) are life-threatening, invasive and aggressive soft tissue infections.

3. NECROTIZING SOFT TISSUE INFECTIONS

NSTI is an inclusive term intended to describe all infections with a necrotizing component involving any or all of the layers of the soft tissue compartment, from the superficial dermis and subcutaneous tissue to the deeper fascia and muscle. NSTIs include necrotizing cellulitis, necrotizing fasciitis, Fournier gangrene and necrotizing myositis.

Microbiologically, NSTIs may be classified into three types: type 1 (polymicrobial), type 2 (monomicrobial), and type 3 (monomicrobial infections initiated by a variety of virulent Gram-positive or Gram-negative bacilli such as *Clostridia*, *Vibrio*, *Aeromonas*, *Eikenella*, and *Bacillus* species).[3]

NSTIs rank among the more difficult disease processes encountered by physicians and surgeons and should be clearly separated from nonnecrotizing infections. The WSES classification[3] stresses the differentiation between nonnecrotizing infections and necrotizing infections. However, both guidelines focus on the necessity for a prompt diagnosis of these infections.[1,3]

In practice, clinical judgment is the most important element to diagnose necrotizing fasciitis. Clinical findings that suggest involvement of necrotizing fasciitis include: severe pain disproportional to the clinical findings; failure to respond to initial antibiotic therapy; evolving systemic toxicity; edema or tenderness extending beyond the cutaneous erythema; crepitus, indicating gas in the tissues; and in a subsequent phase skin necrosis or ecchymoses.

NSTIs are typically caused by toxin-producing bacteria and manifest a very rapid progression of disease with significant local tissue destruction. Systemic toxicity depends on the strain of bacteria and toxins produced. In most cases the progression of disease is typically measured in hours and an early diagnosis and treatment are crucial to survival.[6] The key to a successful outcome is a high index of suspicion and an aggressive and prompt management. Delay in diagnosing and inappropriate treating increases the risk of mortality.

Both computed tomography (CT) and magnetic resonance imaging (MRI) may be useful for diagnosing necrotizing soft tissue infections. MRI has been the imaging modality of choice for necrotizing fasciitis. However, MRI may be difficult to perform under emergency situations.

In these infections a multidisciplinary approach, including prompt aggressive source control, appropriate antimicrobial therapy and supportive management, are crucial for the patient's outcome.

Most patients require multiple surgical debridements and often have large and complex wounds requiring soft tissue coverage and prolonged hospitalizations. Early appropriate empiric antibiotic therapy against suspected pathogens should be started as soon as possible. Both guidelines suggest antibiotics suppressing streptococcal toxin and cytokine production (such as clindamycin). Since it is impossible to exclude with certainty a polymicrobial infection, an aggressive broad-spectrum empiric antimicrobial therapy should initially be started covering Gram-positive, Gram-negative, and anaerobic organisms, until culture-specific results and sensitivities are available.

IDSA guidelines suggest[1] vancomycin, linezolid, or daptomycin combined with one of the following options: piperacillin-tazobactam, a carbapenem (imipenem-cilastatin, meropenem, and ertapenem), ceftriaxone plus metronidazole, or a fluoroquinolone plus metronidazole.

WSES suggestions for antimicrobial therapy of necrotizing fasciitis are very similar[3]: linezolid and piperacillin/tazobactam or daptomycin + piperacillin/tazobactam + clindamycin.

In Fournier gangrene, WSES guidelines[3] suggest piperacillin/tazobactam + clindamycin if signs and symptoms of severe sepsis are not present or meropenem + linezolid if signs and symptoms of severe sepsis are present.

An appropriate deescalation of antimicrobial therapy is suggested by both guidelines once culture results return.

The efficacy of intravenous immunoglobulin (IVIG) in treating streptococcal toxic shock syndrome has not been defined. However, as extracellular streptococcal toxins have a role in organ failure, shock, and tissue destruction, neutralization of these toxins theoretically could be beneficial. WSES guidelines suggest[3] (with low quality of evidence and weak recommendation) IVIG in all patients with NSTI and evidence of organ dysfunction. On the other hand, IDSA guidelines[1] state that additional studies of the efficacy of IVIG are necessary before a recommendation can be made supporting its use in this setting.

4. CONCLUSION

SSTIs encompass a variety of pathological conditions ranging from simple superficial infections to severe necrotizing soft tissue infections.

Among skin and soft tissue infections, NSTIs are a rapidly progressive form representing potentially life-threatening infections.

Both WSES and IDSA guidelines stress the high suspicion that clinicians should always have when confronted with a case or suspected case of necrotizing infections, where the key for a successful outcome is an aggressive and prompt management.

REFERENCES

1. Stevens DL, Bisno AL, Chambers HF, Dellinger EP, Goldstein EJ, Gorbach SL, et al. Practice guidelines for the diagnosis and management of skin and soft tissue infections: 2014 update by the Infectious Diseases Society of America. *Clin Infect Dis* 2014;**59**(2):e10–52.
2. Stevens DL, Bisno AL, Chambers HF, Everett ED, Dellinger P, Goldstein EJ, et al. Practice guidelines for the diagnosis and management of skin and soft-tissue infections. *Clin Infect Dis* 2005;**41**(10):1373–406. Epub 2005 Oct 14. Erratum in: *Clin Infect Dis* 2006;42(8):1219.
3. Sartelli M, Malangoni MA, May AK, Viale P, Kao LS, Catena F, et al. World Society of Emergency Surgery (WSES) guidelines for management of skin and soft tissue infections. *World J Emerg Surg* 2014;**9**(1):57.
4. Klein E, Smith DL, Laxminarayan R. Hospitalizations and deaths caused by methicillin-resistant *Staphylococcus aureus*, United States, 1999–2005. *Emerg Infect Dis* 2007;**13**:1840–6.
5. Sowash MG, Uhlemann AC. Community-associated methicillin-resistant *Staphylococcus aureus* case studies. *Methods Mol Biol* 2014;**1085**:25–69.
6. Hakkarainen TW, Kopari NM, Pham TN, Evans HL. Necrotizing soft tissue infections: Review and current concepts in treatment, systems of care, and outcomes. *Curr Probl Surg* 2014;**51**(8):344–62.

CHAPTER 2

Evidence-Informed Approach to Pilonidal Sinus Disease and Related Disorders

C.L. Harris*, K. Laforet*, R.G. Sibbald†, R. Somayaji‡
*Western University, London, ON, Canada
†University of Toronto, Toronto, ON, Canada
‡University of Calgary, Calgary, AB, Canada

1. INTRODUCTION

Sacrococcygeal pilonidal sinus disease (PSD) is a chronic, painful soft tissue infectious disorder that affects mostly young people. The disease was first described by Mayo in 1833,[1] and Anderson[2] in 1847, with Hodges[3] assigning the name in 1880, using the Latin words *pilus* (hair) and *nidus* (nest). PSD occurs in the intertriginous natal cleft or buttock crease where moisture, friction, and bacteria abound. Similar abscesses can occur elsewhere on the body (axilla, umbilicus, interdigital, breast, etc.) although these are relatively rare.[4–6] For the purposes of this chapter, the focus will be on care of sacrococcygeal/natal cleft pilonidal sinus disease and the resultant wounds.

Often presenting as acute abscesses, the all-too-frequent resultant chronic wounds cause significant upset, being socially disabling, interrupting work and school, and causing physical and emotional suffering. PSD places significant financial burden on the patient, family, and community in terms of days lost, as well as overall costs to the health care system. It can be frustrating for the physicians and clinicians who are involved in the care, and often requires repeated surgeries to achieve healing. Despite more than 100 years of refining surgical and wound care approaches for this etiology, there is still no consensus as to the most effective method of surgery, or wound care, or lifestyle changes, reflected in the large number of articles that exist in the literature; more than 1000 since 2005 were identified.

This chapter reflects on the evidence published in the past 10 years focusing on pathophysiology, known risk factors, differential diagnosis, and the multiple components of care required for this condition: type of surgery, management of bacterial balance, positioning, periwound skin care, type of dressing, level of physical activity, pain management, nutrition, self-care and psychosocial considerations. Successful PSD management requires a systematic interprofessional approach that addresses these principles.

The Microbiology of Skin, Soft Tissue, Bone and Joint Infections
ISSN 2451-9006, http://dx.doi.org/10.1016/B978-0-12-811079-9.00002-1

2. PATHOPHYSIOLOGY

PSD has been considered as one of a tetrad of follicle occlusion disorders described in 1975[7] that also includes hidradenitis suppurativa (HS, formerly termed *acne inversa*), dissecting cellulitis of the scalp, and severe acne congoblata.[8] In fact, HS is present in 20% of patients with PSD.[9] Earlier literature considered PSD as a congenital disorder. More recent theories suggest the disorder is an acquired acute or chronic exacerbating, inflammatory reaction,[10,11] caused by the anatomical location, friction, shearing and pressure forces. Natal cleft PSD can present as acute, chronic or silent. The abscesses or tracks often contain a bundle or nest of hairs, as well as keratin scales from dead keratinocytes, fibrosis, and granulation tissue. Surrounding tissue may contain chronic inflammatory changes.

3. THEORIES OF FORMATION

Bascom[10] first identified that hair follicles become irritated and associated with bacterial damage due to a build-up of keratin and inverted hairs that create an inflammatory response.[10] After studying 6545 cases over 35 years, Karydakis[12] formulated a pathogenic formula citing that three primary variables—loose hair (H), applied with force (F) to vulnerable local skin and tissues (V)—results in insertion of the hair at the depth of the natal cleft (G), leading to local infection and deep and surrounding abscess formation. He noted that the insertion never occurred in the sides of the cleft. A deep natal cleft retains moisture from sweating or bathing that subsequently softens the epidermal layer, making it vulnerable to anaerobic bacterial invasion and accumulation of trapped hair with associated debris.[13] Folliculitis may also occur in this environment, with similar consequences. Hair is trapped due to the friction forces caused by movement of the buttocks, referred to as a "cigarette-rolling" effect of the skin surfaces.[14] Sharp hair ends (from split and broken hair shafts originating from the natal cleft, back, head, or neck) or chisel-like root ends facilitate traumatic insertion[12,13] into an enlarged, inflamed hair follicle or into a preexisting sinus track.[8,9,15,16] Barbs on the shaft of the hair prevent expulsion.[16] Once one hair has entered, others can easily follow.[12] Early PSD is associated with high midline or lateral pits 2–5 cm from the midline due to hyperkeratosis of the enlarged hair follicle ostium.[16] These pits are the secondary exit point of the hairs. Although they are not the cause of the process, they can be an entry point for subsequent hair insertions.[9,12,15,16] Pits are different from sacrococcygeal dimpling in the natal cleft, seen at birth in 0.5% of individuals as a benign variant.[17] Another theory proposes an influence of increased sex hormones during puberty, which coincides with the peak incidence of PSD between the ages of 15 and 20. The hormonal sequelae are enlarged hair follicles that have potential for hair penetration, follicular inflammation, infection, and abscess formation. The process is illustrated in Fig. 1.

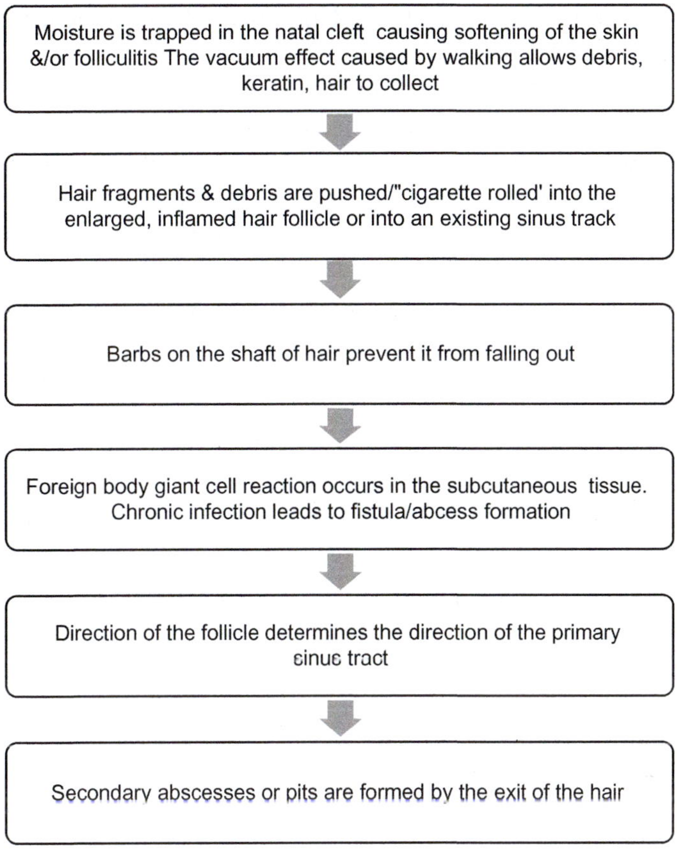

Fig. 1 Formation of pilonidal sinus disease.

4. EPIDEMIOLOGY AND RISK FACTORS

The prevalence of PSD is estimated at 0.5% to 1% of the general population, affecting predominantly young hairy males of Mediterranean descent. Thompson et al.[18] identified that PSD is rare in East Asia, Oceania and sub-Saharan Africa. The sex ratio is estimated to be 2.2 to 3 males to one female.[9,19] The symptom onset is in the late teens: the average age is 21 years for men and 19 for women; then new lesions decrease after age 25.[20] Doll et al.[13] confirmed this in their research on the military population, where older soldiers presenting with the same risk factors as their younger counterparts had significantly lower incidence. In the United States, symptomatic PSD incidence is 26 per 100,000 population,[21] or 70,000 cases per year.[22] This is a consistent finding with other countries, including Norway[21] and France[23] where the average PSD incidence is 0.7%.

There are several risk factors commonly associated with PSD. People with a triad of extensive body hair, sedentary occupations (i.e., sitting >6 hours), and low frequency of bathing ($\leq$2 times per week) are at a 219-fold increased risk for sacrococcygeal PSD than those without these risk factors.[24]

Occupation or lifestyle has been demonstrated as a causal factor in PSD. During the Second World War, PSD was nicknamed "jeep driver disease."[25] Akinci et al.[19] in their study of Turkish soldiers detected 58/88 (65.9%) persons with PSD were vehicle drivers, where rigid suspensions and hard seats may cause minor but repetitive trauma.[21] Benign dimpling of the natal cleft does not pose a risk unless it becomes a source of recurrent infection. A correlation between obesity and pilonidal sinus incidence and risk of recurrence has been studied several times, with varying results. While not all studies have demonstrated causality, there is sufficient evidence to support the need to modify risk factors where possible to improve outcomes (Table 1). Often a combination of risk factors contributes to PSD.

5. TYPES OF PSD

5.1 Acute, Infected Abscesses

Fifty-four percent of cases present with an acute abscess,[21] that may rupture spontaneously or require intervention. Abscess formation may occur due to involvement of multiple hair follicles (inflammation → infection → abscess formation) and require drainage and possible systemic antimicrobial therapies. Acute PSD has fewer sinuses, although edema and inflammation may prevent identification of these sinuses.[13]

5.2 Chronic Fistula Cavity

A consecutive foreign body giant cell reaction to the hair, partial epithelization of tracts (due to epithelial cells from the original hair follicle) and chronic infection were thought to be the key tissue responses which lead to chronic fistulas or abscess-forming disease.[32] However, it is now believed that multiple sinus formation originates before or at the beginning of symptomatic PSD and a younger age of onset is related to more sinuses present at the time of surgery.[13] The findings demonstrate that chronic sinus formation appears only when the onset is between the ages of 21.5 and 22.5 years. While all pilonidal sinus wounds (PSWs) contain chronic bacterial infection, additional sinuses or tracks are most likely unrelated to chronic infection. As with most chronic wounds, individuals who have chronic PSD are at risk for developing squamous cell carcinoma (approximately 0.1% of chronic PSD patients) and need to be monitored closely.[33]

5.3 Asymptomatic PSD

In autopsies on 100 children, aged 1 day to 10 years, Chamberlain and Vowter[34] detected that seven had asymptomatic PSD, with no clinical symptoms, and summarized that the incidence of asymptomatic PSD was double that of clinical acute or chronic PSD.

Table 1 Risk factor analysis and statistics

Nonmodifiable risk factor	Statistics	Reference
Male Gender	Male:Female ratio 2.2:1	Søndenaa et al.[21]
Pubescent endocrine changes/age between 15–29 years	Earlier onset of puberty in women = earlier age of onset of PSD/ 61% of men and 81% of women were in this age range	Søndenaa et al.[21]
Stiff Body Hair	OR 9.23 ($P<.0001$)	Harlak et al.[24]
Family history	20% ($P<.0001$); 38% 71% 50% long-term recurrence rate with + ve family history + earlier onset Family history refuted—Unable to find familial link	Akinci et al.[19] Søndenaa et al[21] Bolandparvaz et al.[26] Doll et al.[27] Harlak et al.[24] Duman et al.[28]
Dark Hair Color	28%	Søndenaa et al.[21]
Local Trauma (bicycling, train rides, falls, physical exercise)	34%	Søndenaa et al.[21]
Modifiable risk factor	**Statistics**	**Reference**
Obesity	$P<.0001$ OR:1.3 ($P<.001$) 37% overweight 71% BMI > 25 Obese patients with a high BMI (mean of 29.35) had a higher recurrence of pilonidal disease Refuted—Lack of evidence that obesity is a cause of pilonidal sinus disease	Akinci et al.[19] Harlak et al.[24] Søndenaa et al.[21] Bendewald & Cima[29] Bolandparvaz et al.[26] and deParades et al.[23] Çubukçu et al.[30] Çubukçu et al.[31]
Sedentary occupation or lifestyle	66% ($P<.0001$) Adjusted OR: 4.03 ($P<.001$) 44% 71% of patient group ($P=.001$)	Akinci et al.[19] Harlak et al.[24] Søndenaa et al.[21] Bendewald and Cima[29] Bolandparvaz et al.[26]
Excessive body hair and/or stiff body hair	OR 9.23 ($P<.0001$)	Harlak et al.[24] Søndenaa et al.[21]
Hygiene (irregular; <2 ×/week) of the sacrococcygeal region	75% bathed <3 ×/week ($P<.000$) Adjusted OR: 6.33 ($P<.001$) 7% 1–2 ×/week, 30% 3–7 ×/week	Bolandparvaz et al.[26] Harlak et al.[24] Søndenaa et al.[21]

Table 2 Type of PSD and percentage of cases

Diagnosis	Clinical criteria
Acute abscess-forming PSD (29.6%)	• Acute inflammation; acute retention • sinus or infected abscess with or without cellulitis • Swelling in the natal cleft skin with or without an opening on the skin surface
Chronically fistulating PSD (58.8%)	• Chronic fistula cavity with one or more sinus tracts communicating with the skin with no symptoms and signs of acute retention • intermittent or chronic draining of pus • local pressure may produce pain or expose a sinus
Chronic remitting PSD (8.3%)	• Previously experienced sensations of pressure or pain but never had perforations, secretion or a combination of these
Asymptomatic or Incidental PSD (3%–8%)	• No history of draining or symptoms • subclinically inflamed with hair, sinus and chronic infection present discovered incidentally during routine physical examination or on autopsy

Doll et al.[35] described "incidental" PSD sinus(es) revealed during investigation on persons without history of exudate or symptoms. By performing three biopsies in the natal cleft on each of 432 adult bodies during autopsy, Aysan et al.[36] discovered that 20 (4.6%) had at least one visualized sinus tract external tip, and considered that to be a positive clinical sign of silent or incidental PSD. By checking for five additional histopathological features, they determined that an additional 16 (3.7%) cases had >3 features, with no clinical signs of natal cleft PSD. The histological markers included:

- enlarged hair follicle bulb, with or without inflammation
- intrafollicular keratinous debris, inflammatory cell infiltration of natal cleft tissue
- one or more hair shafts at any depth of the subepidermal area and/or in the sinuses
- diffuse spreading of collagen fibers causing fibrosis.

The authors considered these histological features characteristic of "silent" or asymptomatic PSD.

Table 2 outlines the possible variations that Doll et al.[13,35] have identified in their review of 1731 patients presenting for PSD surgery.

6. DIFFERENTIAL DIAGNOSIS

A differential diagnosis is critical to ensure a correct diagnosis. People with PSD experience an average of 3–5 years of pain and suffering before being appropriately diagnosed.[2,15,21,37]

i. *Abscesses or sinuses*: These lesions extend beyond the natal cleft, or elsewhere on the body, such as anterior groin, axillae or under the breasts. A history or presence of additional abscesses or sinus may indicate coexisting follicular occlusion disorders (e.g., HS). In this situation, PSD is associated with HS requiring systemic treatment that may include antibiotics and prednisone. Destructive therapies for HS include surgical excision with skin grafting or occasionally radiation with laser epilation of hair follicles for dissecting cellulitis.[8]

ii. *Furuncles/boils involving the deep hair follicle system* may be associated with PSD but usually appear elsewhere on the body and *Staphylococcus aureus* (including methicillin-resistant strains) is the most common organism associated with this skin lesion.

iii. *Anal fistula*: This is a tract traveling from the skin to the anal canal. Identify the tract connection with the anus or the opening using rectal digital exam. Treatment options are dependent on whether the anal sphincter is involved. These modalities include a fistulotomy where there is no risk of fecal incontinence, a Seton Suture, Advancement flap procedure, bioprosthetic plug or ligation of the intersphincteric fistula tract (LIFT) procedures.[38]Although an anal fissure is not a pilonidal sinus, some of the self-care and hygiene aspects of the Harris Pilonidal Sinus Disease Protocol[39] can be applied.

iv. *Perianal ulcers*: These ulcers are often associated with herpes simplex with herpetiform (weeping vesicles in clusters) present on an erythematous base or perianal syphilis, which may appear as a solitary ulcer or two ulcers that may be opposite each other in a "kissing" configuration. The Centers for Disease Control[40] advise that evaluation of unusual genital, anal, or perianal ulcers include:

- syphilis serology, darkfield examination, or PCR testing if available
- culture or PCR testing for genital herpes
- serologic testing for type-specific HSV antibody.[40]

Genital herpes, syphilis, and chancroid have been associated with an increased risk for HIV acquisition and transmission. Perianal tuberculosis may mimic pilonidal sinus disease but would also have inguinal lymphadenopathy (higher incidence with HIV infection).

v. *Unusual thickened or persistent lesions*: There is a need to rule out verrucous (condylomata or genital warts) caused by sexually transmitted viruses, the human papilloma virus (HPV), types 6 and 11. As mentioned, an area of persistent inflammation may develop a squamous cell carcinoma (keratotic papule or nodule).

vi. *Perianal Crohn's disease (PCD)*: is defined as inflammation at or near the anus, including tags, fissures, fistulae, abscesses, or stenosis that is often present in up to 80% of individuals with Crohn's disease.[41]

7. SURGICAL TREATMENT

Although, this chapter does not focus on surgical techniques, Lee et al.[42] summarized the different surgical techniques for the treatment of PSD: Open wound (healing by planned

Table 3 Classification of PSD and recommended interventions adapted from Tezel[43]

Type of PSD and description	**Interventions**
Type I: Asymptomatic pit(s) without a history of abscess and/or drainage	Requires no surgical therapy; consider local hair removal and good personal daily hygiene
Type II: Acute pilonidal abscess	Lateral incision and drainage (I&D), consideration of future surgery
Type III: Pit(s) with a history of abscess and/or previous drainage	Surgery is recommended using the Bascom technique (off midline incision with rotational flap over the gluteal cleft resulting in a modified shallow or lifted gluteal cleft)[44]
Type IV: Extensive disease where one or more sinus opening lies outside the navicular area. Such patients usually have a history of multiple abscess formation and drainages without definitive pilonidal surgery	See Table 4 for other surgical options with associated healing times and results
Type V: Recurrent pilonidal sinus following any surgical treatment	

secondary intention), fistulotomy and curettage, marsupialization, excision and closure: either midline or asymmetric/oblique (e.g., Karydakis, Bascom procedure, cleft closure) or flaps (e.g., rhomboid, V-Y advancement, Z-plasty, or gluteal-myocutaneous). Tezel[43] describes a five-level classification for PSD to inform treatment decision making, based on severity of disease, and defined by the lateral edges and posterior extent of the natal cleft (Table 3).

The 2011 Cochrane review of the published literature determined that patients who had primary closure healed faster and returned to work earlier than patients whose wounds were left open to heal by secondary intention.[45] The review also found that primary closure was associated with a higher degree of recurrence than those healed by secondary intention. Each approach has its advantages and disadvantages, and the decision regarding which approach to select needs to be guided by the patient's own desired goals for treatment. Illustrating the ongoing search for an ideal surgical technique, three separate 2016 papers describe minimally invasive procedures. First, a prospective randomized controlled trial (RCT)[49] with 145 patients compared video-assisted ablation of pilonidal sinus, in which the entire sinus was removed with a minimal surgical wound, to Bascom's cleft lift procedure. There was little difference in complication rates, but patients having the endoscopic treatment had statistically significant less pain at 1 hour, 1 day, 1 week and 1 month postsurgery ($P = .001$ for all), and greater satisfaction at 1 and 6 months ($P = .001$ for both). A second prospective study[50] on endoscopic treatment with 250 patients tracked them for 12 months postop, and found that the failure to close and recurrence rates were both 5%. Third, in a study of 31 patients,[51] a rotating Farrell applicator was used to ablate the sinus tracts, removing hair and debris trapped along the spiral thread each time it is removed. The procedure finished with scraping of the cutaneous orifices. If not healed at 15 days, the procedure was repeated, with an 87% heal rate overall. All three groups need longer follow-up to determine the long-term effects, but are promising. Table 4 provides an overview of information regarding the types of surgical interventions.

Table 4 Surgical techniques for pilonidal sinus disease

Surgical technique	Closure rate/time of closure	Adverse events	Comments
Simple Cystectomy Open Technique (Secondary Closure)	Variable 25.1–41 days Healing may take 41–79 days or longer[44–47]	Maceration Wound dehiscence Hematoma Early suture removal Surgical site infection	– Open technique: wounds left open and allowed to heal via secondary intention – Open healing is associated with a lower risk of recurrence as compared to primary closure; however, involves longer wait time before return to work as compared to primary closure
Simple Cystectomy Primary Closure	Variable[48]	Maceration Wound dehiscence Hematoma Early suture removal Surgical site infection[45]	– Primary closure: after cystectomy, wound is closed via nylon stitches along the midline; higher rate of recurrence as compared to open technique/Secondary closure
Endoscopic Ablation	Variable	Failure to close[49,50]	– Reduced pain, faster return to work, minimally invasive – Requires further study and extended follow-up to fully compare benefits
Farrell Brush	Variable	Failure to close[51]	– Small sample size – Requires further study and extended follow-up to fully compare benefits
Karydakis flap	Median wound healing time = 21 days[52]	Hematoma Wound site infection (high) Fluid collection Flap edema Hematoma Partial wound dehiscence[53,54]	– Procedure designed to move wound away from the midline – Involves elliptical incisions lateral to midline – Low recurrence rate due to a shallow midline furrow free from scar or suture holes

Continued

Table 4 Surgical techniques for pilonidal sinus disease—cont'd

Surgical technique	Closure rate/time of closure	Adverse events	Comments
Bascom procedure/ Cleft Closure	11.3 days[54]	Wound infection Wound drainage Wound dehiscence[54,55]	– Involves lateral incision, removal of hair and excision of sinus tract – Avoids midline incision and involves minimal tissue removal – Recommended for refractory pilonidal disease[13] Very low recurrence rate[44,56]
Limberg Flap	15.3 days[57]	Wound Infection Fluid collection Wound infection Flap edema Hematoma Partial Wound Dehiscence[53,54]	– Involves a rhomboid flap that is designed to keep healing wound away from the midline – Rhomboid incision is made over the affected region, and the tissue is removed – A second rhomboid-shaped flap is dissected from one side and rotated to cover the wound defect from the initial incision – Low recurrence rate. Lower risk of complications, shorter hospital stays, and quicker return to physical activity[53]
Z-plasty	15.4 days[58]	Hematoma Infection Recurrence Hematoma[38]	– Procedure decreases tension on wound after repair – Elliptical lesions are used to remove the cyst and Z-incisions are made approximately 30 degrees to midline – Fast wound healing and return to normal activity – Low recurrence rates[59]
Y-V Plasty	15.3 days[47,60]	Rare Wound dehiscence Seroma Severe wound pain[61]	– Designed for larger wounds – Elliptical incisions used to remove all affected tissue – "V-shaped" incisions made lateral to wound and advanced over the wound – Useful technique for recurrent and refractory pilonidal disease[61,62]

Lifestyle changes, which will be further discussed in the treatment section, and attention to patient-centered concerns may also improve wound closure rates and reduce recurrence, especially with discussion and counseling during the preoperative period. Al-Khayat et al.[63] found that, while excision and primary closure is an acceptable treatment in nonobese and nonsmoking patients with PSD, the infection rate in obese patients and smokers is unacceptably high. They recommended a program of preoperative weight loss and smoking cessation. If this could not be achieved, the planned approach would be excision and secondary closure.

8. OTHER TREATMENT CONSIDERATIONS

The modified Delphi technique used to create the Harris Pilonidal Sinus Wound protocol[39,64] for open and infected pilonidal sinus wounds demonstrated the complexity of caring for and healing these wounds, further supported in 2012 and 2016.[65,66] Delays in healing for this patient population are multifactorial and include:

- Inadequate positioning for wound assessment
- Hair entrapment in the wound cavity leading to inflammation with potential for infection with a need for source control (i.e., hair removal)
- Wound-related exudate with resultant moisture associated skin damage (i.e., maceration)
- Use of dressings that donate moisture to the wound surface aggravating the problem
- Suboptimal wound and periwound skin hygiene
- Hypergranulation or fragile epidermal/dermal components caused by excessive inflammation and/or bacterial damage in the wound or periwound skin (may be associated with pocketing and undermining)
- Unrecognized critical colonization or deep/surrounding infection (slough may facilitate biofilm formation)
- Inadequate nutrition (i.e., protein)
- Physical activities leading to friction or shear with subsequent tissue trauma
- Lack of adherence to meticulous treatment protocols (may be multifactorial).

The following section reviews and updates the 2010 Harris pilonidal sinus work.[39,64] The protocol is principle-focused rather than prescriptive. It does not differentiate between PSD wounds healing by secondary intention, those treated with a local incision and drainage (I&D), or those where primary surgical intervention has failed, resulting in an open wound, because similar issues exist in all three situations.

8.1 Pain Management

The clinician should assess not only procedural-related, postprocedural and in-between procedure pain, but physiological and behavioral indicators of pain. Encourage the patient to take analgesia as prescribed; if ineffective, there is a need to communicate with the physician or prescribing healthcare professional.[67]

A detailed pain assessment is critical to ensure appropriate pain management. Pain can interfere with any and all activities of a person's daily living (e.g., attending work, school or nonadherence to the wound management plan of care, including the wound packing or dressing regime).[68] Price et al.[69] performed a quality-of-life survey of individuals with acute wounds ($n=80$), 62 of which were PSD: 49% identified discomfort significant enough to interfere with sleep and 20% had diminished appetite due to the pain. Some patients are hesitant to use pain medication as prescribed because it impairs activities of daily living: e.g., their ability to drive, work, cook or eat. The result is ineffective pain control. Pain related to pilonidal sinus wounds is like other wounds and could be nociceptive (stimulus-related), neuropathic or a combination.

The World Health Organization's analgesic ladder is a structured approach to stimulus-related pain.[70] Patients presenting with gnawing, aching, tender, or throbbing (nociceptive stimulus-related pain) can be treated with nonsteroidal anti-inflammatory drugs (NSAIDs), with or without adjuvants. Issues of tolerance, addiction and side effects with the use of narcotics need to be discussed with the patient.

Neuropathic pain presents a different picture clinically. Patients may report spontaneous burning, shooting, stabbing, or stinging pain. Tricyclic agents (TCA) have been successful in treating neuropathic pain inhibiting presynaptic reuptake of serotonin and norepinephrine and blocking cholinergic, adrenergic, histaminergic and sodium channels.[71] Nortriptyline and desipramine are useful tricyclics with the antiepileptic agents used in neuropathic pain control, including gabapentin, pregabalin, and carbamazepine.

Patients often present with both neuropathic and nociceptive pain, requiring a combination of therapies. The use of tricyclics for pain management also may have a positive effect on insomnia, but the depression often associated with chronic wounds requires a higher tricyclic dose than needed for pain control in most patients.[72] Although uncertain of the cause, Braungart et al.[73] reported chronic pain in 6% and numbness in 28% of their pediatric patients, and recommended that this phenomenon should be evaluated in a structured way.

To minimize pain associated with dressing changes, prepare the patient with oral premedication a minimum of 1 hour prior to dressing change; use gentle wound care and local dressings that are easy to remove. The use of topical amide local anesthetics may also help prior to wound debridement.

8.2 Optimal Positioning for Wound Assessment and Care

The examination should be conducted with a source of good quality overhead lightening. The traditional prone or side-lying patient position used for assessing and providing wound care is problematic for wounds in the natal cleft. The body's natural response to anticipated pain results in tensed muscles, impeding access to the full wound area and increases patient discomfort. Placing one or two pillows under the anterior hips with the patient in the prone position (modified jackknife position) will slightly flex the hips, improving visualization and access. In the modified jackknife position, the patient finds it harder to tense the buttocks, and can assist in holding them apart to facilitate visual assessment for embedded hairs, bridging or pocketing of hypergranulation tissue or fragile new epithelium. This position also facilitates dressing application.

8.3 Assess for Signs of Localized and Deep Infection

Pilonidal sinus wounds are located close to the rectum and in the deep natal cleft; therefore, the greatest challenges are to reduce wound fecal contamination, and prevent trapping of perspiration and moisture. Both lead to bacterial proliferation and infection. Infection results in friable granulation tissue, premature bridging of epithelial tissue, pocketing of granulation tissue in the base, and/or abscess formation.[69] The wound and surrounding tissue may be in a chronic inflammatory state due to the presence of hair ends, the result of the hair follicle eliciting a foreign-body reaction. Untreated, any of these responses can delay healing. Deep wound infection may or may not present with overt features of inflammation. Even with surgical closure, infection is a risk. One study of 94 patients undergoing excision and primary closure identified PSD wound infection in 12.8% postoperatively.[63] A 2013 systematic review by Mavros et al.[74] reviewed attempts to prevent postoperative infection with the use of prophylactic systemic antibiotics or gentamicin collagen sponges (GCS). They concluded that there were no significant differences in rates of healing or surgical site infections with either approach. However, the authors cautioned that the clinical heterogeneity and paucity of evidence of the existing research precluded any safe conclusions.

The Harris modified Delphi[39,64] utilized the NERDS and STONEES mnemonic as a method of identifying clinical signs suggestive of superficial/local infection (NERDS) and deep tissue infection (STONEES). This tool was described by Sibbald et al. in 2006,[75] and validated by Woo and Sibbald in 2009 (Table 5).[76]

Superficial or localized infection occurs following colonization, most commonly due to a polymicrobial community, and this may lead to the development of biofilm on the wound surface. Biofilm is commonly seen on interfacing wounds such as PSDs.[77] These types of infection result in delayed wound closure. If three or more NERDS signs are present, local debridement and topical antiseptic/antimicrobial dressings (for specified predetermined time periods) are indicated. Topical therapies can include: silver, iodine-containing preparations, polyhexanide (PHMB), chlorhexidine, methylene blue/gentian violet, and other products (Table 6).

8.4 Antimicrobial Dressings

The choice of dressing will depend on the amount of exudate, the desired frequency of dressing changes, and the ability of the patient to manage self-care. Woo et al.[78] completed a scoping literature review on topical agents and dressings for PSD. While silver

Table 5 NERDS and STONEES

Superficial (localized) infection (NERDS)	
N	*Non-healing*: The wound size has not decreased ≥30% within the last 4 weeks
E	*Exudate* has increased
R	*Red* friable granulation tissue presents as abnormal bright red color
D	*Debris*: dead yellow slough or slimy, translucent and confluent film seen on the wound surface
S	*Smell* indicates Gram-negative or anaerobic organisms
Deep tissue infection (STONEES)	
S	*Size* of the wound is increasing
T	*Temperature*: Infrared thermometry demonstrated ≥3°F or more increase in temperature compared to the contralateral side
O	Os: Latin for bone refers to any expose bone or wound depth that probes to bone
N	New satellite areas of wound extension(s) developing in the wound margins often joining with the main wound
E	*Erythema* and *Edema* around the wound margins and periwound (cellulitis)
E	*Exudate*: an increase in the host's response to injury
S	*Smell*: indicative of Gram-negative or anaerobic bacteria present in the wound

and polyhexanide dressings showed superior results in reducing bacterial burden compared to other topical antimicrobials, the evidence is inconclusive due to study biases and methodological limitations.[78] There may be a role for topical foam in a tube dressing format, with surface active agents. This could remove foreign material from the skin surface and help remove fecal contamination and unwanted bacteria.

8.5 Systemic Antimicrobial Selection

If a patient presents with three or more of the seven STONEES signs, treatment of infection with systemic antimicrobial therapies are recommended with debridement as necessary. Selection of systemic antimicrobial therapies must be based on patient clinical status, infection severity, and local bacterial epidemiology/resistance patterns. Although the microbiology of PSD has not been extensively studied, it is thought to be polymicrobial based on anatomic location and tendency for chronicity. Brook et al.[79] examined cultures of aspirated pus in 75 patients with PSD. Their results supported earlier work by Marks et al.[80] that showed a predominance of polymicrobial anaerobic bacteria. Ardelt et al.[81] studied primary and recurrent sacrococcygeal PSD in 96 patients, also demonstrating that PSD wounds were polymicrobial, with both aerobic and anaerobic organisms.

Once a clinical diagnosis of infection is confirmed, a bacterial swab using the Levine technique[82] is recommended to identify the bacterial species and antimicrobial resistance profile. Empiric antibiotic therapies that include coverage for Gram-positives, Gram-negatives (excluding *Pseudomonas aeruginosa*), and anaerobic bacteria may be initiated pending culture results (Table 7).

Table 6 Topical antimicrobials appropriate for PSD wounds after debridement

Agent	Effective antimicrobial spectrum	Other actions	Forms
Silver	Gm +ves, Gm −ves MRSA, VRE viruses & fungi	• Antiinflammatory • Effective in a moist ionized environment only • Not for dry wounds	Hydrogel, polyester film, gauze, nanocrystaline cloth Alginate, rope, sheet Hydrofiber, foam
Chlorhexidine	Gm +ve, Gm −ve MRSA Inactive against spores, fungi, viruses Limited sensitivity to some *Pseudomonas* species and *Proteus*	• Some tissue toxicity if released • May be proinflammatory if released • Neutral in nonreleased formats	Tulle Gauze Solution
Polyhexanide; polyhexamethylene biquanide (PHMB)	Gm +ve, Gm −ve MRSA, VRE, E. coli, CRE, etc. yeasts, fungi, parasites, some viruses and molds	• Nonrelease forms are neutral (gauze, foam) and do not cause tissue toxicity	Gauze, foam Cleanser, Gel
Iodine Formulations	Povidone Iodine: Cadexomer iodine and povidone iodine (PI) Due to small molecular size, able to penetrate biofilm May promote inflammation	• Proinflammatory • Penetrates glycocalyx of biofilms better than other agents • Caution if used on large areas/long period for thyroid abnormalities	Povidone iodine 10% has highest tissue toxicity Cadexomer Iodine—absorbs exudate, autolytic debridement and moisture balance for healable wounds Inadine—slow release in a polyethylene glycol coated fabric—all wounds
Redox dyes	Methylene blue and gentian violet Disinfectant, bacteriostatic and antiseptic	• Relatively nonrelease • Sterilizes above the wound	Foam as a nonrelease: Polyvinyl alcohol—needs to be hydrated with water and will autolytically debride Polyurethane can be put on without hydration but will not perform autolytic rebridement
Gauze	Not antibacterial	• Will foster the growth of biofilms above the wound	• Moist gauze hydrates the wound • Dry gauze will absorb but if the wound is dry, it may stick and cause mechanical injury and pain

Table 7 Selection of oral antimicrobial agents

Antibiotic	Action
Amoxicillin-clavulanic acid	• Gm +ve (anti-staphylococcal)
Cephalexin	• Gm +ve (anti-staphylococcal)
Clindamycin	• Gm +ve, anaerobes • Antiinflammatory action
Cloxacillin	• Gm +ve (antistaphylococcal)
Doxycycline	• May treat MRSA strains • Most antiinflammatory of the tetracyclines • (Tetracycline)
Metronidazole	• Treats anaerobes • Antiinflammatory action
Quinolones (e.g., levofloxacin/ moxifloxacin)	• "Broad spectrum" Gm −ve/Gm +ve • Treats anaeorobes
Sulfonamides	• Gm +ve (antistaphylococcal) (may treat MRSA strains) • Antiinflammatory action

Most antibiotics listed in Table 7 provide coverage for Gram-positive and Gram-negative (certain species) bacterium. Routine coverage for *P. aeruginosa* is not indicated. Infectious disease consultation should be considered for postsurgical, complicated, or unusual infections (i.e., mycobacterial species or unknown organism).

8.6 Periwound Decontamination

The Harris protocol[39,64] recommends the use of topical antiseptic chlorhexidine for the cleansing part of each dressing change. The intent in the protocol's original design was to reduce bacterial contamination and subsequent inflammation. The periwound skin extending 5 cm around the wound is to be cleansed with a topical agent such as 0.5% chlorhexidine (without an alcohol agent often found in mouthwash preparations). Since the introduction of the protocol, additional antiseptics, such as 0.5% acetic acid or polyexanide/betaine 0.25% have been found to be effective for decontamination without harming the wound surface. The advice for chlorhexidine is to leave it in place for 1 min, making it effective against most organisms except *Pseudomonas aeruginosa.* If this organism is present, the compress is to be left in place for 5 min, followed by a saline or potable water cleanse to avoid contact dermatitis. If using other antiseptics, follow the manufacturer's recommendations for length of exposure. The protocol also recommended irrigating or compressing the wound bed using 100 to 150 mL of 0.9% NaCl or sterile water at room or body temperature using the equivalent of 7–12 psi (e.g., 30 cc syringe with a 19-gauge cannula or prefilled bottle of either 0.9% NaCl or sterile H_20 designed for wound irrigation). The NICE guidelines for prevention and treatment of surgical site infection[83] recommend tap water after 48 hours, so the choice of using sterile saline versus a handheld shower should be dependent on the situation and available resources. When patients are doing self-care, showering is usually sufficient. Shaving or using iris scissors or small blade-type scissors in the area (see Section 8.7, "Hair Removal") is to be done following the antiseptic compress.

Pilonidal sinus wounds need to be kept clean and dry. Patients are encouraged to shower once or twice daily (depending on level of activity) and after each bowel movement. Liquid antimicrobial soap substitutes and handheld shower heads are preferred for cleansing. If showering after a bowel movement is not possible, teach the patient to use moist alcohol-free towelettes to cleanse the area, wiping down towards the anal opening, then pat dry using nonwoven gauze.[65] Nonwoven gauze is recommended as the surface area of the gauze is smoother than woven gauze, and will decrease the risk of irritation and subsequent inflammation. A recent Japanese study[84] examined patients who were instructed to shower at home and apply petrolatum ointment to the wound, following lay-open procedures under local anesthesia. Although it was a small sample of nine patients, there was no recurrence, and it highlights the important role of showering and simple self-care of the PSD wound.

Baths are to be avoided for cleansing or as a form of sitz bath, as they increase bacterial contamination. Sitz baths may be used only if specifically ordered. While they provide comfort and improve hygiene, sitz baths should not be used or encouraged as a form of wound cleansing.[85] If sitz baths are ordered, instruct the patient as follows:

- A specialized kit, designed specifically for sitz bath procedure, is to be used
- No more than 2–3 cm of water is to be in the basin
- Sit in the specialized basin for no longer than 5 min.
- Remove fecal contamination before using a sitz bath, preferably using a handheld shower
- Shower following each sitz bath.

8.7 Hair Removal

As mentioned previously, hair is a nidus for infection and chronic inflammation (Images 1 and 2). The clinician performing the dressing change needs to be trained to observe for and remove embedded hair. Periwound skin in the natal cleft often contains bacteria, most commonly staphylococci, and the hairs trap feces that contain bacteria and other debris that needs to be removed promptly. The Harris protocol[39,64] recommends shaving the natal cleft at least weekly in a 5-cm wide strip extending at least 2.5 cm from all edges of the wound, from the anal verge to the presacrum, to remove all hair in the area between the distal wound and the anus. Anecdotally, the authors have seen significantly improved patient outcomes when providing a hair-free wound and wound margin. The need for this step in the protocol cannot be overstated. Razors designed for women's

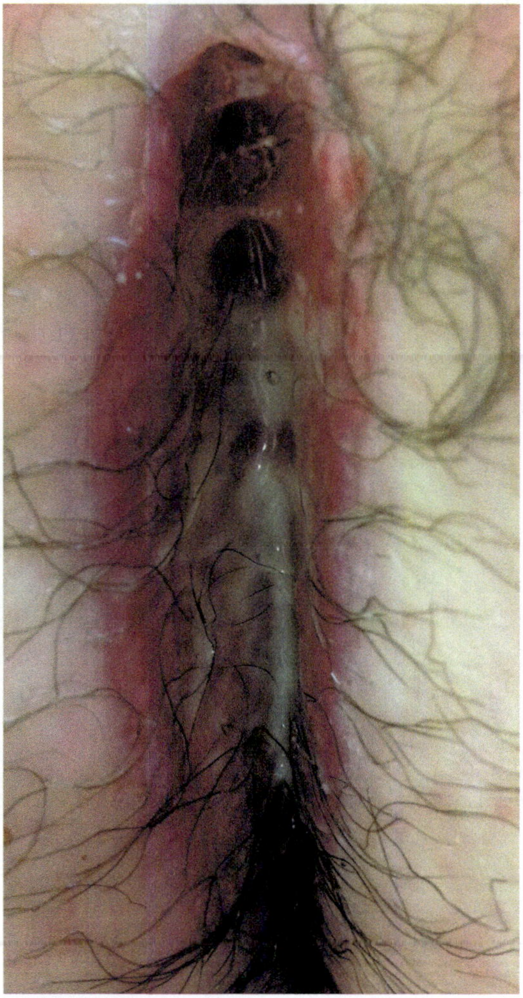

Image 1 Hair as a nidus of infection.

bikini area and razors with a swivel head work best in providing a close shave in the contour areas. The shaving technique should follow manufacturer's recommendations. First, the decontamination procedure should be completed. Next, the area needs to be moist—not too wet. Use a mild soap or shaving cream, being careful not to get this substance into the wound bed. Always shave in the direction of hair growth (follow the way the hair lies on the skin). Shave over each area only once to avoid irritation or skin scraping. In some cases, where there is risk of local trauma, manicure-style scissors or surgical iris-type scissors may be used. For hair trapped in the wound surface, use forceps for extraction.

Laser hair removal is another option for patients whose body hair color is different from that of their skin. Laser has a longer period of being hair-free than other methods,

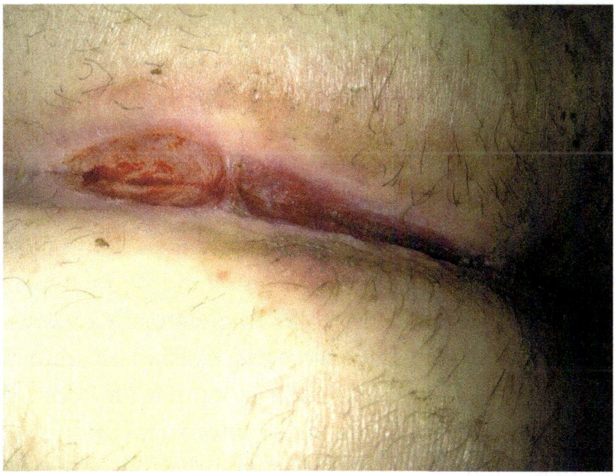

Image 2 Improper shaving technique.

BOX 1 Laser Protocol for PSD

- Plan an average of six treatments at intervals of 6–8 weeks
- The type of laser used should consider the patient's skin type, hair coarseness and pain tolerance
- Shave the area before treatment, extending to 2–5 cm margins from the midline on either side of the gluteal crease (or from the wound margins) from the apex to the anal opening and the hairy area above the buttocks' crease
- Apply topical anesthetic cream to the area 1 h before the procedure
- Use cooling packs after treatment
- Assess the area annually and recommend top-up treatments if needed

Adapted from Marza L. Reducing the recurrence of pilonidal sinus disease. Nurs Times *2013;**109**(25):22–24.*

but it may be expensive, and can also be painful. Laser centers that are cosmetically oriented may be reluctant to carry out procedures adjacent to open wounds. Requiring multiple sessions, the hair removal is not permanent but does require a less frequent maintenance program. Marza[86] suggested a laser protocol for PSD (Box 1).

To prevent recurrences, it is recommended that hair removal should continue up to the age of 40, when it is thought that sex hormone production begins to decline.[62] Four studies using laser depilation postoperatively found recurrence rates of 0 to 0.5% for 1 to 3 years,[87–90] compared to a recurrence rate without laser of up to 22% over 4 years.[91] It is not known what the results are for laser depilation for longer than 3 years, and further research is needed. In a study of pediatric patients, when laser was used for 7 dark-haired patients,[73] the long-term recurrence rate was 0%, prompting the authors to call for further research to evaluate the long-term results, so important with this young age group. Conversely, Petersen et al. [92] found that razor depilation to prevent recurrence actually resulted in a statistically significant ($P = .01$) increase in recurrence when compared with individuals who did not shave.

8.8 Debridement

The Harris protocol[39,64] stressed the need to promptly treat signs of premature epithelial bridging (Image 3), pocketing in the base (Images 3 and 4), or hypergranulation tissue (Image 3) caused by one or more of the following factors: excessive moisture on the wound surface (leads to increased metalloproteases, causing inflammation and development of friable granulation tissue),[93] increased bioburden, friction, shear or irritation from activity or dressing, debris, inflammation or infection. Clinically, hypergranulation tissue presents as edematous granulation buds, tissue that bleeds easily and readily pulls apart, and/or bridging (strands of friable hypergranulation tissue "meshing" together from side to side) or pocketing of granulation tissue. Abnormal bright red friable granulation contains an increased number of blood vessels with minimal mature collagen fibers that will not support healthy reepithelialization. The exudate from this type of hypergranulation tissue is often sanguineous. If patients are changing their own dressings,

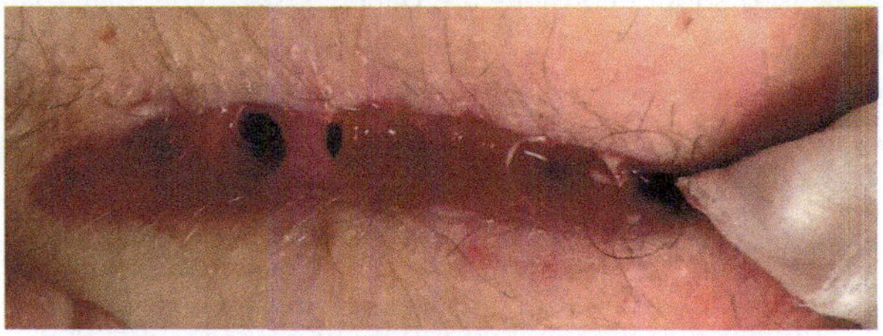

Image 3 Premature epithelialization, pocketing in base, hypergranulation.

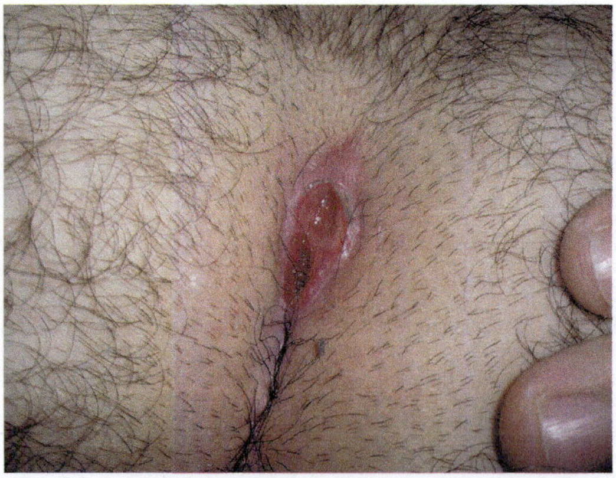

Image 4 Pocketing in base.

ask them to bring the old dressing, contained in an enclosable clear plastic bag, to their next visit for inspection. At the time of dressing change, examine the wound surface for any bleeding points caused by the dressing's removal. Superficial bridging appears as thin, bluish-colored new epithelial tissue (usually 1–2 mm in depth) with a sinus track(s) underneath, at the distal or proximal end of the wound, or as in Image 2 and 3, mid-wound. While bridging may appear to be wound closure, the tissue is unhealthy and needs to be removed. If allowed to close, it will abscess and cause recurrence.

To remove newly formed bridging, spread the buttocks apart with the hands, providing enough force to split the superficial bridge spontaneously. If this is not possible due to the thickness of the bridge, use a local anesthetic and a scalpel to open the superficial undermining distally or proximally.

The Harris protocol[39,64] also recommended cauterizing superficial bridging and hypergranulation tissue with silver nitrate. At the time (2010), it was normal practice in the management of hypergranulation or friable granulation in PSD wounds.[94] This has since been disputed by Widgerow and Leak[95] and is worthy of consideration. Silver nitrate *will* oxidize organic material, coagulate and destroy unwanted tissue and kill bacteria, but is also highly inflammatory. Chronic inflammation has a detrimental effect on wound healing. Therefore, the use of silver nitrate as a primary treatment results in a rebound effect: increased inflammation, increased exudate, and increased patient discomfort. If the causes are not also resolved, the hypergranulation will return.[95]

Debridement is an essential first step to remove debris and slough quickly. The following wound debridement methods should be considered, keeping in mind that they may be used concomitantly.

- *Conservative sharp wound debridement* using either curette or iris scissors and forceps for removal of friable granulation from the surface of the wound. Appropriate pain control: orally, or local anesthetic, or both are needed prior to the procedure. Options for local anesthetic include: lidocaine (xylocaine) with or without adrenaline or topical anesthetic agents containing amine (prilocane, pramoxine).
- *Autolytic debridement*: due to wound location and the propensity of the area to retain perspiration and moisture, autolytic debridement using hydrogels or moisture retentive dressings is to be used with caution. Dressings containing calcium alginates will control bleeding as well as debride; hypertonic sodium chloride (NaCl) impregnated gauze is another option that is absorbent and bacteriostatic.
- *Mechanical debridement* using irrigation with a volume of 100 to 150 mL at 7–12 psi and having the patient use a handheld shower to cleanse prior to dressing change.[91]

8.9 Dressing Options

A wide variety of dressings are available in the market and most provide an active element of wound management. When selecting a dressing, consider the current phase of wound healing, level of moisture, pain, infection, patient activity and preference as well as appropriateness for the setting.

Dressings appropriate for a person with PSD need to be able to:

- Manage exudate: consider primary dressings that laterally wick moisture into the dressing away from the wound surface.
- Conform well to the body surface: select secondary or cover dressings that fit the area, provide adequate insulation to the wound bed, and do not permit strike-through (where exudate soaks through to the outside of the dressing). Strike-through compromises the barrier function of the dressing, potentiating bacterial contamination. Dressings that have a waterproof and breathable barrier top layer and adhesive borders fit these criteria; however, they are meant to stay in place for more than one day. The authors have not found that increasing time between dressing changes is successful with PSD wounds. Often wound deterioration has been associated with dressings holding too much moisture over 2–3 days. Interactive dressings such as hydrophilic polyurethane foams are designed to absorb or wick fluid from the wound bed into the dressing and manage excess moisture through vapor transmission, known as the moisture vapor transmission rate (MVTR).
- Adequately secure: a secure and conformable dressing eliminates friction or shearing forces. Whichever dressing is selected needs to seal the skin in the natal cleft between the distal wound edge and anus, as well as prevent migration of hair and fecal matter into the wound bed. It may be possible during the acute phase of healing to choose dressings that may remain in situ for 2–3 days rather than daily or twice daily dressing changes, but this will be dependent on the amount of exudate, the depth of the natal cleft, the amount of intact skin between the wound edge and the anal verge, along with patient comfort. Watch carefully for signs and symptoms of increased bacterial activity.
- Easy to apply, to promote self-care: for patients who can participate in their own care, dressings need to be simple, i.e., can be applied one-handed and require minimal steps. Self-management is appropriate when the wound is a cavity without tunnels or during the granulation phase of healing. Dressings in this category would be gauze-based dressings. While these dressing options may require more frequent dressing changes than the self-adhesive dressings, they are easier for self-care.
- Manage bioburden or infection: use antimicrobial dressings for wounds that are at high risk for infection or are infected. Pilonidal sinus wounds are at great risk of fecal contamination due to the location. The choice of antimicrobial dressing needs to factor in the exudate control, securement, and ease of application. Encouraging self-management whenever appropriate for the patient will help to reduce infection risk: patients who actively participate in their care have improved hygiene and change dressings as needed to keep the area clean and the dressing dry. Refer to Table 6 for details on antimicrobial options. In the authors' experience, the PSD patient's self-management abilities are facilitated with the use of a topical antiseptic (povidone

iodine, chlorhexidine or polyhexamethylenebiguanide (PHMB)) as prophylaxis and when needed for superficial or localized infection, using soft nonwoven gauze as the secondary dressing. Patients can then change the dressing as needed throughout the day.

- Avoid dressings that form a hard, adherent mass, as they will contribute to friction and inflammation. When choosing dressings for packing, choose a product that will remain soft when moist. Alginates are an excellent choice for exudate management; however, certain alginates have limited gelling capacity, resulting in a formed mass in the wound cavity.
- Prevent migration of hairs and fecal matter into the wound bed to decrease risk of contamination.
- Be comfortable and cost-effective. Patients will not wear a dressing that isn't comfortable. Choosing the right dressing for the patient may be "trial and error."
- The need for regular and persistent personal hygiene needs to be aligned with the local wound care plan. The wound location necessitates frequent hygiene; therefore, dressings need to be easy for the patient to apply as well as fit with their lifestyle, provide the ability to perform self-care and be economically sustainable. A person who normally has two to three bowel movements per day will be resistant to dressings that are too complex to self-apply.

Questions to consider when designing a local wound care plan: what will support patient adherence (concordance) and independence? Soft gauze secondary dressings or superabsorbent dressings with diaper bead polymers provide patients with independence and at the same time adhere to the key principles of preventing contamination, fitting snugly (maintaining contact with the wound surface) and promoting wound closure. For heavily exudative wounds, the authors looked to other forms of moisture control devices, such as superabsorbent panty liners, foam packing strips (longitudinally to make a flat surface) or foam dressings with silver and/or detergents, to be used for daily changes especially after bowel movements. Table 8 reviews the types of dressings available for PSD wounds.

8.10 Negative Pressure Wound Therapy

The Harris protocol,[39,64] based on expert opinion after reviewing the current evidence, suggested considering negative pressure wound therapy (NPWT) for patients with PSD and open wounds experiencing:

- excessive exudate, affecting care and/or skin integrity and/or quality of life;
- lowered quality of life: e.g., students or employees who need to return to school or work and are unable to return without achieving healing;
- a need to promote rapid decrease in the size of the wound bed for large pilonidal sinus wounds;

Table 8 Wound dressing considerations for PSDs

Category	Description	Function	Properties
Calcium Alginates and hydrofibers	Polymers of alginic acid found in seaweed providing a source of mannuronic and guluronic acid residues	Absorbs up to 20 × in weight in exudate Locks fluid within fibers	Comes in sheet or rope. When exposed to exudate fibers turns into hydrophilic gel Absorbent; hemostatic, bacteriostatic Biodegradable
Foams	Polyurethane open cell polymer dressing comes in multitude of shapes, sizes and thickness: nonadhesive, adhesive (acrylic or silicone)	Laterally wicks exudate into dressing Depending on type of foam, able to absorb small to large amounts of exudate	Polyurethane base; adhesive borders or/ and adhesive over foam to secure to area Barrier film on external surface to prevent strike-through Conformable
Absorbent Dressings	Nonocclusive, permeable dressings	Absorbent: small to moderate amounts depending on composition; allow moisture absorbed into the dressing to evaporate into the atmosphere	Soft viscose polyester bonded pads (e.g., gauze-based + film), may have film barrier to prevent strike-through May contain absorbent polymers ("super absorbers")
Gauze	Dry nonwoven Normal saline soaked Povidone iodine soaked PHMB	Absorptive Donates moisture Donates in wound Absorptive	Not antibacterial Not antibacterial Antibacterial, some toxicity Antibacterial, nonrelease but not antibacterial in the wound

- a situation where expected healing times without the NPWT are longer than 6–8 weeks and local wound care is not achieving the expected rate of healing;
- a case involving obese or thin individuals with PSD.

Since that time, there has been interest in applying NPWT to primary closure incisions to provide extra stability, edema reduction and a dressing protected from contamination. Case studies continue to demonstrate some benefit for secondary closure as well, including the use of more portable versions of NPWT, which are less cumbersome for the patient to manage.[96] Biter et al.[97] conducted a small RCT comparing NPWT to conventional dressings for PSD wounds healing by secondary intention. There was no difference in time to complete wound healing and time to resume daily life activities between the groups, although there was a positive effect on wound size reduction in the first 2 weeks in the NPWT group (0.30 vs. 0.57, $P = .02$).

8.11 Optimize Health

8.11.1 Nutrition Needed for Wound Healing

The Harris protocol[39,64] identified the importance of nutrition in wound healing: protein, zinc and vitamin C being the main three substances. Protein is vital for wound healing; therefore when a person with PSD has a highly exudative wound, it results in dehydration due to fluid loss, hypoproteinemia and shifts in capillary hydrostatic pressure. Continued loss of protein and fluids will stall or delay wound closure.

Zinc assists in the proper structure of proteins and cell membranes and has antiinflammatory effects. The best sources for zinc are shellfish, red meat and liver. Vitamin C has a crucial role in the regulation of collagen, which is necessary for the extracellular stability of the skin.

While four servings of meat, dairy products or other sources of protein per day are recommended, the patient population with PSD are young adults, known for diets high in fats and simple carbohydrates and low in protein. It is a challenge to shift younger patients' diets.

In the authors' experience, the current trend of drinking "smoothies" (liquid drinks containing protein powder, milk, yogurt or combinations of the three, along with vegetables and fruits) has been beneficial in improving the diet to include the nutrients necessary for healing. For patients who are reluctant to change their eating habits, it is recommended that they take a multivitamin and mineral supplements that include zinc (minimum of 40 mg/day), along with oral supplement of Vitamin C (1000 mg) daily.

Smoking cessation (or reduction) and minimizing alcohol intake is necessary for wound healing. Smoking rapidly depletes Vitamin C and alcohol intake affects uptake and synthesis of Vitamin A and zinc and production of albumin.

8.12 Assess and Teach re participation in Work, Sports and Exercise

It is important for the health professional to educate patients with PSD on their disease (it isn't "just a wound"), how wounds heal, reasons why wound healing is delayed, and causes of repetitive trauma and risks for certain activities. Examples of preoperative and patient self-management instructions, based on the Harris protocol, were developed at the Children's Hospital of Eastern Ontario (CHEO), and are provided in English and French (Image 5), and postoperative instructions by Calea are provided (Image 6).

Together with the patient, evaluate activities of daily living: how many hours do they sit, what are the requirements of their occupation? For example, a long-haul driver will be sitting for long periods; a nurse or food server will be walking and bending for 8 or

CHEO

Nous aidons les enfants et les familles à vivre en pleine santé

Chirurgie du sinus pilonidal

Qu'est-ce qu'un sinus pilonidal?

Le mot « pilonidal » signifie un « nid de poils ». Un sinus pilonidal est un petit tunnel sous la peau dans le sillon inter-fessier (aussi appelé « fossette du coccyx »). Ce tunnel (ou sinus) s'ouvre près de l'anus, mais parfois comporte plus d'une sortie. Un sinus pilonidal contient souvent des poils, et peut aussi s'infecter. Quand une infection s'installe dans ce sinus, du pus s'accumule et forme un abcès. L'infection peut être douloureuse et peut provoquer une enflure dans le sillon inter-fessier ou le bas du dos.

Quelle est la cause du sinus pilonidal?

La raison n'est pas claire, mais nous pensons que cette maladie est provoquée par l'irritation puis l'infection des follicules pileux. Les follicules pileux sont des pores avec des poils. Pendant l'activité physique, les fesses peuvent frotter l'une contre l'autre et cela peut faire empirer le problème. Le sinus pilonidal apparaît le plus souvent chez l'adolescent et l'adulte de moins de 40 ans. Il est plus fréquent chez les garçons et les hommes.

Quel est le traitement du sinus pilonidal?

Le traitement du sinus pilonidal consiste à ouvrir, nettoyer et laisser guérir. Une intervention chirurgicale est nécessaire en cas de sinus pilonidal qui ne se ferme pas. Votre chirurgien utilisera l'une de ces 2 interventions pour traiter le sinus pilonidal :

Grande excision et guérison par seconde intention

Cette opération consiste à enlever le sinus et tout le tissu enflammé qui l'entoure.

- Ensuite la plaie sera laissée ouverte et guérira de l'intérieur vers l'extérieur. (Cela prend habituellement moins de 12 semaines).
- Il faudra garder la plaie sans poil.
- Il faudra changer le pansement régulièrement.

Excision et fermeture primaire

Cette opération consiste à enlever le sinus et tout le tissu enflammé qui l'entoure. Ensuite, les bords de la plaie sont cousus ensemble. Généralement, la plaie guérit très vite. Mais parfois la plaie s'ouvre. Dans ce cas, la plaie sera laissée ouverte et guérira d'elle-même de l'intérieur vers l'extérieur. Il faudra mettre un pansement jusqu'à la cicatrisation complète.

Prendre soins de votre enfant ou ado après chirurgie

Antibiotiques

Certains patients devront prendre des antibiotiques après la chirurgie. Suivez attentivement les instructions de votre chirurgien.

Les soins à apporter à la plaie

Voir notre fiche d'information « Les plaies du sinus pilonidal » sur les soins à apporter à la plaie.

Maintenir propre

Votre chirurgien vous avertira quand votre enfant ou adolescent pourra prendre une douche.
Voir notre fiche d'information « Les plaies du sinus pilonidal » pour obtenir des informations supplémentaires sur le bain. Il faudra installer une douche à main (vous en trouverez dans une quincaillerie). Votre enfant ou adolescent ne doit pas prendre de bain dans une baignoire tant que la plaie n'est pas cicatrisée. Encouragez votre enfant ou adolescent à se doucher après être allé à la selle (faire caca). Il est aussi conseillé de prendre une douche juste avant un changement de pansement.

P5656F novembre 2013 www.cheo.on.ca

Image 5 CHEO Preop and self-management booklets.

We help children and families be their healthiest

Pilonidal Sinus Wounds

What is a pilonidal sinus?

The word 'pilonidal' means a 'nest of hairs'. A pilonidal sinus is a small tunnel under the skin in the fold between the buttocks (also called the 'natal cleft'). This tunnel (or sinus) opens near the anus, but sometimes there is more than one opening. A pilonidal sinus often contains hairs, and it might also be infected. When there is an infection in this sinus, pus collects in one place, forming an abscess. The infection can cause pain and swelling in the natal cleft or lower back.

What causes pilonidal sinuses?

It's not clear, but we think that they might be caused by hair follicles becoming irritated and then infected. Hair follicles are pores that have hairs growing out of them. During physical activity, the buttocks can rub together, and this may add to the problem. Pilonidal sinuses happen more often in teens and adults under 40, and they are more common in boys and men.

How are pilonidal sinuses treated?

Pilonidal sinuses are often treated first in the Emergency Department, where doctors will:

- Open the abscess, allowing the pus to drain out;
- Clean the wound;
- 'Pack' the wound with special gauze (these wounds have to heal from 'the bottom up');
- Usually prescribe antibiotics to treat the infection.

New skin covering the closed wound will never be as strong as before the sinus developed. It's important to take extra care to prevent the wound from opening again. If the pilonidal sinus does not close, your child or teen may need surgery. If surgery is needed, you'll meet with a surgeon to discuss things further.

Taking care of your child or teen at home

Medications

Follow your doctor's directions for pain medication. It's very important to keep pain under control.

Wound Care

We will arrange to have a nurse come to your home to change the dressing. The nurse may be able to teach you to change the dressing for your child or teen. It will be important to:

- Keep the wound clean, dry and covered with a dressing;
- Change the dressing right away if it becomes soiled with drainage or stool (poop!) ;
- Remove the dressing before your child or teen showers or has a sitz bath (see next page for more information);
- Use a hand held shower to gently flush out the inside of the wound (available at most hardware stores);
- Keep soap, shampoo and loose hair out of the wound;
- After showering, clean the wound as the nurse has taught you;
- Pat the area dry;
- Apply a new dressing, as the nurse has taught you;
- Have your child or teen wear cotton underwear and loose fitting pants.

Silver nitrate treatment

Sometimes the doctor or nurse may need to treat the unhealthy tissue in the pilonidal sinus wound with silver nitrate. It comes on a stick that looks a little like a match. Your child or teen may notice some burning pain for a short time after treatment with silver nitrate. The drainage from the wound increases after it is used, and may be a gray colour - this is normal.

Form # P5657E November 2013 www.cheo.on.ca

Image 5—cont'd

more hours a day. The first contributes to sweating in the natal cleft, the second to friction forces. What sports or activities do they enjoy? Sports that require multiple stops and starts, bending and twisting or movement of the buttocks include soccer, hockey, baseball, gymnastics, bicycling, gardening and shoveling snow. Any activities that cause the buttocks to rub together causing friction—sometimes in multiple directions—result in increased trauma to the wound and periwound area.

If physical activities are important to the person with PSD, eliminating them can negatively impact their quality of life.[39,64,68] In the same vein, a person's job is something one cannot dismiss or change easily. Encourage the patient to start with modified activities, build in pain management and use the parameters of pain levels, watching for any trauma to the wound area and wound healing rates to monitor effects.[64] If pain is controlled and wound closure rates continue to reduce, encourage patients to continue with their activities, increasing participation as tolerated. They need to use caution as to how much they will contribute, balancing their need to resume activities of daily living versus the potential for increasing trauma to the area and/or delayed healing.

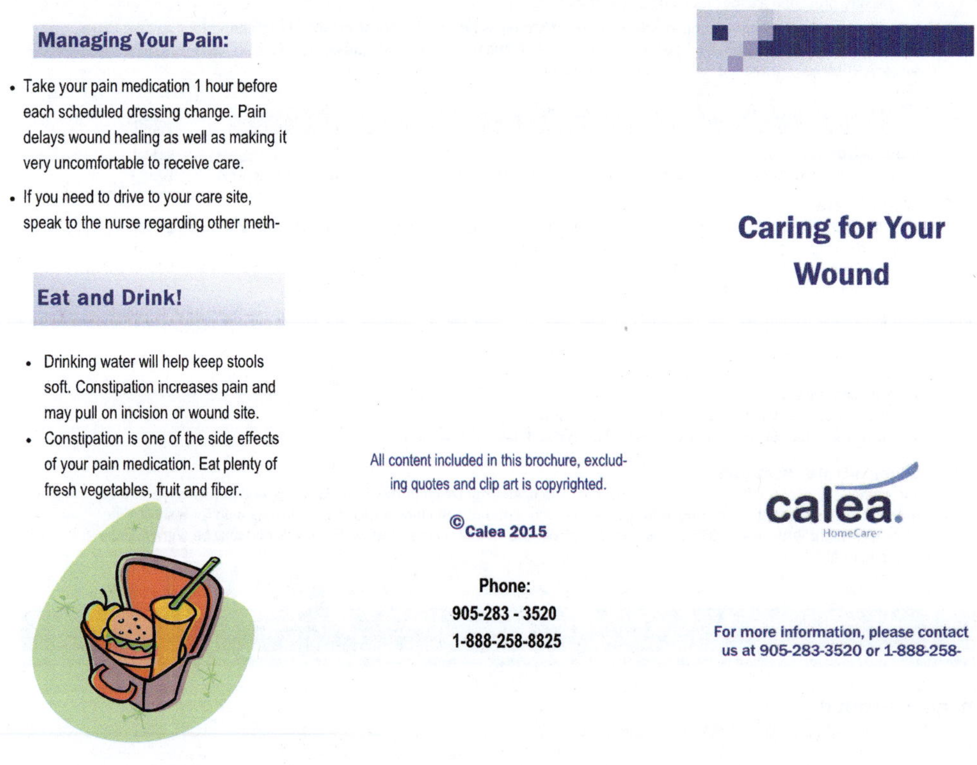

Managing Your Pain:

- Take your pain medication 1 hour before each scheduled dressing change. Pain delays wound healing as well as making it very uncomfortable to receive care.
- If you need to drive to your care site, speak to the nurse regarding other meth-

Eat and Drink!

- Drinking water will help keep stools soft. Constipation increases pain and may pull on incision or wound site.
- Constipation is one of the side effects of your pain medication. Eat plenty of fresh vegetables, fruit and fiber.

All content included in this brochure, excluding quotes and clip art is copyrighted.

©Calea 2015

Phone:
905-283 - 3520
1-888-258-8825

Caring for Your Wound

calea. HomeCare

For more information, please contact us at 905-283-3520 or 1-888-258-

Image 6 Calea Postop teaching handout.

What is a Pilonidal Abscess:

"A Pilonidal (Abscess, Cyst, Sinus, Dimple) is an abscess in the natal cleft (more commonly referred to as your butt crack) that tends to become infected and cause intense pain and drainage. Sometimes a Pilonidal contains hair and sometimes they don't" (www.pilonidal.org)

Please bring the following to each visit:

- Wear loose pants or bring a pair of boxer shorts to change into following your treatment. If wearing hip-hop jeans, place the belt above the area where your wound is.
- Razor—bikini line razors are the best. (e.g. Schick™ or Noxema™, etc.)
- Day-time incontinent briefs or panty-liners

Supply Care Kit:

- Dressing supplies will be provided to you along with instructions for use.
- Inform the nurse when your supplies are low.

Personal Hygiene:

- Your dressing needs to be kept clean AND dry all the time to protect your wound
- If your dressing gets soiled or wet, Change it.
- Clean the wound as shown, pat the area dry and apply a new clean dressing
- Shower after each bowel movement (if possible) and at least daily
- Remove soiled dressing before showering or Sitz bath.
- Use a hand held sprayer to gently flush out the inside of the wound and direct soap, shampoo and loose hair away from open area.
- Baths are not permitted. If your doctor has recommended a Sitz bath, take one following a bowel movement. A Sitz bath consists of 2 – 3 inches of warm water in the bath tub. Sit in the tub with knees to chest and stay for no longer than 5 min. or purchase a Sitz Bath Kit.
- Pat area and wound dry before applying dressing as demonstrated by the nurse.

Activities:

- Refrain from picking up heavy objects (e.g. young child) for at least the first week following surgery.
- It is important to change positions frequently to relieve stress on the area.
- Sit for short periods: get up, slowly walk around and stand.
- It is recommended NOT to drive a car for the first 5 – 7 days following surgery
- Gentle yoga will help manage pain as well as support wound healing. **NOTE:** Gentle yoga only.
- Check with your doctor before resuming any physical activities

At Home:

- If at all possible, attach a hand-held shower head to your shower stall.
- Purchase a fragrance-free liquid soap (like Spectro-Gel™) for cleansing in the shower.
- Keep moistened towelettes near the toilet ; these may be used for extra-clean up after each bowel movement

Image 6—cont'd

9. CONCLUSION

PSD requires all the components of wound bed preparation to optimize healing. While management of localized and deep/surrounding tissue infection is a vital component of the care for PSD, the challenges to success are complex. Effective treatment of PSD requires a comprehensive assessment and management plan that incorporates local, systemic, and patient factors. Appropriate surgical intervention, including newer surgical techniques (especially with off-midline incisions) have been associated with improved outcomes, and should be considered. Teaching the patient self-care strategies for the wound and periwound environment and to make life-style changes engages and empowers them to be successful, and decreases their risk of recurrence.

REFERENCES

1. Mayo OH. *Observations on injuries and diseases of the rectum.* London: Burgess and Hill; 1833. pp. 45–46.
2. Anderson AW. Hair extracted from an ulcer. *Boston Med Surg J* 1847;**36**:74–6.
3. Hodges RM. Pilonidal sinus. *Boston Med Surg J* 1880;**103**:485–6.

4. Eryilmaz R, Okan I, Ozkan OV, et al. Interdigital pilonidal sinus: a case report and literature review. *Dermatol Surg* 2012;**38**:1400–3.
5. Eryilmaz R, Sahin M, Okan I, et al. Umbilical pilonidal sinus disease: predisposing factors and treatment. *World J Surg* 2005;**29**(9):1158–60.
6. Lahiri R, Mullen R, Ashton MA, et al. Pilonidal abcess in the breast: a case report. *J Surg Case Rep* 2014;**2014**(6):rju061. http://dx.doi.org/10.1093/jscr/rju061. Published online 2014 Jun 20.
7. Plewig G, Kligman AM. *Acne: morphogenesis and treatment*. Berlin: Springer-Verlag; 1975. pp. 192–193.
8. Vasanth V, Chandrashekar BS. *Follicular occlusion tetrad. Ind Dermatol Online J* 2014;**5**(4):491–3. http://www.idoj.in. IP: 184.69.53.98 [accessed 12.07.16].
9. Hartschuh W. Chapter 5.4. Pilonidal Sinus. In: Arnold W, Ganzer U, Herold A, Lehur PA, Matzel KE, O'Connell PR, editors. *Coloproctology (European Manual of Medicine)*. Springer Science & Business Media. Verlag Berlin Heidelberg; 2008. p. 77.
10. Bascom J. Pilonidal disease: origin from follicles of hairs and results of follicle removal as treatment. *Surgery* 1980;**87**:567–72.
11. von Laffert M, Stadie V, Ulrich J, et al. Morphology of Pilonidal Sinus Disease: some evidence of its being a unilocalized type of hidradenitis suppurativa. *Dermatology* 2012;**223**:349–55.
12. Karydakis GE. Easy and successful treatment of pilonidal sinus after explanation of its causative process. *Aust N Z J Surg* 1992;**62**:385–9.
13. Doll D, Friederichs J, Dettmann H, Boulesteix A-L, Duesel W, Petersen S. Time and rate of sinus formation in pilonidal sinus disease. *Int J Color Dis* 2008;**23**(4):359–64.
14. Brearley R. Pilonidal Sinus: A new theory of origin. *Br J Surg* 1955;**43**:62–6.
15. Khanna A, Rombeau JL. Pilonidal disease. *Clinics Colon Rectal Surg* 2011;**24**(1):46–53.
16. Jones DJ. ABC of colorectal diseases. Pilonidal sinus. *BMJ* 1992;**305**:410–2.
17. Lee ACW, Kwong NS, Wong YC. *Management of sacral dimples detected on routine newborn examination: a case series and review. HK J Paediatr* 2007;**12**:93–5. http://www.hkjpaed.org/pdf/2007;12;93-95.pdf [accessed 27.08.16].
18. Thompson MR, Senapati A, Kitchen P. Simple day-case surgery for pilonidal sinus disease. *Br J Surg* 2011;**98**(2):198–209.
19. Akinci OF, Bozer M, Uzunköy A, Düzgün SA, Coşkun A. Incidence and aetiological factors in pilonidal sinus among Turkish soldiers. *Eur J Surg* 1999;**165**(4):339–42.
20. Hull TL, Wu J. Pilonidal disease. *Surg Clin North Am* 2002;**82**:1169–85.
21. Søndenaa K, Andersen E, Nesvik I, Søreide JA. Patient characteristics and symptoms in chronic pilonidal sinus disease. *Int J Colorectal Dis* 1995;**10**(1):39–42.
22. Fitzpatrick EB, et al. Pilonidal disease in a military population: how far have we really come? *Am J Surg* 2014;**207**:907–14.
23. de Parades V, Bouchard M, Janier M, Berger A. Pilonidal sinus disease. *J Vis Surg* 2013;**150**(4):237–47.
24. Harlak A, Mentes O, Kilic S, Coskun K, Duman K, Yilmaz F. Sacrococcygeal pilonidal disease: analysis of previously proposed risk factors. *Clinics (Sao Paulo, Brazil)* 2010;**65**(2):125–31.
25. Leak K, Widgerow A, Johnson S. Pilonidal sinus (Jeep's disease): an improved model of care. *Wounds UK* 2011;**7**(2):72–7.
26. Bolandparvaz S, Moghadam Dizaj P, Salahi R, Paydar S, Bananzadeh M, Abbasi HR, et al. Evaluation of the risk factors of pilonidal sinus; A single centre experience. *Turk J Gastroenterol* 2012;**23**(5):535–7.
27. Doll D, Matevossian E, Wietelmann K, Evers T, Kriner M, Petersen S. Family history of pilonidal sinus predisposes to earlier onset of disease and a 50% long-term recurrence rate. *Dis Colon Rectum* 2009;**52**(9):1610–5.
28. Duman K, Mentes O, Kilic S, Coskun K, Duman K, Yilmaz F. Sacrococcygeal pilonidal disease: analysis of previously proposed risk factors. *Clinics* 2010;**65**(92):125–31.
29. Bendewald F, Cima R. Pilonidal disease. *Clinics Col Rec Surg* 2007;**20**(2):86–95.
30. Çubukçu A, Gönüllü NN, Paksoy M, Alponat A, Kuru M, Özbay O. The role of obesity on the reoccurrence of pilonidal sinus disease in patients, who were treated by excision and Limberg flap transposition. *Int J Colorectal Dis* 2000;**15**:173–5.
31. Çubukçu A, Çarkman S, Gönüllü NN, Alponat A, Kayabasi B, Eyüboglu E. Lack of evidence that obesity is a cause of pilonidal sinus disease. *Eur J Surg* 2001;**167**:297–8.

32. Franckowiak JJ, Jackman RJ. The etiology of pilonidal sinus. *Dis Colon Rectum* 1962;**5**:28–36.
33. Irkoren S, Sivrioglu N, Karaca H. Rare complication after VAC therapy in the treatment of tissue defect due to surgical excision of pilonidal sinus. *J WOCN* 2013;**40**(6):641–3.
34. Chamberlain JW, Vawter GF. The congenital origin of pilonidal sinus. *J Pediatr Surg* 1974;**9**:441–4.
35. Doll D, Friederichs J, Düsel W, Fend F, Petersen S. Surgery for asymptomatic pilonidal sinus disease. *Int J Colorectal Dis* 2008;**23**:839–44.
36. Aysan E, Ilhan M, Bektas H, Kaya EA, Sam B, Buyukpinarbasili N, Muslumanoglu M. Prevalence of sacrococcygeal pilonidal sinus as a silent disease. *Surg Today* 2013;**43**:1286–9.
37. Kooistra HP. Pilonidal sinuses: review of the literature and report of three hundred fifty cases. *Am J Surg* 1942;**55**(1):3–17.
38. Treatments for an anal fistula. NHS choices: your health, your choices. http://www.nhs.uk/Conditions/Anal-fistula/Pages/Treatment.aspx [accessed 05.09.16].
39. Harris C, Holloway S. Development of an evidence-based protocol for care of pilonidal sinus wounds healing by secondary intent using a modified reactive Delphi procedure. Part two: methodology, analysis and results. *Int Wound J* 2012;**9**(2):173–88.
40. Diseases characterized by genital, anal, or perianal ulcers. 2015 Sexually Transmitted Diseases Treatment Guidelines. Centers for Disease Control and Prevention. http://www.cdc.gov/std/tg2015/genital-ulcers.htm [accessed 05.09.16].
41. de Zoeten EF, Pasternak BA, Mattei P, Kramer R, Kader HA. Diagnosis and treatment of perianal Crohn disease: NASPGHAN clinical report and consensus statement. *J Pediatr Gastroenterol Nutr* 2013;**57**:401–12.
42. Lee PJ, Raniga S, Biyani DK, Watson AJM, Faragher IG, Frizelle FA. Sacrococcygeal pilonidal disease. *Color Dis* 2008;**10**:639–52.
43. Tezel E. A new classification according to navicular area concept for sacrococcygeal pilonidal disease. *Color Dis* 2007;**9**(6):575–6.
44. Bascom J, Bascom T. Utility of the cleft lift procedure in refractory pilonidal disease. *Am J Surg* 2007;**193**(5):606–9.
45. AL-Khamis A, McCallum I, King PM, Bruce J. Healing by primary versus secondary intention after surgical treatment for pilonidal sinus. *Cochrane Database Syst Rev* 2010;(1)CD006213. http://dx.doi.org/10.1002/14651858.CD006213.pub3.
46. Bascom J. Pilonidal disease: long-term results of follicle removal. *Dis Colon Rectum* 1983;**26**(12):800–7.
47. McCallum IJ, King PM, Bruce J. Healing by primary closure versus open healing after surgery for pilonidal sinus: systematic review and meta-analysis. *BMJ* 2008;**336**(7649):868–71.
48. Mahdy T. Surgical treatment of the pilonidal disease: primary closure or flap reconstruction after excision. *Dis Colon Rectum* 2008;**51**(12):1816–22.
49. Milone M, Fernandez LM, Musella M, Milone F. Safety and efficacy of minimally invasive video-assisted ablation of pilonidal sinus: a randomized clinical trial. *JAMA Surg* 2016;**151**(6):547–53.
50. Meinero P, Stazi A, Carbone A, Fasolini F, Regusci L, La Torre M. Endoscopic pilonidal sinus treatment: a prospective multicentre trial. *Color Dis* 2016;**18**(5):O164–70.
51. Neola B, Capasso S, Caruso L, Falato A, Ferulano GP. Scarless outpatient ablation of pilonidal sinus: a pilot study of a new minimally invasive treatment. *Int Wound J* 2016;**13**:705–8.
52. Anderson JH, Yip CO, Nagabhushan JS, Connelly SJ. Day-case Karydakis flap for pilonidal sinus. *Dis Colon Rectum* 2008;**51**(1):134–8.
53. Ersoy E, Devay AO, Aktimur R, Doganay B, Özdoğan M, Gündoğdu RH. Comparison of the short-term results after Limberg and Karydakis procedures for pilonidal disease: randomized prospective analysis of 100 patients. *Color Dis* 2009;**11**(7):705–10.
54. Bali İ, Aziret M, Sözen S, Emir S, Erdem H, Çetinkünar S, İrkörücü O. Effectiveness of Limberg and Karydakis flap in recurrent pilonidal sinus disease. *Clinics* 2015;**70**(5):350–5.
55. Tezel E, Bostanci H, Anadol AZ, Kurukahvecioglu O. Cleft lift procedure for sacrococcygeal pilonidal disease. *Dis Colon Rectum* 2009;**52**(1):135–9.
56. Bascom J, Bascom T. Failed pilonidal surgery: new paradigm and new operation leading to cures. *Arch Surg* 2002;**137**(10):1146–50.

57. Tavassoli A, Noorshafiee S, Nazarzadeh R. Comparison of excision with primary repair versus Limberg flap. *Int J Surg* 2011;**9**(4):343–6.
58. Fazeli MS, Adel MG, Lebaschi AH. Comparison of outcomes in Z-plasty and delayed healing by secondary intention of the wound after excision of the sacral pilonidal sinus: results of a randomized, clinical trial. *Dis Colon Rectum* 2006;**49**(12):1831–6.
59. Dogra S, Rao PP. Z-plasty a rational and effective technique for pilonidal sinus. *Hell J Surg* 2014;**86**(4):208–11.
60. Schoeller T, Wechselberger G, Otto A, Papp C. Definite surgical treatment of complicated recurrent pilonidal disease with a modified fasciocutaneous VY advancement flap. *Surgery* 1997;**121**(3):258–63.
61. Dýlek ON, Bekereciodlu M. Role of simple VY advancement flap in the treatment of complicated pilonidal sinus. *Eur J Surg* 1998;**164**(12):961–4.
62. Khatri VP, Espinosa MH, Amin AK. Management of recurrent pilonidal sinus by simple VY fasciocutaneous flap. *Dis Colon Rectum* 1994;**37**(12):1232–5.
63. Al-Khayat H, Al-Khayat H, Sadeq A, Groof A, Haider HH, Hayati H, et al. Risk factors for wound complication in pilonidal sinus procedures. *J Am Coll Surg* 2007;**205**(3):439–44.
64. Harris C, Holloway S. Development of an evidence-based protocol for care of pilonidal sinus wounds healing by secondary intent using a modified reactive Delphi procedure. Part one: the literature review. *Int Wound J* 2012;**9**(2):156–72.
65. Harris CL, Laforet K, Sibbald RG, Bishop R. Twelve common mistakes in pilonidal sinus care. *Adv Skin Wound Care* 2012;**25**(7):324–31.
66. Harris C, Sibbald RG, Mufti A, Somayaji R. Pilonidal sinus disease: 10 steps to optimize care. *Adv Skin Wound Care* 2016;**29**(10):469–78.
67. Howell D, Beadle M, Brignell A, Deachman M, Lackenbaruer H, Palozzi L, Preston S, Struthers C, Velji K, Watt-Watson J. Assessment and management of pain RNAO Nursing Best Practice Guidelines Project, Toronto, ON; 2002. p. 71.
68. Bradley L. The lived experience of young adults with chronic pilonidal sinus disease: a phenomenological approach. In: Poster presentation: World Union of Wound Healing Societies Conference, July 8, 2004, Paris; 2004.
69. Price PE, Butterworth RJ, Bale S, Harding KG. Measuring quality of life in patients with granulating wounds. *J Wound Care* 1994;**3**(1):49–50.
70. World Health Organization. WHO's pain ladder. 2009; www.who.int/cancer/palliative/painladder/en/ [accessed 09.07.16].
71. Jensen TS, Madsen CS, Finnerup NB. Pharmacology and treatment of neuropathic pain. *Curr Opin Neurol* 2009;**22**(5):467–74.
72. Saarto T, Wiffen PJ. *Antidepressants for neuropathic pain. Cochrane Database Syst Rev* 2007;(4)CD005454. http://www.cochrane.org/CD005454/SYMPT_antidepressants-for-treating-neuropathic-pain [accessed 09.07.16].
73. Braungart S, Powis M, Sutcliffe JR, Sugarman JD. Improving outcomes in pilonidal sinus disease. *J Ped Surg* 2016;**51**(2):282–4.
74. Mavros MN, Mitsikostas PK, Alexiou VG, Peppas G, Falagas ME. Antimicrobials as an adjunct to pilonidal disease surgery: a systematic review of the literature. *Eur J Clin Microbiol Infect Dis* 2013;**32** (7):851–8. http://dx.doi.org/10.1007/s10096-013-1830-z. Epub 2013 Feb 5.
75. Sibbald RG, Woo K, Ayello EA. Increased bacterial burden and infection: the story of NERDS and STONES. *Adv Skin Wound Care* 2006;**19**(8):447–61.
76. Woo KY, Sibbald RG. A cross-sectional validation study of using NERDS and STONEES to assess bacterial burden. *Ostomy Wound Manage* 2009;**55**(8):40–8.
77. Wolcott RD, Gontcharova V, Sun Y, Zischakau A, Dowd SE. Bacterial diversity in surgical site infections: not just aerobic cocci any more. *J Wound Care* 2009;**18**(8):317–23.
78. Woo KY, Kwong EW, Jimenez C, Bishop R. Topical agents and dressings for pilonidal sinus wound healing by secondary intention: a scoping review. *Surg Technol Int* 2015;**26**:57–63.
79. Brook I. Microbiology of infected pilonidal sinuses. *J Clin Path* 1989;**42**(11):1140–2.
80. Marks J, Harding KG, Hughes LE, Ribeiro CD. Pilonidal sinus excision—healing by open granulation. *Br J Surg* 1985;**72**(8):447–61.
81. Ardelt M, Dittmar Y, Kocijan R, Rödel J, Schulz B, Scheuelein H, et al. Microbiology of the infected recurrent sacrococcygeal pilonidal sinus. *Int Wound* 2016;**13**(2):231–7. Epub 2014 Apr 24. [accessed 08.09.16].

82. Levine NS, Lindberg RB, Mason AD, Pruitt BA. The quantitative swab culture and smear; a quick, simple method for determining the number of viable aerobic bacteria on open wounds. *J Trauma* 1976;**16**:89–94.
83. NICE surgical site infection prevention and treatment: postoperative cleansing. 2008; https://www.nice.org.uk/guidance/CG74/chapter/1-Guidance#postoperative-phase [accessed 08.09.16].
84. Yamashita Y, Nagae H, Hashimoto I. Ambulatory surgery for pilonidal sinus: tract excision and open treatment followed by at-home irrigation. *J Med Invest* 2016;**63**(3-4):216–8. http://dx.doi.org/10.2152/jmi.63.216.
85. Tejiran T, Abbas MA. Sitz bath: where is the evidence? Scientific basis of a common practice. *Dis Colon Rectum* 2005;**48**(12):336–40.
86. Marza L. Reducing the recurrence of pilonidal sinus disease. *Nurs Times* 2013;**109**(25):22–4.
87. Conroy FJ, Kandamany N, Mahaffey PJ. Laser depilation and hygiene: preventing recurrent pilonidal sinus disease. *J Plast Reconstr Aesthet Surg* 2008;**61**(9):1069–72. Epub 2007 Jul 30.
88. Badawy EA, Kanawati MN. Effect of hair removal by Nd YAG laser on recurrence of pilonidal sinus. *J Eur Acad Dermatol Venereol* 2009;**23**(8):883–6.
89. Lukish JR, Kindelan T, Marmon LM, Pennington M, Norwood C. Laser epilation is a safe and effective therapy for teenagers with pilonidal disease. *J Pediatr Surg* 2009;**44**(1):282–5.
90. Ghnnam WM, Hafez DM. Laser hair removal as adjunct to surgery for pilonidal sinus: our initial experience. *J Cutan Aesthet Surg* 2011;**4**:192–5.
91. Doll D, Krueger CM, Schrank S, Dettmann H, Petersen S, Duesel W. Timeline of recurrence after primary and secondary pilonidal sinus surgery. *Dis Colon Rectum* 2007;**50**:1928–34.
92. Petersen S, Wietelmann K, Evers T, Hűser N, Matevossian E, Doll D. Long-term effects of postoperative razor epilation in pilonidal sinus disease. *Dis Colon Rectum* 2009;**52**:131–4.
93. Borkowski S. G tube care: managing hypergranulation tissue. *Nursing* 2005;**35**(8):24.
94. Torkington J. Pilonidal sinus disease. In: Téot L, Banwell P, Ziegler UE, editors. *Surgery in wounds*. Berlin and Heidelberg: Springer-Verlag GmbH & Co. K; 2004. p. 317–24.
95. Widgerow AD, Leak K. *Hypergranulation tissue: evolution, control and potential elimination. Wound Healing South Africa* 2010;**3**(2):7–9. http://www.woundhealingsa.co.za/index.php/WHSA/article/view/87 [accessed 07.07.16].
96. Murphy S, Pwell G. Pilonidal sinus wounds: Successful use of the novel negative pressure wound therapy device PICO™. *Wounds UK* 2013;**9**(4):80–3.
97. Biter LU, Beck GM, Mannaerts GH, Stok MM, van der Ham AC, Grotenhuis BA. The use of negative-pressure wound therapy in pilonidal sinus disease: a randomized controlled trial comparing negative-pressure wound therapy versus standard open wound care after surgical excision. *Dis Colon Rectum* 2014;**57**(12):1406–11.

FURTHER READING

1. Ince M, Atilgan H, Ozcelik F, Arslan E. *Smoking and lipid profile in pilonidal sinus patients. Eur J Gen Med* 2014;**11**(2):90–3. http://sabad.dergipark.gov.tr/download/article-file/61206 [accessed 25.09.16].
2. Bowling FL, Stickings DS, Edwards-Jones V, Armstrong DG, Boulton AJ. Hydrodebridement of wounds: effectiveness in reducing wound bacterial contamination and potential for air bacterial contamination. *J Foot Ankle Res* 2009;**2**(1):1.

CHAPTER 3

The Microbiome of Diabetic Foot Ulcers and the Role of Biofilms

M. Malone
High Risk Foot Service, Liverpool Hospital, South West Sydney LHD, Sydney, Australia and Liverpool Diabetes Collaborative Research Unit, Ingham Institute

1. DIABETIC FOOT INFECTION

The most common pathway leading to a foot infection in a person with diabetes is through a physical break in the protective barrier of the skin, in the form of a diabetic foot ulcer (DFU).[1] Once the skin is breached, a DFU exposes the underlying soft tissues to potential bacterial colonization with infection arising if certain conditions favor bacterial replication. Further contiguous extension to deeper structures may produce scenarios where failure in controlling the spread of infection leads to extensive damage of host tissue and bone, and this in part may be driven by several aspects of altered immunologic function.[2] Of particular focus is the reduction in polymorphonuclear leukocyte response to bacterial infection and the role of hyperglycemia, all being extensively reviewed.[3] This scenario optimizes the pathway to lower extremity amputation in a person with diabetes.

Bacteria that contaminate and colonize wounds likely originate from the surrounding skin flora but other sources include the environment and endogenous mucous membranes, such as the gastrointestinal tract or nares.[4] One reason for this is that DFUs present an ideal environment for harboring microorganisms since they offer a warm, moist, nutritive home, especially if devitalized tissue is present in the wound bed.[4] The longer a wound remains open, the greater the chances of a more diverse and abundant bacterial colonization, with the type, depth, location, level of perfusion and the efficacy of the host immune response dictating the niche of colonizing bacteria.[4,5]

1.1 Colonization Versus Infection?

Bacterial colonization of wounds versus infection is an area that must be greatly appreciated by treating clinicians, so that the appropriate use of antimicrobials and adjunct therapies can be made effectively. While all wounds contain bacteria, colonization refers to a specific scenario where bacteria are presently multiplying but the sum of their actions are not enough to elicit an immune response.[4] Diabetic Foot Infections (DFIs) have been

The Microbiology of Skin, Soft Tissue, Bone and Joint Infections
ISSN 2451-9006, http://dx.doi.org/10.1016/B978-0-12-811079-9.00003-3

described as occurring when bacterial organisms proliferate within a wound (i.e., infra-malleolar) and, in the course of doing so, cause substantial tissue damage that induces a host response accompanied by inflammation, which is clinical infection.[6] For this reason, the diagnosis of DFI has been promoted by expert groups as a "clinical diagnosis" using more than two clinical signs of infection: inflammation, erythema, local tenderness or pain, warmth and purulent discharge.[6]

In some people with diabetes, the overt clinical signs of infection are diminished or absent, and this may be due to the failure to exhibit an inflammatory response[7] with reports identifying that people with diabetes and infected chronic wounds express erythema less frequently than people without diabetes.[8] This has led to a clinical perspective that some chronic DFUs may have "secondary signs" of infection that include but are not limited to malodor, delayed wound closure and poor quality wound bed tissue.[7,9]

The fine line between colonization and infection can be clinically challenging and some clinicians have adopted more quantitative measures to differentiate potential "healthy" colonization from pathogenic infection by relying on the density of bacteria present per gram of tissue. Greater than 10^5 colony forming units (cfu) of bacteria per gram of tissue has been widely used as a key indicator of potential "bioburden" as the causative factor associated with delayed wound healing. This numerical indicator is based largely on early evidence from various wound aetiologies[10] and further incorporated by others,[11] yet controversy persists over whether a burden of $>10^5$ cfu of bacteria per gram of tissue is required to cause wound infection.

Kingsley[12] proposed a wound infection continuum model that placed an emphasis on the progression from colonization of bacteria within a wound through to infection (Fig. 1).[12] An important component of the wound continuum model is the concept of "critical colonization" which refers to the multiplication of organisms within a wound without invasion or interfering with wound healing. While the concept of critical colonization is still the center of much debate, it is often used by clinicians to explain delayed wound healing in the absence of any overt clinical signs of infection and other wound-delaying variables. This concept "if true" may be of importance for clinicians, as chronic wounds with critical colonization may benefit from local and/or topical treatments such as antimicrobial wound dressings and wound debridement, rather than systemic management with the use of antibiotics; however, no quantifiable data exists to prove or disprove this theory.

1.2 How to Obtain Adequate Cultures for Infected DFUs

The current method to identify potential pathogens of infection from DFUs is by obtaining a wound culture following the removal of devitalized tissue (either by debridement or curettage) after cleansing. The most widely available method employed by many clinicians is the use of a cotton or nylon swab of the superficial tissue using the Levine

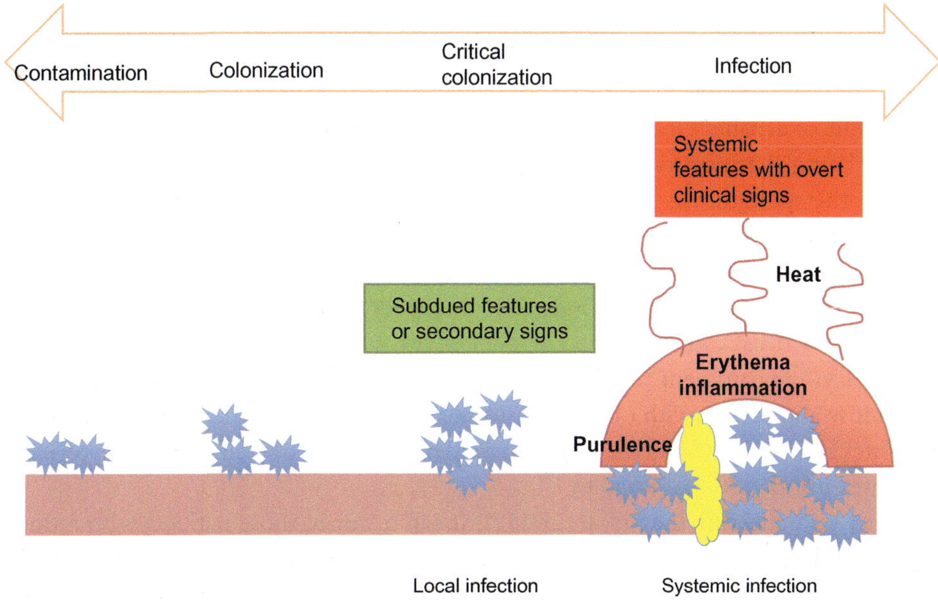

Fig. 1 A wound infection continuum model.

technique[13]; however, the concordance between superficial swabbing using the Levine technique and tissue biopsy has been reported at 78% providing only a "reasonably accurate measurement"[13] and much debate still surrounds this collection method.

The Infectious Disease Society of America (IDSA) currently propose that tissue samples are the gold standard method of collection for identifying potential bacterial pathogens from infected DFUs, as they yield the most accurate results.[6] This is exemplified by recent studies employing molecular DNA-based techniques where tissue biopsy samples from both the superficial and deep tissues of a DFU identified that the pathogens of infection were from microorganisms residing in deeper tissues.[14–16] In all instances the pathogens of infection were situated within deeper tissues and were typically fastidious aerobes along with anaerobes that were more frequently underrecognized from superficial cultures. Importantly, this supports the requirement in both a clinical scenario and for studies exploring the microbiome of wounds to utilize samples of tissue that capture both the superficial and deeper tissues through the use of a punch biopsy.

The importance in managing DFIs is underpinned by the requirement to identify any pathogen/s of infection so that antimicrobial therapy can be directed. Traditional culture-dependent methods have been utilized to identify planktonic organisms subjected to controlled laboratory conditions. The limitations of such are that culture-dependent methods select for species that flourish under the controlled conditions of the diagnostic microbiology lab and this may not necessarily reflect the most abundant or clinically important organisms in DFIs.[17]

1.3 Cultivation-Based Approaches to Defining the Bacteriology of Infected DFUs

Current culture-dependent studies on the bacteriology of DFIs have widely reported the involvement of aerobic Gram-positive cocci, mainly Staphylococcal and Streptococcal species as being the most common etiological agents that acutely infect DFUs.[18,19] In particular, the spotlight has centered on the well-known pathogen *Staphylococcus aureus* and its role in DFI has been well reported.[1,18–20]

In two large culture-dependent studies on the bacteriology of infected DFUs[18,19] samples from over 1266 patients indicated the predominant pathogens of infection as being Staphylococcal and Streptococcal species with *Enterococcus* spp. and *Corynebacterium* spp. also playing prominent roles. In both studies, anaerobes were generally reported as being of low abundance with one of the studies identifying the ratio of aerobe to anaerobe isolates being ~7:1[18] and the other study identifying the overall percentage of anaerobe isolates in only 28% of samples.[19]

Further complicating the picture of DFI, it can occur in those patients with reduced immune-mediated responses, whereby the potential involvement of low-virulence colonizers of devitalized tissue or bone such as coagulase-negative *Staphylococci* and *Corynebacterium* spp. may assume a more pathogenic role.[21] Additionally, while it is generally regarded that DFUs of short duration presenting with acute infection are monomicrobial,[1] chronic infected DFUs seem to harbor a much more complex polymicrobial flora including *Pseudomonas* spp., *Enterococci*, Enterobacteriaceae and other obligate anaerobes.[22]

1.4 Current Studies Employing Molecular Microbiology for Diabetic Foot Infections

The advent of new molecular DNA-based techniques that are culture-independent have identified limitations of traditional culture-dependent methods, acknowledging their capacity to identify a limited number of known culturable bacteria (<1%).[23]

DNA-based techniques, supported by ever-growing gene reference libraries, allow for the characterization of microbial communities or microbiomes that encompass the sum of all microorganisms residing on and within the host—bacterial, fungal, and viral.[24]

Characterizing the microbiome of infected and uninfected DFUs using molecular methods is in its infancy, and only four studies with comparable data have been published in recent years[5,25–27] (Fig. 2).

In addition to the limited work in the area of DFUs, composite evidence by authors employing molecular techniques exist for chronic wounds, where studies have pooled multiple wound etiologies to provide an overview of the chronic wound microbiome.[28–31] Most chronic wound microbiome studies to date have employed pyrosequencing centered approaches that amplify and sequence the small subunit ribosomal

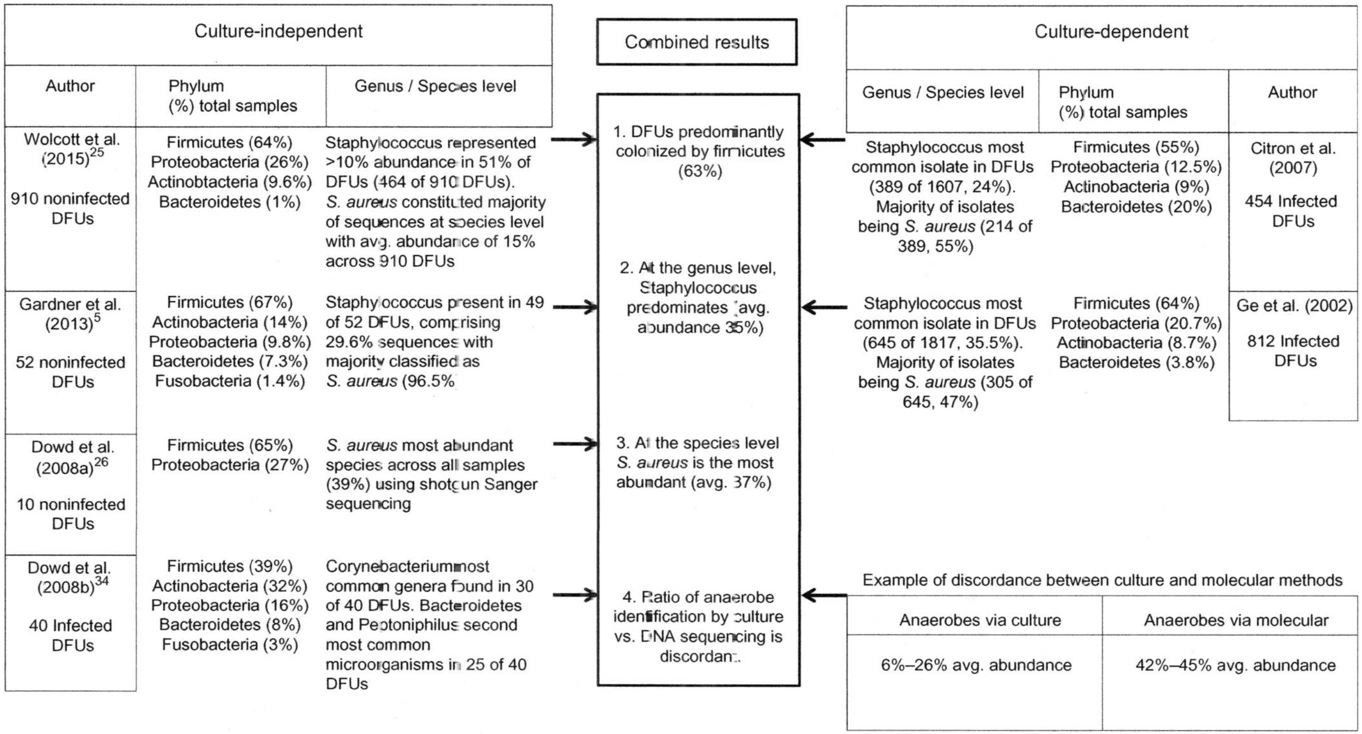

Fig. 2 Microbiome of infected and uninfected diabetic foot ulcers. *Source: used with permission from Malone M, Gosbell IB, Dickson HG, Vickery K, Espedido BA, Jensen SO. Can molecular DNA-based techniques unravel the truth about diabetic foot infections?* Diabetes Metab Res Rev *2017;* ***33****:e2834. http://dx.doi.org/10.1002/dmrr.2834. Wiley.*

RNA (16S rRNA) gene, a highly conserved gene present in all prokaryotes (bacteria) but not eukaryotes (humans). This has revealed a vastly more complex array of bacterial communities than those identified by traditional culture-dependent methods. The 16S rRNA gene is an ideal target for bacterial analysis given that it possesses nine hypervariable regions that have considerable sequence diversity between bacterial taxa.[32] The hypervariable regions are bordered by stretches of sequences that are highly conserved between bacteria, ideal for designing universal and species-specific primers to amplify the hypervariable regions by Polymerase Chain Reaction (PCR).[33]

The first report in the literature on the microbiome of DFUs was undertaken by Dowd et al.[34] reporting on 10 chronic DFUs using multiple genomic approaches that included: partial ribosomal amplification and pyrosequencing (PRAPS), full ribosomal amplification, cloning and Sanger sequencing (FRACS), density gel electrophoresis (DGGE), and Sanger sequencing (PRADS). Identification of species was undertaken using operational taxonomic units (OTUs); these are clusters of single DNA sequences grouped together based on their similarity and used to define microorganisms to a species or genus-level. DNA reads from each wound were compiled to create a pooled DNA data set from multiple DFUs. Facultative and strict anaerobic Gram-positive cocci formed the majority of sequences with genus-level identification highlighting the predominance of *Staphylococcus* spp. (PRAPS = 10,874 of 36,508 sequences, 30%; FRACS = 70 of 178 sequences, 39%) in addition to *Peptoniphilus*, *Anaerococcus*, *Rhodopseudomonas*, *Enterococcus*, *Veillonella*, *Bacterioides*, *Clostridium*, and *Finegoldia* spp., each contributing to >5% of the microbial diversity.

The second study to explore the DFU microbiome was undertaken by Dowd et al.[26] who employed bacterial tag-encoded FLX amplicon pyrosequencing (bTEFAP) to sample 40 chronically infected DFUs from a range of locations on the foot and ankle. Dowd et al.[26] hypothesized that a single major pathogen would be associated with all wounds; therefore DNA reads for each DFU were reported individually and not pooled. This allowed the compilation of community profiles for each DFU, including the accurate identification of the number of samples of each genera that were detected, and the average percentage each genus contributed to its positive sample.

Results identified that each DFU contained a rich diversity of microorganisms with *Corynebacterium* (30 of 40 DFUs, avg abundance = 14.4%), *Bacteroides* (25 of 40 DFUs, avg abundance = 24.2%), *Peptoniphilus* (25 of 40 DFUs, avg abundance = 13.6%), *Fingoldia* spp. (23 of 40 DFUs, avg abundance = 6.7%) *Anaerococcus* (22 of 40 DFUs, avg abundance = 7.7%), and *Streptococcus* (21 of 40 DFUs, avg abundance = 36.5%) being present in all samples. Surprisingly, *Staphylococcus* was only present in 13 of 40 (avg abundance = 8.3%) DFUs compared to a previous study by the same authors that identified *Staphylococcus* as the major bacterial genus in 10 infected DFUs.[34] The location of ulcers originating from multiple sites of the foot and ankle may have increased the heterogeneity observed between samples in the latter study. Some DFUs may have been purely ischemic or neuroischemic in origin, increasing the likelihood of the presence of devitalized, hypoxic tissue and favoritism towards low-virulent colonizers and anaerobic microorganisms.

No single genus of bacteria was present in all 40 DFUs and this led Dowd et al.[26] to hypothesize that the microbiome of DFUs (and chronic wounds) is dominated by multiple species that cooperate in a biofilm, a concept Dowd et al. "coin" as functionally equivalent pathogroups (FEP). This centers on genotypically distinct bacteria working cooperatively to induce pathogenesis, similar to what is observed from a known single pathogen, such as *S. aureus*; however, there is currently limited data available to support this.

Gardner and colleagues[5] profiled the microbiome of 52 individuals with noninfected DFUs using 16S rRNA gene sequence analysis and were the first to restrict the sampling of patients to a homogenous sample of DFUs (neuropathic DFUs only). The group proposed that previous studies within the area[26,28,29,34] were tainted by heterogeneous sampling (e.g., pooled samples of differing chronic wounds) and of all three DFU etiology types (neuropathic, ischemic, and neuroischemic) and, as such, these pathophysiologically distinct lesions likely harbored their own unique microenvironments and thus microbiota.

In characterizing the microbiome of 52 noninfected DFUs, a total of 13 phyla were reported from 300,660 DNA sequences with the majority being classified into Firmicutes (67%), Actinobacteria (14%), Proteobacteria (9.8%), Bacteroidetes (7.3%), and Fusobacteria (1.4%). *Staphylococcus* was identified as the most common and abundant genus in 49 of 52 DFU samples (DNA sequences—88,995 of 300,660, abundance—29.6%). At a species-level the majority of sequences belonged to the common pathogen *S. aureus* (*S. aureus* sequences = 85,880 of 88,995, 96.5%), an unsurprising finding considering the highly documented role of this microorganism in diabetes-related foot infection.[6]

Further analysis of microbial diversity in 52 DFUs reported on average 30 OTUs per DFU (range 7–64) in comparison to culture-dependent analysis that detected on average 4 OTUs per DFU ($P < .0001$). Comparisons of the relative bacterial abundances within each DFU using culture-dependent analysis identified the overestimation in the abundance of *Staphylococcus* spp. (0.47 vs 0.32, $P = .0001$) and underestimation of anaerobes (0.11 vs 0.18, $P = .0063$) in comparison to 16S rRNA gene sequence analysis. By culture, anaerobes were identified as the predominant organisms in only 6 of 52 DFUs (12%), a finding consistent with the known limitations of culture-dependent methods, particularly in the identification of slow-growing, fastidious anaerobic organisms.[5] In contrast, 16S rRNA gene sequence analysis identified twice this amount (12 DFUs, 23%); this is of importance at a clinical level where the role of anaerobes in DFI and wounds is often underestimated. Their significance as highly virulent pathogens, detrimental to the wound environment or as pathogens of DFI, is only becoming appreciated through advances in new molecular techniques that are culture-independent.

Gardner et al.[5] also depicted bacterial community structure using a statistical approach of Euclidean distances to determine how similar DFU attributes were to the indicators of wound bioburden (microbial load, diversity, and presence of pathogens). The partitioning of data identified three clusters, which Gardener et al.[5] referred to as EUC1, EUC2, and EUC3. EUC1 contained significantly greater OTU richness ($P=.006$), diversity ($P=.02$), and bacterial load ($P=.02$), whereas EUC2 contained a greater abundance of *Staphylococcus* and lower abundance of anaerobes ($P=.0003$), and EUC3 contained the highest abundance of *Streptococcus* ($P=.0002$).

Euclidean clusters were also mapped against clinical markers that included HbA1c, mean tissue oxygenation, ulcer duration, ulcer depth, ulcer surface area, and necrotic tissue. EUC1 containing greater OTU richness, diversity, and bacterial load, were associated with DFUs of longer duration ($P=.02$) and increased ulcer depth ($P=.01$). The EUC1 cluster also contained a greater abundance of anaerobes ($P=.01$) and Proteobacteria ($P=.005$), a likely association with wound longevity and deeper tissue involvement. EUC2 and EUC3 had the highest proportions of both *Staphylococci* and *Streptococci* with laboratory markers from these participants recording the highest levels of glycosylated hemoglobin (HbA1c %) (Median EUC 2 HbA1c=9.2% versus median EUC 3 HbA1c=9.4%).

Assumptions on infections in people with diabetes have repeatedly associated *Staphylococci* and *Streptococci* as major pathogens, noting higher colonization rates in those with suboptimal glycemic control.[35] A plethora of studies have also identified that hyperglycemia causes immunosuppression, and a reduction in glucose by a variety of means reverses immune function deficits.[36–38] However, no evidence exists identifying direct links between hyperglycemia, infection and potential deficits in the immune response to specific microorganisms.[6,39]

Gardner et al.[5] invested a significant proportion of their analysis on undertaking a statistical approach to partitioning DFUs into meaningful clusters based on their associations to the three dimensions of bioburden (i.e., microbial load, microbial diversity, and presence of pathogens). Interpretation and validation of cluster analysis occurred through a silhouette score (SI), a method used to report the graphical representation of how well each object lies within its cluster in comparison to other clusters. An average SI score of 0.42 was reported by Gardner et al.,[5] but an SI score of <0.5 signifies only modest support for the proposed cluster algorithms.[40,41] Wu et al.[42], who explored gut microbial enterotypes using molecular approaches, recommended using a high SI threshold (≥0.75) for validating clusters.

The statistical significance of partitioning 52 DFUs into similar clusters given the moderate SI score should be aired on the side of caution, as a low SI score suggests that clustering could be due to chance. Further studies aiming to increase cluster validity through higher SI scores may be required in supporting the partitioning of the DFU microbiome to clinical indicators similar to those observed by Gardner et al.

Wolcott et al.[25] retrospectively reported 910 noninfected DFUs using 16S rRNA pyrosequencing and currently represents the largest body of evidence surveying the microbiome of DFUs. Wound samples were collected from superficial debridement material obtained from the wound bed and relative abundances were calculated from species-level OTUs constituting >10% of all DNA reads per sample. Given the large sample size from multiple wound etiologies, individual wound microbiomes were not reported and DNA sequences were pooled. Of 910 DFUs, the most abundant phylum was the Firmicutes (64%) followed by Proteobacteria (26%), Actinobacteria (9.6%), and Bacteroidetes (1%). *Staphylococcus* (31%) was the most abundant OTU with *S. aureus* (48%) and *S. epidermidis* (35%) predominating at the species-level.

In review of the studies employing amplification and sequence analysis of the 16S rRNA gene to characterize the microorganisms involved in DFI, few have sampled participants with overt clinical signs of infection. It is generally regarded that clinically uninfected wounds should not be treated with antimicrobials and therefore should not be cultured.[6,43] The clinical relevance of culturing uninfected DFUs to characterize the microbiome is thus debatable, and worth expanding upon further. In patients with diabetes presenting with overt clinical signs of infection, the decision to obtain a culture from a wound to identify a pathogen of infection is clinically justified. It has been well addressed in the literature; however, many patients with diabetes do not display overt clinical signs of infection in the presence of a nonhealing DFU.[7,44]

Some clinical researchers have proposed under these circumstances to utilize secondary signs and symptoms (i.e., malodor, increased exudate, and poor quality tissue) to diagnose infections that may be more pertinent to guiding clinicians facing wounds that are nonhealing in the presence of standard care.[45] Therefore, obtaining cultures to examine the microbiome of patients with "masked" signs of infection in DFUs, that are nonhealing despite standard care, may prove greatly beneficial. With the exception of Gardner et al.,[5] who restricted sampling to a homogenous group of DFUs (neuropathic), all other microbiome studies have included the sampling of different DFU etiologies. The extent to which community diversity is affected through variances in wound etiologies is currently the focus of debate. It is generally accepted that DFUs of ischemic origin may contain a wound bed environment of devitalized, slough tissue. This may perpetuate very different bacteria from those of the healthy vascularized tissue common in neuropathic DFUs.[1,46] Further studies comparing each DFU etiology individually may clarify this matter further.

Lastly, it has been cited in the literature that certain aerobic species are often overestimated as the primary pathogens in DFI. When comparing the evidence on the bacteriology of DFIs through both culture-dependent and -independent approaches, both corroborate *S. aureus* as being the predominant pathogen colonizing DFUs. In tandem with this, reports have also suggested that the abundance and role of anaerobes are greatly underestimated. The significance, however, at a clinical level of identifying additional "hidden" anaerobes that often form part of a polymicrobial infection is yet undefined. Many antimicrobial agents commonly used in the treatment of DFIs have a broad spectrum of activity to cover most anaerobes. Furthermore, no published studies are available reporting improved clinical outcomes when adding additional antianaerobic directed therapy to concurrent regimes.

2. BIOFILMS

During most of the history and development of microbiology, the general understanding of the role microbes play in human health and disease has been that they exist as planktonic or free-floating single-cell organisms. Seminal works by Louis Pasteur and Robert Koch in the mid-to-late 1800s paved the way in the field of microbiology, and laboratories still use the 150-year-old methods developed by these pioneers. These techniques postulate that microbial cells act in a planktonic state: that is, they disperse in a liquid environment.

However, emerging evidence from the previous century, based on microbial studies of aquatic environments and dental plaque, provided insights that microorganisms have a natural tendency to associate surfaces, preferring a sessile lifestyle.[47,48] This early work, which focused predominantly on environmental samples, later provided a platform for the contemporary medical models that we have come to understand as "microbial biofilms." Biofilms are frequently defined based on in vitro observations through methods such as scanning electron microscopy (SEM) or peptide nucleic acid-flourescent in situ hybridization (FISH) (Fig. 3A and B).

The classic definitions often describe biofilms as bacteria attached to surfaces, encapsulated in a self-produced matrix, and tolerant to antimicrobials.[49,50] The biofilm concept and its involvement in human health and disease were first described in the medical literature in 1978,[48] with much debate following. This was followed by an explosion of biofilm analysis in human health and disease in the 1980s, which was driven largely

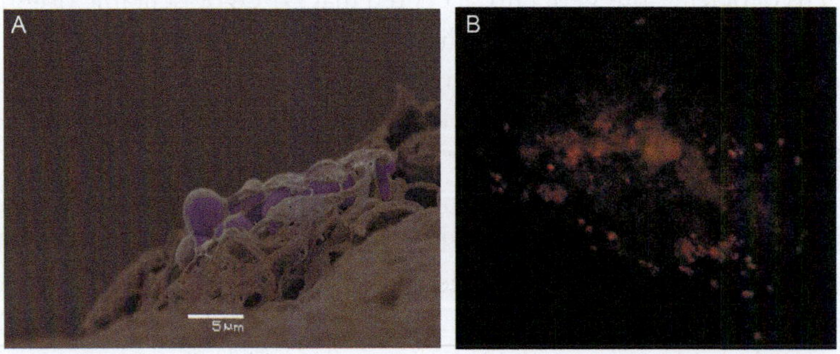

Fig. 3 Scanning electron microscopy (SEM) and peptide nucleic acid fluorescent in situ hybridization (FISH) methods for detection of biofilms (A and B, respectively).

by advancements in emerging technologies and techniques applicable to the study of bacterial populations in situ. Using light and electron microscopy, in combination with specific probes to define cell surface structures, William Costerton (1934–2012), a microbial ecologist, was credited with advancing the understanding of the role biofilms play in human health and disease.

Costerton's early enterprising works were fixed on environmental models, but this quickly transcended to the medical arena with thoughts of "how does a bacterium know whether it is in a urinary catheter or an alpine stream?" hypothesizing that the bacterium would grow as a biofilm on both surfaces regardless.[51] Costerton and colleagues were also the first to propose the role of *Pseudomonas aeruginosa* biofilms in the sputum of cystic fibrosis patients,[52] much to the energetic refutation of medical peers at the time. The concept of biofilms in human health and disease is now universally accepted in tuberculosis,[53] periodontal disease and dental caries,[54] cystic fibrosis,[52,55,56] in-dwelling medical device infections,[57] otitis media and other upper respiratory infections,[58,59] and chronic wounds.[60,61]

Early work by Costerton and colleagues[62] identified that bacteria growing on medical devices existed within a biofilm, and that they exhibited a remarkable tolerance to both host defenses and antimicrobial therapy. A plethora of in vitro biofilm models have elucidated that bacterial biofilms can withstand antimicrobial concentrations 100–1000 times higher than those of planktonic counterparts.[63–65] This would suggest that biofilm cells harbor the capability to switch on multidrug resistance mechanisms that they only express during biofilm growth. If this were true, then mutants expressing the same resistance would have been seen in rapidly multiplying planktonic cultures. This has not been the case, which has therefore led to the riddle of how a biofilm can be resistant to killing by many antimicrobials in the absence of specific resistance mechanisms.[66]

In spite of the wealth of research undertaken to identify biofilm resistance and/or tolerance to antimicrobials, no single causative mechanism has been identified. Instead, it has been suggested that a likely combination of factors contributes to biofilm recalcitrance, with several areas of interest including but not limited to: slow or incomplete permeation of antimicrobials through extracellular polymeric substance (EPS)[63] altered microenvironment and niches within biofilms promoting slow growth rates and adaptive stress response,[67,68] efflux pumps,[69] and the role of "persisters."[70]

2.1 Clinical Significance of Bacterial Biofilms in Chronic Wounds

2.1.1 How Do Biofilms Impede Wound Healing?

The exact mechanisms of biofilm impairment on the healing processes of wounds remain ambiguous. Existing data suggest a chronic wound is kept in either a vicious inflammatory state, or subject to localized low oxygen tensions. The pathways behind this are not clear, but several systemic and local factors may contribute to the occurrence and maintenance of the wound chronicity.

2.1.2 Biofilms Sustain Hyperinflammation

In a review article, Wolcott and colleagues[71] present a detailed hypothesis summarizing that once a biofilm community becomes established, its stubbornness and frequent failure to many treatments propagates hyperinflammation. Specifically, they propose that biofilm phenotype bacteria produce proteases that inhibit and destroy extracellular matrix. In addition, the chronic wound environment also possesses host-derived proteases. Together this may overfill a chronic wound with a proteolytic mix of proteases, elastases, and gelatinases (these are commonly referred to as matrix metalloproteinases, or MMPs). In parallel, biofilm adherence to the wound bed may also potentially inhibit the release of the natural suppressor of MMPs—tissue inhibitors of MMPs (TIMPS). This scenario may therefore sustain a perpetual state of hyperinflammation.

Walcott and colleagues also put forward the potential of biofilms to "bait" the immune system through releasing planktonic bacteria. Walcott and colleagues suggest that the presence of anaerobic bacteria play a key role throughout, releasing a cell wall constituent lipopolysaccharide, a potent inflammatory inducer. In an animal model on cystic fibrosis patients, biofilms forming *P. aeruginosa* were shown to undergo lipopolysaccharide modifications that induced greater inflammatory responses in mice.[72] No human in vivo data exist to support this theory for chronic wounds.

2.1.3 Biofilms May Contribute to Localized Areas of Low Oxygen Tension Within a Wound

Early microelectrode studies of aerobic in vitro biofilm models found discrete areas within biofilm that had significant oxygen depletion.[67] This suggested that areas of biofilm housed microniches favoring differing microorganisms, and may explain how the presence of anaerobes in mixed-species biofilms exist, contribute and cooperate with aerobic neighbors.

Further studies employing microelectrodes with Confocal Laser Scanning Microscopy (CLSM) have identified microdomains with different biochemical environments including alterations in pH and oxygen.[73] Recent data by James and colleagues[68] have provided further evidence to support a concept of localized low oxygen tensions contributing to wound chronicity. Using oxygen microsensors and transcriptomics (examining microbial metabolic activities) to study in situ biofilms, steep oxygen gradients and induced oxygen-limitation stress responses were identified from bacteria. Additionally, through transcriptomics, it was established that metabolic activities of the biofilm and the recruitment of cells that consume oxygen for host defensive processes were the primary pathways of oxygen depletion. Taken collectively, this data supports the concept of a biofilm establishing and maintaining localized low oxygen tensions in a wound, thus contributing to chronicity.

3. SUMMARY

The progression of infection in a person with diabetes can lead to the devastating requirement for a lower extremity amputation. Therefore the clinical identification of infection and the grading of its severity is key. To direct targeted therapy, clinicians need to understand the microorganisms involved in the process, but it is clear that molecular DNA-based approaches identify many "hidden" microorganisms. Both the progress and application of current DNA sequencing continues to rapidly evolve, but thus far we have only been enlightened with a broader view of "which microorganisms are there." The interpretation and any clinical implications of additional bacteria within samples remain unclear, but as DNA-based evidence increases we may be enlightened. Importantly, the concept of biofilms and their involvement as contributors to chronic nonhealing wounds and in the pathogenesis of chronic infection poses a significant shift in therapeutic paradigms. This needs to be approached cautiously, with input from antimicrobial stewardships, wound-care clinicians and the broader medical community.

REFERENCES

1. Lipsky BA, Berendt AR, Deery HG, Embil JM, Joseph WS, Karchmer AW, et al. Diagnosis and treatment of diabetic foot infections. *Clin Infect Dis* 2004;**39**:885–910.
2. Joshi N, Caputo GM, Weitekamp MR, Karchmer AW. Infections in patients with diabetes mellitus. *N Engl J Med* 1999;**341**:1906–12.
3. Geerlings SE, Hoepelman AIM. Immune dysfunction in patients with diabetes mellitus (DM). *FEMS Immunol Med Microbiol* 1999;**26**:259–65.
4. Bowler PG, Armstrong DG. Wound microbiology and associated approaches to wound management. *Clin Microbiol Rev* 2001;**14**:244–69.
5. Gardner SE, Hillis SL, Heilmann K, Segre JA, Grice EA. The neuropathic diabetic foot ulcer microbiome is associated with clinical factors. *Diabetes* 2013;**62**:923–30.
6. Lipsky BA, Berendt AR, Cornia PB, Pile JC, Peters EJG, Armstrong DG, et al. Infectious Diseases Society of America clinical practice guideline for the diagnosis and treatment of diabetic foot infections. *Clin Infect Dis* 2012;**54**:e132–73.
7. Gardner SE, Hillis SL, Frantz RA. Clinical signs of infection in diabetic foot ulcers with high microbial load. *Biol Res Nurs* 2009;**11**:119–28.
8. Gardner SE, Saltzman CL. Diabetes and inflammation in infected chronic wounds. *Wounds* 2005;**7**(8):203–5.
9. Cutting KF, White R. Defined and refined: criteria for identifying wound infection revisited. *Br J Community Nurs* 2004;**9**:S6–S15.
10. Robson MC. Wound infection: a failure of wound healing caused by an imbalance of bacteria. *Surg Clin N Am* 1997;**77**:637–50.
11. Browne AC, Sibbald RG. High bacterial load in asymptomatic diabetic patients with neurotrophic ulcers retards wound healing after application of dermagraft. *Ostomy Wound Manage* 2001;**10**:44–9.
12. Kingsley A. The wound infection continuum and its application to clinical practice. *Ostomy Wound Manage* 2003;**49**:1–7.
13. Gardner SE, Frantz RA, Saltzman CL, Hillis SL, Park H, Scherubel M. Diagnostic validity of three swab techniques for identifying chronic wound infection. *Wound Repair Regen* 2006;**14**:548–57.

14. Sotto A, Lina G, Richard JL, Combescure C, Bourg G, Vidal L, et al. Virulence potential of *Staphylococcus aureus* strains isolated from diabetic foot ulcers: a new paradigm. *Diabetes Care* 2008;**31**(12):2318–24.
15. Imirzalioglu C, Sethi S, Schneider C, Hain T, Chakraborty T, Mayser P, et al. Distinct polymicrobial populations in a chronic foot ulcer with implications for diagnostics and anti-infective therapy. *BMC Res Notes* 2014;**7**:196.
16. Dunyach-Remy C, Schuldiner S, Bayle S, Roig B, Sotto A, Lavigne JP. Polymerase chain reaction-denaturing gradient gel electrophoresis (PCR-DGGE): a promising tool to diagnose bacterial infections in diabetic foot ulcers. *Diabetes Metab* 2014;**40**:476–80.
17. Grice EA, Segre JA. The skin microbiome. *Nat Rev Microbiol* 2011;**9**(4):244–53.
18. MacDonald YGD, Hait H, Lipsky BA, Zasloff M, Holroyd K. Microbiological profile of infected diabetic foot ulcers. *Diabet Med* 2002;**19**:1032–4.
19. Citron DM, Goldstein EJ, Merriam CV, Lipsky BA, Abramson MA. Bacteriology of moderate-to-severe diabetic foot infections and in vitro activity of antimicrobial agents. *J Clin Microbiol* 2007;**45**(9):2819–28.
20. Malone M, Bowling FL, Gannass A, Jude EB, Boulton AJM. Deep wound cultures correlate well with bone biopsy culture in diabetic foot osteomyelitis. *Diabetes Metab Res Rev* 2013;**29**:546–50.
21. Bessman AN, Geiger PJ, Canawati H. Prevalence of Corynebacteria in diabetic foot infections. *Diabetes Care* 1992;**15**:1531–3.
22. Gerding DN. Foot infections in diabetic patients: the role of anaerobes. *Clin Infect Dis* 1995;**20**:S283–8.
23. Grice EA, Kong HH, Renaud G, Young AC, Bouffard GG, Blakesley RW, et al. A diversity profile of the human skin microbiota. *Genome Res* 2008;**18**:1043–50.
24. Peterson J, Garges S, Giovanni M, McInnes P, Wang L, et al. The NIH human microbiome project. *Genome Res* 2009;**19**:2317–23.
25. Wolcott RD, Hanson JD, Rees EJ, Koenig LD, Phillips CD, Wolcott RA, et al. Analysis of the chronic wound microbiota of 2,963 patients by 16S rDNA pyrosequencing. *Wound Repair Regen* 2016;**24**(1):163–74.
26. Dowd SE, Sun WR, McKeehan T, Smith E, Rhoads D. Polymicrobial nature of chronic diabetic foot ulcer biofilm infections determined using bacterial tag encoded FLX amplicon pyrosequencing (bTEFAP). *PLoS One* 2008;**3**:326.
27. Malone M, Gosbell IB, Dickson HG, Vickery K, Espedido BA, Jensen SO. Can molecular DNA-based techniques unravel the truth about diabetic foot infections? *Diabetes Metab Res Rev* 2017;**33**:e2834. http://dx.doi.org/10.1002/dmrr.2834.
28. Price LB, Liu CM, Melendez JH, Frankel YM, Engelthaler D, Aziz M, et al. Community analysis of chronic wound bacteria using 16S rRNA gene-based pyrosequencing: impact of diabetes and antibiotics on chronic wound microbiota. *PLoS One* 2009;**4**:e6462.
29. Rhoads DD, Wolcott RD, Sun Y, Dowd SE. Comparison of culture and molecular identification of bacteria in chronic wounds. *Int J Mol Sci* 2012;**13**:2535–50.
30. Ammons MCB, Morrissey K, Tripet BP, Van Leuven JT, Han A, Lazarus GS, et al. Biochemical association of metabolic profile and microbiome in chronic pressure ulcer wounds. *PLoS One* 2015;**10**:e0126735.
31. Han A, Zenilman JM, Melendez JH, Shirtliff ME, Agostinho A, James G, et al. The importance of a multifaceted approach to characterizing the microbial flora of chronic wounds. *Wound Repair Regen* 2011;**19**:532–41.
32. Van de Peer Y, Chapelle S, De Wachter RA. Quantitative map of nucleotide substitution rates in bacterial rRNA. *Nucleic Acids Res* 1996;**24**:3381–91.
33. Baker GC, Smith JJ, Cowan DA. Review and re-analysis of domain-specific 16S primers. *J Microbiol Methods* 2003;**55**:541–55.
34. Dowd S, Sun Y, Secor P, Rhoads D, Wolcott R, James G, et al. Survey of bacterial diversity in chronic wounds using Pyrosequencing, DGGE, and full ribosome shotgun sequencing. *BMC Microbiol* 2008;**8**:43.
35. Lipsky BA, Pecoraro RE, Chen MS, Koepsell TD. Factors affecting staphylococcal colonization among NIDDM outpatients. *Diabetes Care* 1987;**10**:483–6.

36. Bouter KP, Hoekstra JB, Masurel N, Erkelens DW, Diepersloot RJ. Influence of blood glucose levels on peripheral lymphocytes in patients with diabetes mellitus. *Diabetes Res* 1992;**19**:77–8.
37. von Känel R, Mills PJ, Dimsdale JE. Short-term hyperglycemia induces lymphopenia and lymphocyte subset redistribution. *Life Sci* 2001;**69**:255–62.
38. Clement S, Braithwaite SS, Magee MF, Ahmann A, Smith EP, Schafer RG, et al. Management of diabetes and hyperglycemia in hospitals. *Diabetes Care* 2004;**27**:553–91.
39. Park S, Rich J, Hanses F, Lee JC. Defects in innate immunity predispose C57BL/6J-Leprdb/Leprdb mice to infection by *Staphylococcus aureus*. *Infect Immun* 2009;**77**:1008–14.
40. Rousseeuw PJ. Silhouettes: a graphical aid to the interpretation and validation of cluster analysis. *J Comput Appl Math* 1987;**20**:53–65.
41. Koren O, Knights D, Gonzalez A, Waldron L, Segata N, Knight R, et al. A guide to enterotypes across the human body: meta-analysis of microbial community structures in human microbiome datasets. *PLoS Comput Biol* 2013;**9**:e1002863.
42. Wu GD, Chen J, Hoffmann C, Bittinger K, Chen YY, Keilbaugh SA, et al. Linking long-term dietary patterns with gut microbial enterotypes. *Science* 2011;**334**:105–8.
43. Gardner SE, Haleem A, Jao YL, Hillis SL, Femino JE, Phisitkul P, et al. Cultures of diabetic foot ulcers without clinical signs of infection do not predict outcomes. *Diabetes Care* 2014;**37**:2693–701.
44. Gardner S, Saltzman CL. Diabetes and inflammation in infected chronic wounds. *Wounds* 2005; **17**:203–5.
45. Gardner SE, Troia C, Eastman S, MacDonald M, Buresh K, Healy D. A tool to assess clinical signs and symptoms of localized infection in chronic wounds: development and reliability. *Ostomy Wound Manage* 2001;**47**:40–7.
46. Edmonds ME, Foster AVM. Diabetic foot ulcers. *Br Med J* 2006;**332**:407–10.
47. Zobell CE, Anderson DQ. Observations on the multiplication of bacteria in different volumes of stored sea water and the influence of oxygen tension and solid surfaces. *Biol Bull* 1936;**71**:324–42.
48. Geesey GG, Mutch R, Costerton JW, Green RB. Sessile bacteria: an important component of the microbial population in small mountain streams. *Limnol Oceanogr* 1978;**23**:1214–23.
49. Carpentier B, Cerf O. Biofilms and their consequences, with particular reference to hygiene in the food industry. *J Appl Bacteriol* 1993;**75**:499–511.
50. Costerton JW, Cheng KJ, Geesey GG, Ladd TI, Nickel JC, Dasgupta M, et al. Bacterial biofilms in nature and disease. *Annu Rev Microbiol* 1987;**41**:435–64.
51. Costerton JW, Irvin RT, Cheng KJ, Sutherland IW. The role of bacterial surface structures in pathogenesis. *Crit Rev Microbiol* 1981;**8**:303–38.
52. Lam J, Chan R, Lam K, Costerton JW. Production of mucoid microcolonies by *Pseudomonas aeruginosa* within infected lungs in cystic fibrosis. *Infect Immun* 1980;**28**:546–56.
53. Ojha AK, Baughn AD, Sambandan D, Hsu T, Trivelli X, Guerardel Y. Growth of *Mycobacterium tuberculosis* biofilms containing free mycolic acids and harbouring drug-tolerant bacteria. *Mol Microbiol* 2008;**69**:164–74.
54. Marsh PD, Bradshaw DJ. Dental plaque as a biofilm. *J Ind Microbiol* 1995;**15**:169–75.
55. Costerton JW. Cystic fibrosis pathogenesis and the role of biofilms in persistent infection. *Trends Microbiol* 2001;**9**:50–2.
56. Bjarnsholt T, Jensen PØ, Fiandaca MJ, Pedersen J, Hansen CR, Andersen CB. *Pseudomonas aeruginosa* biofilms in the respiratory tract of cystic fibrosis patients. *Pediatr Pulmonol* 2009;**44**:547–58.
57. Donlan RM. Biofilm formation: a clinically relevant microbiological process. *Clin Infect Dis* 2001;**33**:1387–92.
58. Hall-Stoodley L, Hu FZ, Gieseke A, Nistico L, Nguyen D, Hayes J. Direct detection of bacterial biofilms on the middle-ear mucosa of children with chronic otitis media. *JAMA* 2006;**296**:202–11.
59. Boase S, Foreman A, Cleland E, Tan L, Melton-Kreft R, Pant H. The microbiome of chronic rhinosinusitis: culture, molecular diagnostics and biofilm detection. *BMC Infect Dis* 2013;**13**:1–9.
60. James G, Swogger E, Wolcott R, Pulcini E, Secor J, Sestrich J. Biofilms in chronic wounds. *Wound Repair Regen* 2008;**16**:37–44.
61. Bjarnsholt T, Kirketerp-Møller K, Jensen PØ, Madsen KG, Phipps R, Krogfelt K. Why chronic wounds will not heal: a novel hypothesis. *Wound Repair Regen* 2008;**16**:2–10.

62. Costerton JW, Irvin RT, Cheng KJ. The bacterial glycocalyx in nature and disease. *Annu Rev Microbiol* 1981;**35**:299–324.
63. Walters MC, Roe F, Bugnicourt A, Franklin MJ. Contributions of antibiotic penetration, oxygen limitation, and low metabolic activity to tolerance of *Pseudomonas aeruginosa* biofilms to ciprofloxacin and tobramycin. *Antimicrob Agents Chemother* 2003;**47**:317–23.
64. Anwar H, Costerton JW. Enhanced activity of combination of tobramycin and piperacillin for eradication of sessile biofilm cells of *Pseudomonas aeruginosa. Antimicrob Agents Chemother* 1990;**34**:1666–71.
65. Machado I, Graça J, Lopes H, Lopes S, Pereira MO. Antimicrobial pressure of ciprofloxacin and gentamicin on biofilm development by an endoscope-isolated *Pseudomonas aeruginosa. ISRN Biotechnol* 2012;**2013**:178646.
66. Lewis K. Persister cells and the riddle of biofilm survival. *Biochem Mosc* 2005;**70**:267–74.
67. de Beer D, Stoodley P, Roe F, Lewandowski Z. Effects of biofilm structures on oxygen distribution and mass transport. *Biotechnol Bioeng* 1994;**43**:1131–8.
68. James GA, Zhao AG, Usui M, Underwood RA, Nguyen H, Beyenal H, et al. Microsensor and transcriptomic signatures of oxygen depletion in biofilms associated with chronic wounds. *Wound Repair Regen* 2016;**24**(2):373–83. http://dx.doi.org/10.1111/wrr.12401.
69. Soto SM. Role of efflux pumps in the antibiotic resistance of bacteria embedded in a biofilm. *Virulence* 2013;**4**:223–9.
70. Brooun A, Liu S, Lewis KA. Dose-response study of antibiotic resistance in *Pseudomonas aeruginosa* biofilms. *Antimicrob Agents Chemother* 2000;**44**:640–6.
71. Wolcott RD. Biofilms and chronic wound inflammation. *J Wound Care* 2008;**17**:333–41.
72. Ciornei CD, Novikov A, Beloin C, Fitting C, Caroff M, Ghigo JM. Biofilm-forming *Pseudomonas aeruginosa* bacteria undergo lipopolysaccharide structural modifications and induce enhanced inflammatory cytokine response in human monocytes. *Innate Immun* 2010;**16**:288–301.
73. Lawrence JR, Swerhone GDW, Kuhlicke U, Neu TR. In situ evidence for microdomains in the polymer matrix of bacterial microcolonies. *Can J Microbiol* 2007;**53**:450–8.

CHAPTER 4

Cutaneous and Musculoskeletal Complications in Human Immunodeficiency Virus Infection

A.A. Prasetyo, R. Sariyatun
Universitas Sebelas Maret, Surakarta, Indonesia

1. INTRODUCTION

Cutaneous and musculoskeletal complications are common in the course of human immunodeficiency virus (HIV) infection, occurring in up to 95% and 23.5%, respectively, of people living with HIV/AIDS (PLWH).[1,2] Due to immunodeficiency, the manifestations are generally more severe than those in immunocompetent individuals, leading to a significant morbidity in PLWH.[3] It is worth noting that some skin and musculoskeletal manifestations have been considered signs/symptoms of early or acute HIV infection, including rash (presents in 56% of cases), myalgia (55%), arthralgia (31%), oral ulcers (29%), oral candidiasis (17%), and genital ulcers (7%).[4] Moreover, some atypical appearances are unique to HIV infection.[5] This valuable knowledge could lead to initial suspicion for HIV diagnosis, particularly in areas where routine HIV testing is uncommon. Moreover, a number of skin and musculoskeletal manifestations are related to the patient's degree of immunodeficiency. Thus, some complications have been used as indicators of advanced HIV infection and declined $CD4^+$ T cells, particularly when routine $CD4^+$ T cell calculation is unavailable.[6,7] On the other hand, a number of skin and musculoskeletal complications can resemble other diseases.[8] There is a myriad of possible causes of skin and musculoskeletal complications in PLWH, either infection-related or noninfection-related pathology. This situation hampers the diagnosis and commencement of the appropriate therapies. Therefore, knowledge of the rates and presentations of cutaneous and musculoskeletal complications in PLWH is important. It could help diagnose new HIV cases as well as predict a patient's progression, and more importantly, provide appropriate clinical care for PLWH. Hence, this chapter aims to provide descriptions of the common cutaneous and musculoskeletal complications in HIV infection. Our perspectives for further research in this area and a summary of the complications aforementioned are also provided in this chapter (Table 1).

The Microbiology of Skin, Soft Tissue, Bone and Joint Infections
ISSN 2451-9006, http://dx.doi.org/10.1016/B978-0-12-811079-9.00004-5

Table 1 Cutaneous and musculoskeletal complications in people living with HIV/AIDS

Complications	Rates	Cutaneous manifestations	Musculoskeletal manifestations
Herpes simplex virus type 1 and 2 infections	52%–100%[a9]	✓	–
Varicella zoster virus reactivation	2.9%/year[b10]	✓	✓
Epstein-Barr virus infection	42%[11]	✓	✓
Cytomegalovirus disease	0.75%[b12]	✓	✓
Human herpes virus-8-associated Kaposi's sarcoma	2%[b13]	✓	–
Merkel cell polyomavirus-associated carcinoma	31%[a14]	✓	–
Poxvirus-caused molluscum contagiosum	5%–18%[15]	✓	–
Human papilloma virus infection	7.5%[a16]	✓	–
Hepatitis C virus infection	1.2%–98.5%[a17]	✓	✓
Syphilis	8%[b18]	✓	✓
Staphylococcus aureus infection		✓	✓
Tuberculosis	35–500 higher incidence in AIDS patients[19]	✓	✓
Mycobacterium leprae infection		✓	–
Atypical mycobacterial infection		✓	✓
Oral bacterial infection		✓	✓
Scabies	300 million cases globally/year[20]	✓	–
Leishmaniasis		✓	✓
Candidiasis		✓	–
Dermatophytosis		✓	–
Cryptococcosis	1.7%–6.6%[b] in AIDS patients[21]	✓	–
Histoplasmosis	5%–27%[b] in AIDS patients[22]	✓	–
Aspergillosis	5%–11%[a6]	✓	–
Malignancies		✓	–
Drug reactions		✓	✓
Seborrheic dermatitis	30%–80%[b23]	✓	–
Psoriasis	1.3%–5%[a24]	✓	✓
Pruritic papular eruption	11%–46%[24]	✓	–
Eosinophilic folliculitis		✓	–
Reiter's syndrome	5%–10%[a8,25]	✓	✓
Granuloma annulare		✓	–
HIV-associated vasculitides	1%[a26,27]	✓	✓
Bone diseases		–	✓

[a]Prevalence.
[b]Incidence.

2. INFECTIVE COMPLICATIONS

2.1 Viral Infections

2.1.1 Human Herpes Viruses

Herpes Simplex Virus Type 1 and 2

Herpes simplex virus type 1 (HSV-1) and herpes simplex virus type 2 (HSV-2) are among the most frequent coinfections found in PLWH. Emerging as a lifelong infection, HSV is more common in older than younger PLWH.[28] HSV-1 and HSV-2 are traditionally associated with labial and genital infections, respectively, albeit HSV-1-associated genital infection has been increasingly reported.[9] HSV manifestations frequently appear once the patient becomes severely immunosuppressed; however, they can also occur because of immune restoration after highly active antiretroviral therapies (HAARTs).[29,30] The latter is described in another part of this chapter. HSV manifestations comprise vesicles, erosions, and ulcerations of the mouth, esophagus, perianal areas, genitals, and distal fingers, which are generally atypical, less likely self-resolved and in greater rate, aggressiveness, and duration compared to immunocompetent individuals.[9,23,24] In immunocompromised individuals, the lesions can progress into persistent deep ulcers or coalesce and extend into the mucosal and deep cutaneous layers, occasionally with necrosis, bleeding, and severe pain.[24] Further, atypical lesions sometimes resemble long-standing lichenoid, granulomatous, and pseudolymphomatous HSV reactions.[31] Notably, HSV chronicity and reactivation can still emerge despite suppressive HAART.[31,32]

HHV-3/Varicella Zoster Virus

Primary infection of HHV-3 (varicella zoster virus, VZV), presenting as varicella (chicken pox), is generally followed by viral migration to the spinal and cranial sensory ganglia where VZV becomes dormant.[33] Once reactivated, the dormant VZV causes a secondary infection called herpes zoster (HZ) or shingles.[33] PLWH have greater HZ incidence compared to the general population, regardless of HIV stage.[34] HZ can be found in PLWH with $CD4^+$ T cell counts of 400/μL, and thus is among the earliest manifestations of HIV infection and conceivably predictive for HIV infection.[30] In immunosuppressed PLWH, HZ can appear as large bullae, hemorrhagic and necrotic lesions, or multidermatomal or generalized disease, until destruction of cartilage tissues due to severe necrotic lesions.[10,30] The lesions can also be superinfected with *Streptococcus pyogenes* or *Staphylococcus aureus*, causing impetigo-type lesions.[24] Atypically, such hyperkeratotic papules may unfollow the dermatomal pattern and persist for months or years.[24] Ecthymatous HZ can occur, appearing as multiple ulcers with black eschars in the center surrounded by a rim of vesicles.[24] Notably, HZ in PLWH can appear without rash, namely zoster sine herpete, in which patients describe a deep-seated burning pain at the involved dermatome and possibly with pain in the chest, back, or abdomen.[24]

HHV-4/Epstein-Barr Virus

Epstein-Barr virus (EBV) has been associated with small muscle tumors (SMTs) in PLWH, particularly among children, where SMTs predominantly occur in the luminal gastrointestinal (GI) tract and the hepatobiliary tree.[35] A specific EBV-caused skin condition in PLWH is oral hairy leukoplakia (OHL), which is also one of the seven cardinal oral findings of HIV infection.[36] OHL appears as nonmalignant white lesions on oral mucosa, tongue, oral mucosa, or gingiva.[36] OHL is correlated with HIV viral load and $CD4^+$ T cell count, thus becoming of prognostic value to AIDS.[11]

HHV-5/Cytomegalovirus

Cytomegalovirus (CMV) disease is the commonest cause of life-threatening opportunistic viral infection in PLWH.[37] It mostly occurs because of reactivation of prior latent CMV infection.[34] Although cutaneous manifestation is rare, it can appear as ulcers, vesicles, purpuric macules, verrucous lesions, prurigo nodularis-like lesions, erythematous and crusted papules, or digital infarctions.[34] Musculoskeletal manifestations are uncommon, although a case of bone marrow dysplasia has been reported.[12]

HHV-6 and HHV-7

HHV-6 and HHV-7 are genetically close, hence posing similar tropisms and disease associations.[38] In immunocompetent individuals, HHV-6 has been linked to some cutaneous and musculoskeletal manifestations, including mononucleosis-like syndrome, Sjögren syndrome, and Hodgkin's and non-Hodgkin's lymphomas, whereas HHV-7 has been associated with exanthema subitum and pityriasis rosea.[38,39] The manifestations in PLWH remain unclear. However, increased HHV-6 and HHV-7 levels in oral cavities of PLWH have been associated with higher levels of periodontopathic bacteria, leading to a hypothesis that HHV-6/HHV-7 and oral bacteria coactively play important roles in periodontal disease.[40]

HHV-8

HHV-8 has been associated with Kaposi sarcoma (KS). KS typically affects the skin or mucosal surfaces as purplish, reddish-blue, or dark-brown/black macules, plaques, or nodules that may bleed and ulcerate easily.[13] KS-related skeletal conditions are uncommon. When they occur, the axial skeleton (skull, vertebrae, pelvis, ribs, and sternum) are typically affected, whereas the joints, appendicular skeleton, and skeletal muscle are rarely involved.[8]

2.1.2 Merkel Cell Polyomavirus

Merkel cell polyomavirus (MCV) has been associated with Merkel cell carcinoma (MCC), a neuroendocrine skin cancer.[41] As HIV is linked to higher MCV prevalence and cutaneous viral load, PLWH are at a greater risk of developing MCC, presenting as painless, rapidly growing, single erythematous nodules.[14] Atypical presentations, like MCC in black people and primary nodal or intraparotid MCC, have also been reported.[14]

2.1.3 Poxvirus

Poxvirus is the etiological agent of molluscum contagiosum, manifesting as skin-colored, dome-shaped papules, or nodules with central umbilication.[23] Presentation in HIV infection can be atypical, extensive, and resembling basal cell carcinoma, keratoacanthoma, or cryptococcosis.[23] Likewise, such atypical molluscum contagiosum could be the first sign of HIV infection.[15]

2.1.4 Human Papilloma Virus

Human papilloma virus (HPV) is highly prevalent globally, and even the infection becomes more persistent in PLWH due to immunosuppression.[42] Warts are common among PLWH, and they tend to be multifocal, recalcitrant, and resistant to treatment.[42] Condyloma acuminata and anogenital tumors are strongly increased with HIV progression.[16]

2.1.5 Hepatitis C Virus

As both hepatitis C virus (HCV) and HIV are transmitted in similar ways and exposed to similar at-risk populations, dual HIV-HCV infections are common.[17] While HCV apparently infects hepatocytes, such extrahepatic manifestations are common, including cryoglobulinemia, porphyria cutanea tarda, lichen planus, pruritus, arthralgia, and bone loss.[43,44] HIV/HCV coinfected people more commonly develop pigmentary disorders, vitamin D deficiency, osteoporosis, and fractures.[44–46]

2.2 Bacterial Infections

2.2.1 Syphilis

Syphilis, caused by *Treponema pallidum*, shares similar transmission routes and risk factors with HIV, hence is highly prevalent in PLWH.[18,23] PLWH with primary syphilis, compared to HIV-negative people, more likely have multiple chancres with larger and deeper ulcers.[47] In secondary syphilis, PLWH commonly suffer simultaneous genital ulcers and can also develop osteoarticular manifestations, most frequently the subacute or chronic arthritis of large joints, arthralgias, osteolytic lesions, or such atypical presentations.[23,30,48]

2.2.2 Staphylococcus aureus

S. aureus is the commonest bacterial pathogen in PLWH, known to cause folliculitis, impetigo, ecthyma, and skin abscesses.[23] It is the commonest cause of septic arthritis in PLWH, which frequently affects large and weight-carrying joints of the lower extremities, knees, and hips.[19] Septic arthritis, pyomyositis, septic bursitis, pyogenic osteomyelitis are other common manifestations associated with *S. aureus*.[19]

2.2.3 Mycobacteria

Tuberculosis (TB), a disease caused by *Mycobacterium tuberculosis* (*M. tuberculosis*), is a leading cause of death in PLWH. On the other hand, HIV increases TB incidence and the risk of reactivation from latent infection.[49] While TB generally affects the pulmonary system, extrapulmonary manifestations are frequent in PLWH and could be found in more than 50% of TB cases.[50] *M. tuberculosis* can infect the skin, musculoskeletal, as well as the GI and nervous systems.[19,23] Typical manifestations include lupus vulgaris and scrofuloderma, tuberculous spondylitis, arthritis, osteomyelitis, tenosynovitis, and bursitis.[19,23,48,49] The most common site of musculoskeletal TB is the vertebra (Pott's disease), mainly in the lower thoracic or upper lumbar segments.[19]

Mycobacterium leprae (*M. leprae*) infection, or leprosy, is another mycobacterial infection prevalent in PLWH. Yet, there is a lack of evidence that it shows worse prognosis or different manifestations in PLWH.[51] Instead, leprosy in PLWH is often diagnosed when the patient's immune system is restored through HAART, thus indicating that leprosy in PLWH is an unmasking immune reconstitution inflammatory syndrome (IRIS) phenomenon.[30,51]

In addition to *M. tuberculosis* and *M. leprae*, coinfection with some atypical mycobacteria (e.g., *M. avium-intracellulare*, *M. chelonae*, *M. fortuitum*, *M. xenopi*, *M. kansasii*, and *M. haemophilum*) have been found in PLWH.[19] Atypical mycobacteria-related conditions usually occur in AIDS patients and often present with other opportunistic infections.[19,49] The bacteria generally disseminate hematogenously and tend to affect multiple sites, such as the metaphyses and diaphyses of long bones, the tendon sheaths of the upper extremities, and the bursae of the wrists and knees.[19,49] The *M. avium-intracellulare complex* (MAC) is the most frequent atypical mycobacterial species found in AIDS patients and is responsible for numerous cases of septic arthritis, osteomyelitis, and tenosynovitis in this immunosuppressed group.[19] Moreover, approximately 50% of atypical mycobacteria-coinfected PLWH show cutaneous lesions, appearing as nodules, ulcers, and draining sinus tracts.[5]

2.2.4 Oral Bacteria

Oral bacteria colonization on the surfaces of the teeth can generate dental biofilm, subsequently induce inflammatory response and cause gingivitis.[52] Continuous inflammation leads to the loosening of the teeth, occasional pain, impaired mastication, and eventual tooth loss.[52,53] This pathological condition is termed periodontal disease and is associated with *Porphyromonas gingivalis*, *Aggregatibacter actinomycetemcomitans*, *Tannerella forsythia*, *Treponema denticola*, *Prevotella intermedia*, *Campylobacter rectus*, *Eubacterium nodatum*, *Peptostreptococcus micros*, *Streptococcus intermedius*, and *Treponema* sp. infections.[52,54]

Notably, periodontal disease considerably has a predictive value to AIDS and treatment failure, while some forms of the disease are among the seven cardinal oral lesions indicative of HIV infection, including linear gingivitis erythema (LGE), necrotizing ulcerative gingivitis (NUG), and necrotizing ulcerative periodontitis (NUP).[36,52] LGE is characterized by a fiery linear erythema along the margin of the gingiva and occasionally with bleeding or petechial-like patches.[36,52] NUG presents as acute and painful gingivitis with limited destruction of the soft tissues, whereas NUP involves the destruction of the periodontal ligament and the alveolar bone.[36,52]

2.3 Parasitic Infections

2.3.1 *Sarcoptes Scabiei*

Scabies is the commonest parasitic infection in PLWH.[20,24] In immunosuppressed PLWH, scabies often appears as atypical and fulminant eruptions with high density of parasites, thus making the patients potential sources of scabies in their surroundings.[23] There are two categories of scabies in PLWH: exaggerated or popular/atypical scabies, which is characterized by highly pruritic and generalized papules topped by a burrow, and Norwegian or crusted scabies, which is less pruritic and presents as thick, white-gray, hyperkeratotic and scaly plaques either generalized or localized.[23,24] Both forms can occur simultaneously.[55] Due to their resemblance to seborrheic dermatitis, psoriasis, and certain drug-induced eruptions, the diagnosis may be difficult.[23]

2.3.2 *Leishmania*

Leishmaniasis is a condition caused by some species of protozoa of the genus *Leishmania* transmitted by sand fly bites endemic in some African, Middle Eastern, European, and Central and South American nations.[56] Classically, leishmaniasis is categorized into cutaneous leishmaniasis (CL), mucocutaneous leishmaniasis (MCL), and visceral leishmaniasis (VL). CL is the least severe type, presenting as singular or nodular lesions in the area of the insect bite, although cases of diffuse CL have also been reported.[56,57] MCL occurs due to the extension or hematogenous/lymphatic spread of primary infection in the mucosal layer, causing lesions in the oral, nasal, laryngeal, or pharyngeal mucosa, which can cause tissue destruction in the nose, pharynx, and palate if not treated properly.[56] CL can be the first symptom of HIV infection.[58]

HIV increases the risk of patients to suffer VL; conversely, VL promotes HIV replication and development of serious symptoms.[59] VL is a fatal and systemic condition as the parasite proliferates within the reticuloendothelial system of the liver, spleen, and bone marrow.[56] Concomitant cutaneous, GI, and respiratory tract involvements can also occur.[60] Cutaneous manifestations are common in PLWH, appearing as psoriasiform eruption, papulonodular lesions, erythrodermic pattern, linear brown macules, dermatomyositis-like eruption, or polymyositis-like syndrome.[61] VL is also associated with a darkening of the skin, the so-called *kala-azar* in Hindi, which means "black fever".[60] VL is considered a condition of immunosuppression; thus its occurrence may be an indicator for HIV testing and HAART initiation.[62] Up to 60% of VL cases can develop post-kala-azar dermal leishmaniasis (PKDL), in which the parasite is present in the skin, allowing the affected patients to act as human reservoirs.[59] PKDL manifests as macular, papular, or nodular lesions around the perinasal areas and it often spreads to the face, trunk, and extremities.[60]

2.4 Fungal Infections

2.4.1 Candida albicans

C. albicans is a well-known fungus causing oral thrush/candidiasis, in addition to angular cheilitis, folliculitis, and intertrigo.[30] Oral candidiasis is thought to be an early predictor of severe immunosuppression and can progress to esophageal or systemic dissemination.[63] It is often recurrent and systemic antifungal is frequently required.[63] Yet, the rate of candidiasis diminishes once PLWH begin HAART.[16]

2.4.2 Dermatophytes

Dermatophytosis is an infectious condition caused by geophilic, zoophilic, or anthropophilic fungi, such as *Epidermophyton floccosum*, *Microsporum canis*, *Trichophyton mentagrophytes*, *Trichophyton rubrum*, *Trichophyton tonsurans*, and *Trichophyton violaceum*, which invade the superficial dead layer of the skin or keratinized tissues like hair and nails.[23] The rate is higher in immunosuppressed subjects, although it becomes lower with HAART.[16] In PLWH, dermatophytosis is generally more extensive and severe, and can affect multiple regions as tinea corporis, cruris, faciei, pedis, manuum, or unguium.[24] Notably, tinea corporis can present as an acute or chronic infection; the acute form presents as classic ringworm with sharply demarcated erythematous and scaly macules, whereas the chronic form presents as lesions coalescing into much larger hyperpigmented plaques.[24] The chronic form is problematic in PLWH as it requires oral antifungals, which potentially interact with HAART.[24] Moreover, tinea versicolor/pityriasis versicolor is also common and can even be seen early in HIV infection, presenting as clearly margined macules that appear hypopigmented on dark skin and hyperpigmented on light skin.[24]

2.4.3 Cryptococcus neoformans

C. neoformans infection, cryptococcosis, is the second commonest fungal infection in AIDS and is an important cause of morbidity and mortality in immunosuppressed individuals.[21] Cryptococcosis usually appears as a systemic infection rather than a primary dermal infection, with predispositions to the skin and central nervous system.[23] Skin manifestations occur in up to 10% of coinfected PLWH, appearing as a variety of nonspecific lesions, including papules, nodules, pustules, ulcers, plaques, KS-like and molluscum contagiosum-like lesions commonly on the face, neck, and scalp.[6,23,24]

2.4.4 Histoplasma capsulatum

H. capsulatum is a dimorphic fungus causing a granulomatous condition termed histoplasmosis, a deep fungal infection associated with AIDS.[6,22,23] Skin lesions occur in 11% of patients, commonly as the result of hematogenous dissemination from pulmonary infection. Clinical presentations typically include papules, patches, nodules, abscesses, plaques, pustules, and ulcers.[23,24]

2.4.5 Aspergillus fumigatus

A. fumigatus infection, aspergillosis, is an important cause of mortality and morbidity in immunosuppressed individuals.[64] It affects multiple areas, as well as the cutaneous and musculoskeletal systems.[6] Skin manifestations include clam-like lesions, verrucous plaques and pustules, ulcerated nodules, and thickened areas of overlying erythema more commonly in the extremities and the head.[6]

3. NONINFECTIVE COMPLICATIONS

3.1 Malignancies

Despite increased life expectancy due to HAART, PLWH continue to carry a high-risk of certain malignancies.[65] The cause of malignancies in PLWH can be either cancer-associated viruses (e.g., HCV, MCV, HPV, and HHV) or the HIV itself.[66] Some cancers can even occur despite suppressive HAART and do not seem to be associated with HIV/AIDS progression, the so-called non-AIDS-defining cancers (NADCs).[65] Melanoma and nonmelanoma skin cancers are among NADCs.[65]

3.2 Drug Reactions

PLWH are exposed to multiple drugs, at least lifelong HAART and medication for opportunistic infections. Consequently, they are at a greater risk of developing adverse drug effects.[67] Drug reactions can affect numerous organs, presenting rapidly or later during medication, ranging from cosmetic symptoms to life-threatening conditions.[68]

3.2.1 Stevens-Johnson Syndrome and Toxic Epidermal Necrosis

Stevens-Johnson syndrome and toxic epidermal necrosis (SJS/TEN) are more frequent in PLWH and their manifestations tend to be severer.[2] Antiretroviral therapies (ARTs) are associated with SJS/TEN incidence, including didanosine, nevirapine, efavirenz, indinavir, amprenavir, fosamprenavir, and darunavir for SJS, and zidovudine and nevirapine for TEN.[1,3,24,68,69] Other drugs, such as anticonvulsants, allopurinol, and antibiotics, can also cause SJS/TEN.[24,69]

3.2.2 Lipodystrophy

Globally 13.1%–52.9% of PLWH develop lipodystrophy syndrome,[70] characterized by selective subcutaneous fat wasting on the face, extremities, and buttocks (lipoatrophy) and/or the accumulation of visceral fat and gynecomastia (lipohypertrophy).[1] Lipodystrophy is associated with stavudine, efavirenz, and indinavir, and often occurs months to years after the regimen intake.[68] Lipodystrophy affects cosmetic appearance and causes metabolic syndromes, hence potentially leads to stigmatization and HAART discontinuation.[1]

3.2.3 Abacavir Hypersensitivity

Abacavir can induce a systemic hypersensitivity syndrome, defined by the appearance of two of the following symptoms: fever, rash, GI disturbances, other constitutional complaints, and respiratory symptoms.[68] Fever and rash are frequent as they occur in more than 60% of patients, while other manifestations like myalgia, arthralgia, pharyngitis, transaminitis, and hypotension can also occur.[68] The rash is typically maculopapular, yet urticaria, diffuse erythema, erythema multiforme, and targetoid skin appearances have also been described.[68]

3.2.4 Nail and Mucocutaneous Hyperpigmentation

Darkening of the nail tissue and changes in the mucocutaneous appearance are common side effects of zidovudine and emtricitabine.[68,71] Careful examination should be performed as hyperpigmentation can also be related to HIV itself.[68] Zidovudine-related hyperpigmentation is dose-dependent; hence dose reduction or withdrawal of the drug would reduce the hyperpigmentation.[30] This evaluation could be performed to distinguish it from HIV-related hyperpigmentation.

3.2.5 Indinavir-Induced Retinoid-Like Effects

Indinavir is the only ART known to induce retinoid-like effects, which appear as cutaneous xerosis (with asteatotic eczema, acquired ichthyosis, pruritus, and burning sensation), cheilitis, certain nail conditions (paronychia, ingrown toenails, hypertrophic granulation tissue deposition, and nail plate dystrophy), hair alterations (body and scalp alopecia, and the growth of curly hair), and hyperlipidemia within 2 months after drug administration.[68,72] Cutaneous xerosis and cheilitis are the most common manifestations.[72] Xerosis most commonly occurs on the hands, feet, and face, and it presents as rough and scaly skin and possibly with excoriation due to scratching.[73] Chronic or recurrent paronychia is another condition strongly related to indinavir, which appears as inflamed tissue around the nail plates, most frequently on the large toe.[72,73] Moreover, indinavir-associated alopecia, presenting as diffuse scalp alopecia, alopecia areata, and localized hair loss on the body surfaces, often strongly affects personal appearance and hence leads to drug discontinuation.[73]

3.2.6 Enfuvirtide-Related Injection-Site Rejection

Enfuvirtide, a fusion inhibitor, is the first ART administered through subcutaneous injection.[74] Unfortunately, most patients (98% of PLWH in the pivotal trials) underwent injection-site reactions, including pain, discomfort, pruritus, erythema, induration, nodules, ecchymosis, cysts, and occasionally cellulitis or abscesses, which generally occurs within the first week of therapy and lasts for an average of 7 days.[24,73]

3.2.7 Drug-Induced Myopathy

Zidovudine has been reported to induce diffuse myopathy, characterized by subtle onset of myalgia, muscle tenderness, and proximal muscle weakness after approximately 11 months of medication.[5] The degree of myopathy tends to be dose-dependent, and the symptoms and defect can be improved after zidovudine is discontinued.[5]

3.3 Seborrheic Dermatitis

Seborrheic dermatitis is a chronic inflammatory condition usually affecting sebaceous gland-rich areas.[23] In PLWH, seborrheic dermatitis is prevalent and often with intense symptoms, and in fact, its severe appearance is indicative for HIV infection.[23] The pathogenesis and clinical appearance can be similar to those in immunocompetent individuals, although atypical and severe manifestations often occur in PLWH.[23] The disease commonly starts in seborrheic areas (e.g., face, scalp, and upper chest) before spreading rapidly to the back, extremities, axillae, and groin during immunosuppression.[23,24] The clinical presentation varies from acute dermatitis and psoriasiform plaques, and in pigmented skin it can present annular hypopigmented patches with petaloid seborrheic dermatitis and less visible erythema.[23] The rash appears yellow-white with greasy scales on hypopigmented and/or erythematous patches.[24]

3.4 Psoriasis

Compared to the general population, the rate of psoriasis in PLWH is just slightly higher; however, exacerbation and worsening of preexisting disease are more common.[24] Psoriasis in PLWH tends to be more severe, progressive, destructive, recalcitrant, and treatment-resistant.[23,75,76] It can occur at any stage of HIV infection, and the severity increases as the immunity declines.[76] Any forms of psoriasis can occur in PLWH; however, erythrodermic, guttae, and inverse psoriasis are the most frequent forms, and multiple forms of psoriasis often appear concurrently in a patient.[23,76] Rupioid psoriasis, presenting as coin-shaped, hyperkeratotic crusted plaques, has also been reported.[23]

3.5 Psoriatic Arthritis

Psoriatic arthritis in PLWH is generally progressive, deforming, erosive, and refractory to conventional treatment.[8,23,76] It tends to manifest as polyarticular and asymmetric arthropathy, mostly affecting the large joints of the lower extremities and often involving the surrounding tendons and fascia.[8,25,49] Pain, swelling, and pitting edema are clinical features representing enthesitis, synovitis, dactylitis, and peritendositis.[8] Enthesopathies of the lower limb can occur, whereas sacroiliitis and spondylitis are rare.[8] Psoriatic arthritis is radiologically indistinguishable from Reiter's syndrome.[25]

3.6 Pruritic Papular Eruption

Pruritic popular eruption (PPE) is among the most frequent pruritus condition in PLWH.[24] The underlying cause remains obscure, but it is thought to be a hypersensitivity reaction to insect bites or stings.[77] PPE is associated with low $CD4^+$ T cell levels and advanced HIV disease; hence it is considered to be a sign of severe immunosuppression and a parameter for antiretroviral (ARV) initiation in resource-limited areas.[23] A new-onset, recurrent or worsening PPE has also been thought to be an indicator of treatment failure.[78] The clinical presentations of PPE include intensely pruritic, symmetric, skin-colored to erythematous papules or pustules on the extremities, which subsequently become widespread to the trunk and face, making it difficult to distinguish from eosinophilic folliculitis (EF).[23] As the disease is naturally very pruritic, the papules are often excoriated and secondarily infected, thus causing inflammation, hyperpigmentation, scarring, and eventual nodular appearance.[23,24]

3.7 Eosinophilic Folliculitis

EF is another pruritic skin condition strongly related to HIV infection, and commonly found in PLWH with $CD4^+$ T cell count below 300 cells/μL, although it can also emerge as an IRIS manifestation.[23,24,30,75] Its cause is also unknown, but it is thought to be the result of hypersensitivity to various antigens in a dysregulated immune system.[24] EF clinically presents as pruritic, erythematous, follicular papules, mainly distributed on the face, neck, upper arms, and torso above the nipple line, whereas the soles and palms are unaffected.[23,24,30,75] As the papules are intensely pruritic, they often become excoriated, inflamed and eventually hyperpigmented, which can be a major concern on the facial areas of patients with pigmented skin.[23] Unlike PPE, EF rarely is associated with sterile pustules, and histological examination shows eosinophilic infiltrates around the affected follicles.[24]

3.8 Reiter's Syndrome (Reactive Arthritis)

Reiter's syndrome occurs more frequently and more severely in PLWH than in the general population, either in the early or late stage of HIV infection.[8,23,25] In PLWH, it rarely presents with the classic triad of urethritis, conjunctivitis, and arthritis.[8] Enthesopathies, flexor tendonitis, rotator cuff tendonitis, medial/lateral epicondylitis, or systemic symptoms like fever, lymphadenopathy, weight loss, and diarrhea may occur.[8,25] The other manifestations of Reiter's syndrome include oral lesions, uveitis, ulcerative vulvitis, annular dermatitis on the glans penis (circinate balanitis), keratoderma hemorrhagica, thick hyperkeratotic plaques on the palms and soles (keratoderma blenorrhagicum), and subungual hyperkeratosis of the nail.[23,25] Clinically, the combination of the mucocutaneous appearances, arthritis, and systemic symptoms can resemble secondary syphilis.[23]

3.9 Granuloma Annulare

Granuloma annulare (GA) has been associated with some infective conditions, including HIV. GA can appear at any stage of the HIV infection, yet the rate is slightly higher in AIDS patients.[79] In PLWH, GA tends to be more generalized than localized.[80] In fact, the development of generalized and atypical forms of GA in PLWH might be a prominent cutaneous marker of AIDS.[79]

3.10 HIV-Associated Vasculitides

HIV itself is linked to vasculitides, the inflammation and necrosis of the blood vessel leading to tissue injury.[81] Before the introduction of ART, vasculitides was reported in as many as 23% of symptomatic PLWH, but the rate declined to approximately 1% in the early ART era.[26,27] Vasculitides in HIV infection can affect all vessels (large, medium, and small vessels), but mostly affects small arteries.[26] The pathogenesis of vascolitides in HIV infection involves the direct and indirect effects of HIV and the resulting immunological responses.[26] Some common types of vasculitides in PLWH are described in the following subsections.

3.10.1 Polyarteritis Nodosa

Polyarteritis nodosa (PAN) is the most common HIV-associated vasculitides, mainly affecting the skin, peripheral nerves, and skeletal muscle, in addition to the central nervous system, lung, GI tract, and kidneys.[82] PAN occurs regardless of $CD4^+$ T cell levels or HIV stage, but tends to be higher in males (78% of reported cases).[26] Compared to classical hepatitis B virus-associated PAN, PAN in PLWH presents acutely with fewer skin, musculoskeletal and systemic symptoms, and overall mild severity.[82]

3.10.2 Cryoglobulinemia

Cryoglobulinemia is common in people with high HIV viral load, which may be the consequence of continuous HIV antigenic stimulation of B lymphocytes.[83,84] Such cryoglobulin-containing immune complexes can precipitate in vessels (mainly small to medium-sized vessels) and lead to vasculitis.[85] Cutaneous purpura is the most characteristic manifestation of cryoglobulinemia-induced vasculitis, appearing in 54%–82% of cases, although systemic symptoms are also common, such as fever, weakness, myalgia, and arthralgia (mainly in the hands and wrists).[86]

3.10.3 Henoch-Schönlein Purpura

Henoch-Schönlein purpura (HSP) is occasionally found in PLWH. The clinical presentations include palpable purpura, arthralgia, glomerulonephritis, and GI symptoms.[87] In PLWH, the manifestations may overlap with HIV-associated nephropathy and secondary IgA nephropathy.[88]

3.11 Bone Disease

PLWH are at increased risk of developing osteopenia, osteoporosis, fractures, and osteonecrosis (bone infarction at the epiphysis or near a joint).[49,89] The risk of osteoporosis increases not only in postmenopausal women but also in men of all ages and younger women.[90] These pathological bone conditions can be a direct consequence of HIV infection or other predisposing factors, like alcohol, corticosteroids, hyperlipidemia, cigarette smoking, antiphospholipid, and anticardiolipin antibodies, and ART (especially protease inhibitors).[49,89,90] Osteonecrosis in PLWH should be a concern as it leads to total joint replacement.[89]

4. IMMUNE RECONSTITUTION INFLAMMATORY SYNDROME

The success of HAART has been hurdled by the occurrence of IRIS, defined as the deterioration of previously recognized and treated diseases (paradoxical IRIS) or the appearance of an unrecognized pathogen (unmasking IRIS) attributable to the immune restoration following ART.[71] IRIS occurs in approximately 15%–35% of PLWH starting ART, and the incidence becomes higher in patients with an active opportunistic infection at the time of ART initiation.[91] The clinical presentations are related to underlying infective diseases.[71] In particular, IRIS is common in patients with CMV retinitis, cryptococcal meningitis, progressive multifocal leukoencephalopathy, tuberculosis, HZ, and Kaposi's sarcoma.[92] Some IRIS-related noninfectious conditions have also been reported, comprising sarcoidosis, Reiter's disease, systemic lupus erythematous, vasculitis, alopecia universalis, acne vulgaris, EF, Kaposi's sarcoma, cutaneous reactions to tattoo pigments, Graves' disease, rheumatoid arthritis, Guillain-Barré syndrome, and lymphoma.[71,93,94]

5. PERSPECTIVES FOR FURTHER RESEARCH

During the course of HIV infection, interactions between HIV and co- or opportunistic infections have resulted in numerous clinical presentations affecting the cutaneous and musculoskeletal systems. However, the host genetic profiles, risk factors, and sexual behaviors could be critical in determining the outcomes of the patients. We previously showed associations of host polymorphisms with the occurrences of some coinfections in HIV patients, such as hepatitis viruses, Torque Teno virus, and *Toxoplasma gondii*[95–99] and multidrug-resistant tuberculosis,[100,101] as well as with the patient's $CD4^+$ T cell level, which is important for HIV management.[95,102] Likewise, we showed that high-risk behavior is associated with HIV and HIV-associated infections.[103] Continuous research in this field is critical to better understand the roles of individual genetic profiles, risk factors, and sexual behaviors to predict their susceptibility to such pathological conditions as well as to predict the general prognosis of HIV patients.

6. CONCLUSION

Appearing with impaired immune systems, PLWH are under greater risk of infections by opportunistic pathogens (viruses, bacteria, fungi, and parasites) as well as HIV-associated pathogens. The infections often result in various to atypical skin and musculoskeletal appearances, which often complicates the diagnosis and therapy. Otherwise, recognition of the unique presentations could lead to suspicion of HIV infection in remaining undiagnosed individuals and prediction of AIDS or declined $CD4^+$ T cells. Moreover, noninfection-related inflammatory conditions, drug reactions, malignancies, bone diseases, and IRIS phenomenon are also prevalent among PLWH. Overall, knowledge in these matters is important for clinicians to provide appropriate treatment.

ACKNOWLEDGMENTS

This work was supported by grants from the Indonesian Ministry of Research, Technology and Higher Education (PUPT No 353/UN 27.21/PN/2016 and INSINAS No 453/UN27.21/LT/2016).

REFERENCES

1. Tschachler E. The dermatologist and the HIV/AIDS pandemic. *Clin Dermatol* 2014;**32**(2):286–9.
2. Ntsiba H, Ngandeu-Singwé M, Ondzala A-E. Musculoskeletal infections in human immunodeficiency virus. *Open J Rheumatol Autoimmune Dis* 2012;**2**(2):32–6.
3. Mankahla A, Mosam A. Common skin conditions in children with HIV/AIDS. *Am J Clin Dermatol* 2012;**13**(3):153–66.
4. Yerly S, Hirschel B. Diagnosing acute HIV infection. *Expert Rev Anti-Infect Ther* 2012;**10**(1):31–41.
5. Reveille JD, Williams FM. Infection and musculoskeletal conditions: rheumatologic complications of HIV infection. *Best Pract Res Clin Rheumatol* 2006;**20**(6):1159–79.
6. Ramos-e-Silva M, Lima CMO, Schechtman RC, Trope BM, Carneiro S. Systemic mycoses in immunodepressed patients (AIDS). *Clin Dermatol* 2012;**30**(6):616–27.
7. Oninla OA. Mucocutaneous manifestations of HIV and the correlation with WHO clinical staging in a tertiary hospital in Nigeria. *AIDS Res Treat* 2014;**2014**(2014):1–6.
8. Booth TC, Chhaya NC, Bell JRG, Holloway BJ. Update on imaging of non-infectious musculoskeletal complications of HIV infection. *Skelet Radiol* 2012;**41**(11):1349–63.
9. Tan DHS, Kaul R, Walsmley S. Left out but not forgotten: should closer attention be paid to coinfection with herpes simplex virus type 1 and HIV? *Can J Infect Dis Med Microbiol* 2009;**20**(1):e1–7.
10. Gnann JW, Whitley RJ. Herpes zoster. *N Engl J Med* 2002;**347**(5):340–6.
11. Di Lernia V, Mansouri Y. Epstein-Barr virus and skin manifestations in childhood. *Int J Dermatol* 2013;**52**(10):1177–84.

12. Majumder S, Mandal SK, Bandyopadhyay D, Chowdhury SR, Chakraborty PP, Mitra K. Multiorgan involvement due to cytomegalovirus infection in AIDS. *Braz J Infect Dis* 2007;**11**(1):176–8.
13. Bhutani M, Polizzotto MN, Uldrick TS, Yarchoan R. Kaposi sarcoma-associated herpesvirus-associated malignancies: epidemiology, pathogenesis, and advances in treatment. *Semin Oncol* 2015;**42**(2):223–46.
14. Wieland U, Kreuter A. Merkel cell polyomavirus infection and Merkel cell carcinoma in HIV-positive individuals. *Curr Opin Oncol* 2011;**23**(5):488–93.
15. Filo-Rogulska M, Pindycka-Piaszczyńska M, Januszewski K, Jarząb J. Disseminated atypical molluscum contagiosum as a presenting symptom of HIV infection. *Postepy Dermatol Alergol* 2013;**30**(1):56–8.
16. Kore SD, Kanwar AJ, Vinay K, Wanchu A. Pattern of mucocutaneous manifestations in human immunodeficiency virus-positive patients in North India. *Indian J Sex Transm Dis* 2013;**34**(1):19–24.
17. Ye S, Pang L, Wang X, Liu Z. Epidemiological implications of HIV-hepatitis C co-infection in South and Southeast Asia. *Curr HIV/AIDS Rep* 2014;**11**(2):128–33.
18. Courjon J, Hubiche T, Dupin N, Grange PA, Del Giudice P. Clinical aspects of syphilis reinfection in HIV-infected patients. *Dermatology* 2015;**230**(4):302–7.
19. Tehranzadeh J, Ter-Oganesyan RR, Steinbach LS. Musculoskeletal disorders associated with HIV infection and AIDS. Part I: infectious musculoskeletal conditions. *Skeletal Radiol* 2004;**33**(5):249–59.
20. Leone PA. Scabies and pediculosis pubis: an update of treatment regimens and general review. *Clin Infect Dis* 2007;**44**(Suppl. 3):S153–9.
21. Ashwini BR, Raghupathi AR, Srinarthan A. Cryptococcososis of bone marrow: a case report with review of literature. *J Clin Diagn Res* 2014;**8**(1):158–9.
22. Bonifaz A, Chang P, Moreno K, Fernández-Fernández V, Montes de Oca G, Araiza J, et al. Disseminated cutaneous histoplasmosis in acquired immunodeficiency syndrome: report of 23 cases. *Clin Exp Dermatol* 2009;**34**(4):481–6.
23. Motswaledi MH, Visser W. The spectrum of HIV-associated infective and inflammatory dermatoses in pigmented skin. *Dermatol Clin* 2014;**32**(2):211–25.
24. Khambaty MM, Hsu SS. Dermatology of the patient with HIV. *Emerg Med Clin North Am* 2010;**28**(2):355–68.
25. Tehranzadeh J, Ter-Oganesyan RR, Steinbach LS. Musculoskeletal disorders associated with HIV infection and AIDS. Part II: non-infectious musculoskeletal conditions. *Skeletal Radiol* 2004;**33**(6):311–20.
26. Ojaimi S, Lin M-W, Singh KP, Woolley I. The two-edged sword: vasculitis associated with HIV and hepatitis C coinfection. *Int J STD AIDS* 2014;**25**(2):77–88.
27. Guillevin L. Infections in vasculitis. *Best Pract Res Clin Rheumatol* 2013;**27**(1):19–31.
28. Looker KJ, Magaret AS, Turner KME, Vickerman P, Gottlieb SL, Newman LM. Global estimates of prevalent and incident herpes simplex virus type 2 infections in 2012. *PLoS One* 2015;**10**(1):e114989.
29. Amerson EH, Maurer TA. Dermatologic manifestations of HIV in Africa. *Top HIV Med* 2010;**18**(1):16–22.
30. Ameen M. The impact of human immunodeficiency virus-related diseases on pigmented skin types. *Br J Dermatol* 2013;**169**(Suppl. 3):11–8.
31. Wauters O, Lebas E, Nikkels AF. Chronic mucocutaneous herpes simplex virus and varicella zoster virus infections. *J Am Acad Dermatol* 2012;**66**(6):e217–27.
32. Tan DHS, Raboud JM, Kaul R, Walmsley SL. Antiretroviral therapy is not associated with reduced herpes simplex virus shedding in HIV coinfected adults: an observational cohort study. *BMJ Open* 2014;**4**:e004210.
33. Johnson RW, Alvarez-Pasquin M-J, Bijl M, Franco E, Gaillat J, Clara JG, et al. Herpes zoster epidemiology, management, and disease and economic burden in Europe: a multidisciplinary perspective. *Ther Adv Vaccines* 2015;**3**(4):109–20.
34. Shiohara T, Kurata M, Mizukawa Y, Kano Y. Recognition of immune reconstitution syndrome necessary for better management of patients with severe drug eruptions and those under immunosuppressive therapy. *Allergol Int* 2010;**59**(4):333–43.

35. Tetzlaff MT, Nosek C, Kovarik CL. Epstein-Barr virus-associated leiomyosarcoma with cutaneous involvement in an African child with human immunodeficiency virus: a case report and review of the literature. *J Cutan Pathol* 2011;**38**(9):731–9.
36. Mataftsi M, Skoura L, Sakellari D. HIV infection and periodontal diseases: an overview of the post-HAART era: HIV infection and periodontal diseases. *Oral Dis* 2011;**17**(1):13–25.
37. Paparone PP, Paparone PA. Cytomegalovirus colitis in a human immunodeficiency virus-positive patient with a normal CD4 count. *Am J Med Sci* 2012;**344**(6):508–10.
38. Black JB, Pellett PE. Human herpesvirus 7. *Rev Med Virol* 1999;**9**(4):245–62.
39. Corti M, Villafañe MF, Trione N, Mamanna L, Bouzas B. Human herpesvirus 6: report of emerging pathogen in five patients with HIV/AIDS and review of the literature. *Rev Soc Bras Med Trop* 2011;**44** (4):522–5.
40. Feller L, Meyerov R, Lemmer J. The association between human herpes viruses and periodontal disease, part 2. *SADJ* 2007;**62**(4)170, 172, 174.
41. Grulich AE, Vajdic CM. The epidemiology of cancers in human immunodeficiency virus infection and after organ transplantation. *Semin Oncol* 2015;**42**(2):247–57.
42. Wieland U, Kreuter A, Pfister H. Human papillomavirus and immunosuppression. *Curr Probl Dermatol* 2014;**45**:154–65.
43. Galossi A, Guarisco R, Bellis L, Puoti C. Extrahepatic manifestations of chronic HCV infection. *J Gastrointest Liver Dis* 2007;**16**(1):65–73.
44. Dong HV, Cortés YI, Shiau S, Yin MT. Osteoporosis and fractures in HIV/hepatitis C virus coinfection: a systematic review and meta-analysis. *AIDS* 2014;**28**(14):2119–31.
45. El-Maouche D, Mehta SH, Sutcliffe CG, Higgins Y, Torbenson MS, Moore RD, et al. Vitamin D deficiency and its relation to bone mineral density and liver fibrosis in HIV/HCV co-infection. *Antivir Ther* 2013;**18**(2):237–42.
46. Cunha VS, Meotti C, Oliveira JH, Sprinz E, Álvares-da-Silva MR, Goldani LZ. Different patterns of dermatological presentations in patients co-infected with human immunodeficiency virus and hepatitis C virus (HCV), and those infected with HCV alone. *Clin Exp Dermatol* 2012;**37**(2):122–7.
47. Zetola NM, Klausner JD. Syphilis and HIV infection: an update. *Clin Infect Dis* 2007;**44**(9):1222–8.
48. Colmegna I, Koehler JW, Garry RF, Espinoza LR. Musculoskeletal and autoimmune manifestations of HIV, syphilis and tuberculosis. *Curr Opin Rheumatol* 2006;**18**(1):88–95.
49. Takhar SS, Hendey GW. Orthopedic illnesses in patients with HIV. *Emerg Med Clin North Am* 2010;**28** (2):335–42.
50. Cortez MV, Oliveira CMC, Monte RL, de Araújo JR, Braga BB, Reis DZ, et al. HIV-associated tuberculous lymphadenitis: the importance of polymerase chain reaction (PCR) as a complementary tool for the diagnosis of tuberculosis—a study of 104 patients. *An Bras Dermatol* 2011;**86**(5):925–31.
51. Pires CAA, Jucá Neto FOM, de Albuquerque NC, Macedo GMM, Batista K de NM, Xavier MB. Leprosy reactions in patients coinfected with HIV: clinical aspects and outcomes in two comparative cohorts in the Amazon region, Brazil. *PLoS Negl Trop Dis* 2015;**9**(6):e0003818.
52. Gonçalves LS, Gonçalves BML, Fontes TV. Periodontal disease in HIV-infected adults in the HAART era: clinical, immunological, and microbiological aspects. *Arch Oral Biol* 2013;**58**(10):1385–96.
53. Imai K, Ochiai K. Role of histone modification on transcriptional regulation and HIV-1 gene expression: possible mechanisms of periodontal diseases in AIDS progression. *J Oral Sci* 2011;**53**(1):1–13.
54. Lovegrove JM. Dental plaque revisited: bacteria associated with periodontal disease. *J N Z Soc Periodontol* 2004;**87**:7–21.
55. Thappa DM, Karthikeyan K. Exaggerated scabies: a marker of HIV infection. *Indian Pediatr* 2002;**39** (9):875–6.
56. Marra F, Chiappetta MC, Vincenti V. Ear, nose and throat manifestations of mucocutaneous Leishmaniasis: a literature review. *Acta Biomed* 2014;**85**(1):3–7.
57. Hooja S, Sharma B, Jindal A, Vyas N. First reported cases of diffuse cutaneous leishmaniasis in human immunodeficiency virus positive patients in Jaipur District of Rajasthan. *India Trop Parasitol* 2014;**4** (1):50–2.
58. Talat H, Attarwala S, Saleem M. Cutaneous leishmaniasis with HIV. *J Coll Physicians Surg Pak* 2014;**24** (Suppl. 2):S93–5.

59. Zijlstra EE. PKDL and other dermal lesions in HIV co-infected patients with leishmaniasis: review of clinical presentation in relation to immune responses. *PLoS Negl Trop Dis* 2014;**8**(11):e3258.
60. van Griensven J, Diro E. Visceral leishmaniasis. *Infect Dis Clin N Am* 2012;**26**(2):309–22.
61. Bari AU. Clinical spectrum of cutaneous leishmaniasis: an overview from Pakistan. *Dermatol Online J* 2012;**18**(2):4.
62. Lindoso JA, Cota GF, da Cruz AM, Goto H, Maia-Elkhoury ANS, Romero GAS, et al. Visceral leishmaniasis and HIV coinfection in Latin America. *PLoS Negl Trop Dis* 2014;**8**(9):e3136.
63. Umoru D, Oviawe O, Ibadin M, Onunu A, Esene H. Mucocutaneous manifestation of pediatric human immunodeficiency virus/acquired immunodeficiency syndrome (HIV/AIDS) in relation to degree of immunosuppression: a study of a West African population. *Int J Dermatol* 2012;**51**(3): 305–12.
64. Walsh TJ, Anaissie EJ, Denning DW, Herbrecht R, Kontoyiannis DP, Marr KA, et al. Treatment of aspergillosis: clinical practice guidelines of the Infectious Diseases Society of America. *Clin Infect Dis* 2008;**46**(3):327–60.
65. Deeken JF, Tjen-A-Looi A, Rudek MA, Okuliar C, Young M, Little RF, et al. The rising challenge of non-AIDS-defining cancers in HIV-infected patients. *Clin Infect Dis* 2012;**55**(9):1228–35.
66. Bouvard V, Baan R, Straif K, Grosse Y, Secretan B, El Ghissassi F, et al. A review of human carcinogens—Part B: biological agents. *Lancet Oncol* 2009;**10**(4):321–2.
67. Zhang J-C, Sun Y-T. Efavirenz-induced exfoliative dermatitis. *Scand J Infect Dis* 2013;**45**(1):70–2.
68. Introcaso CE, Hines JM, Kovarik CL. Cutaneous toxicities of antiretroviral therapy for HIV. *J Am Acad Dermatol* 2010;**63**(4):549–61.
69. Altman K, Vanness E, Westergaard RP. Cutaneous manifestations of human immunodeficiency virus: a clinical update. *Curr Infect Dis Rep* 2015;**17**(3):464.
70. Iwuala SO, Lesi OA, Fasanmade OA, Sabir AA, Olamoyegun MA, Okany CC. Prevalence of and risk factors for lipoatrophy in patients with HIV infection in Nigeria. *AIDS Res Treat* 2015;**2015**:1–7.
71. Ramdial PK. Dermatopathological challenges in the human immunodeficiency virus and acquired immunodeficiency syndrome era. *Histopathology* 2010;**56**:39–56.
72. García-Silva J, Almagro M, Peña-Penabad C, Fonseca E. Indinavir-induced retinoid-like effects: incidence, clinical features and management. *Drug Saf* 2002;**25**(14):993–1003.
73. Luther J, Glesby MJ. Dermatologic adverse effects of antiretroviral therapy: recognition and management. *Am J Clin Dermatol* 2007;**8**(4):221–33.
74. Sidhu HK, Chaffee BH, Tvetenstrand CD, Sidhu JS. Fuzeon-induced collagenophagic granuloma: a peculiar granulomatous injection site reaction to fuzeon—a case report and review of literature. *Int J Surg Pathol* 2010;**18**(5):384–7.
75. Cedeno-Laurent F, Gómez-Flores M, Mendez N, Ancer-Rodríguez J, Bryant JL, Gaspari AA, et al. New insights into HIV-1-primary skin disorders. *J Int AIDS Soc* 2011;**14**:5.
76. Menon K, Van Voorhees AS, Bebo Jr. BF, Gladman DD, Hsu S, Kalb RE, et al. Psoriasis in patients with HIV infection: from the medical board of the National Psoriasis Foundation. *J Am Acad Dermatol* 2010;**62**(2):291–9.
77. Resneck Jr. JS, Van Beek M, Furmanski L, Oyugi J, LeBoit PE, Katabira E, et al. Etiology of pruritic papular eruption with HIV infection in Uganda. *JAMA* 2004;**292**(21):2614–21.
78. Labhardt ND, Lejone T, Setoko M, Poka M, Ehmer J, Pfeiffer K, et al. A clinical prediction score in addition to WHO criteria for anti-retroviral treatment failure in resource-limited settings—experience from Lesotho. *PLoS One* 2012;**7**(10):e47937.
79. Choudhary SV, Singhal KV, Shukla S, Koley S. Generalized granuloma annulare—a rare cutaneous marker of AIDS. *Indian J Sex Transm Dis* 2011;**32**(2):137–8.
80. Lukács J, Schliemann S, Elsner P. Treatment of generalized granuloma annulare—a systematic review. *J Eur Acad Dermatol Venereol* 2015;**29**(8):1467–80.
81. Guillevin L, Dörner T. Vasculitis: mechanisms involved and clinical manifestations. *Arthritis Res Ther* 2007;**9**(Suppl. 2):S9.
82. Forbess L, Bannykh S. Polyarteritis nodosa. *Rheum Dis Clin N Am* 2015;**41**(1):33–46.
83. Ferri C. Mixed cryoglobulinemia. *Orphanet J Rare Dis* 2008;**3**(1):25.

84. Kosmas N, Kontos A, Panayiotakopoulos G, Dimitrakopoulos A, Kordossis T. Decreased prevalence of mixed cryoglobulinemia in the HAART era among HIV-positive, HCV-negative patients. *J Med Virol* 2006;**78**(10):1257–61.
85. Takada S, Shimizu T, Hadano Y, Matsumoto K, Kataoka Y, Arima Y, et al. Cryoglobulinemia (review). *Mol Med Rep* 2012;**6**(1):3–8.
86. Ramos-Casals M, Stone JH, Cid MC, Bosch X. The cryoglobulinaemias. *Lancet* 2012;**379** (9813):348–60.
87. Sugimoto T, Tsuda A, Kito K, Uzu T, Kashiwagi A. Henoch-Schönlein purpura in a patient with human immunodeficiency virus infection. *Rheumatol Int* 2008;**28**(6):615–6.
88. Zaid M, Tan K, Smitasin N, Tambyah PA, Archuleta S. Henoch-Schönlein purpura associated with adult human immunodeficiency virus infection: case report and review of the literature. *Ann Acad Med Singapore* 2013;**42**(7):358–60.
89. Swensen S, Schwarzkopf R. Total joint arthroplasty in human immunodeficiency virus positive patients: joint arthroplasty in HIV positive patients. *Orthop Surg* 2012;**4**(4):211–5.
90. Güerri-Fernández R, Villar-García J, Díez-Pérez A, Prieto-Alhambra D. HIV infection, bone metabolism, and fractures. *Arq Bras Endocrinol Metabol* 2014;**58**(5):478–83.
91. Beatty GW. Immune reconstitution inflammatory syndrome. *Emerg Med Clin North Am* 2010;**28** (2):393–407.
92. Müller M, Wandel S, Colebunders R, Attia S, Furrer H, Egger M. Immune reconstitution inflammatory syndrome in patients starting antiretroviral therapy for HIV infection: a systematic review and meta-analysis. *Lancet Infect Dis* 2010;**10**(4):251–61.
93. Eyer-Silva Wde A, Salgado MC, Pinto JF, Ferry FR, Neves-Motta R, Azevedo MC, et al. Acute gouty arthritis as a manifestation of immune reconstitution inflammatory syndrome after initiation of antiretroviral therapy. *Rev Inst Med Trop Sao Paulo* 2012;**54**(4):231–3.
94. Tsang CS, Samaranayake LP. Immune reconstitution inflammatory syndrome after highly active antiretroviral therapy: a review. *Oral Dis* 2010;**16**(3):248–56.
95. Prasetyo AA, Sariyatun R, Reviono, Sari Y, Hudiyono, Haryati S, et al. The APOBEC3B deletion polymorphism is associated with prevalence of hepatitis B virus, hepatitis C virus, Torque Teno virus, and *Toxoplasma gondii* co-infection among HIV-infected individuals. *J Clin Virol* 2015;**70**:67–71.
96. Prasetyo AA, Nurfitria FB, Adnan ZA, Redhono D, Sari Y, Hudiyono. The -1082 A allele of the interleukin-10 single nucleotide polymorhpism -1082 G/A is associated with hepatitis virus co-infection in Indonesian Javanese HIV patients. *Int J Interferon Cytokine Mediat Res* 2015;**7**:1–7.
97. Haryati S, Prasetyo AA, Sariyatun R, Sari Y, Murkati. Interferon-γ +874A/T polymorphism associated with *Toxoplasma gondii* seropositivity in HIV patients. *Asian Pac J Trop Dis* 2015;**5**(10):798–803.
98. Prasetyo AA, Dharmawan R, Raharjo I, Hudiyono. Human leukocyte antigen-E alleles are associated with hepatitis C virus, Torque Teno virus, and Toxoplasma co-infections but are not associated with hepatitis B virus, hepatitis D virus, and GB virus C co-infections in human immunodeficiency virus patients. *J Global Infect Dis* 2016;**8**(2):75–81.
99. Sari Y, Haryati S, Raharjo I, Prasetyo AA. Toxoplasma and viral antibodies among HIV patients and inmates in Central Java, Indonesia. *Southeast Asian J Trop Med Public Health* 2015;**46**(6):977–85.
100. Marwoto, Lismana UH, Prasetyo AA, Suradi, Reviono, Harsini. Correlation of single nucleotide polymorphism 35-kb upstream of HLA-C and clinical profile of multidrug-resistant tuberculosis. *J Clin Diagn Res* 2015;**9**(9):DC10–3.
101. Marwoto, Prasetyo AA, Sariyatun R, Suradi, Reviono, Harsini. Human leukocyte antigen-G (HLA-G) 14-bp deletion polymorphism is associated with decreased risk of pulmonary fibrosis in Indonesian Javanese patients with multidrug-resistant tuberculosis. *Asian Biomed* 2015;**9**(4):473–9.
102. Prasetyo AA, Zaini KUN. Establishing mean CD4+ T cell values among healthy Javanese adults in Indonesia. *Southeast Asian J Trop Med Public Health* 2015;**46**(4):1–7.
103. Prasetyo AA, Ariapramuda ER, Kindi EA, Dirgahayu P, Sari Y, Dharmawan R, et al. Men having sex with men in Surakarta, Indonesia: demographics, behavioral characteristics and prevalence of blood borne pathogens. *Southeast Asian J Trop Med Public Health* 2014;**45**(5):1032–47.

CHAPTER 5

Diseases Caused by *Malassezia* Species in Human Beings

M. Dolenc-Voljč
University Medical Centre Ljubljana; Faculty of Medicine, University of Ljubljana, Ljubljana, Slovenia

1. INTRODUCTION

Malassezia yeasts are dimorphic facultative pathogenic fungi, belonging to the phylum Basidiomycota, class Malasseziomycetes.[1] In saprophytic form, they are part of normal cutaneous microbial flora in adults. They undergo asexual reproduction and are able to exist in both yeast and mycelial forms, with the yeast form being most commonly isolated in normal skin.[2] Under certain circumstances, they are potentially pathogenic to humans and animals. In humans, they can cause various superficial skin diseases in predisposed individuals. Apart from dermatophytes, they are the most common causative microorganisms for fungal skin diseases worldwide.[3] Diseases can be induced either by direct invasion of the tissue by fungal filaments or indirectly through immunological and metabolic mechanisms induced by the yeast. On very rare occasions, some *Malassezia* species can cause severe life-threatening systemic infections.[1]

First recognition of *Malassezia* yeasts dates back to 1846, when Karl Ferdinand Eichstedt noticed yeast cells and hyphae in patients with pityriasis versicolor.[4] In 1873, Sebastiano Rivolta observed lipophilic yeasts in patients with psoriasis.[5] In 1874, Louis-Charles Malassez recognized oval spores in scaly lesions of the scalp, which were in 1884 named *Sacharomyces ovalis* and *Sacharomyes sphaericus* by Giulio Bizzozero.[4,6] In 1889 Henri Baillon suggested the name *Malassezia furfur* for yeasts he had observed in patients with pityriasis versicolor. Raymond Sabouraud assigned the yeast forms found in dandruff to a new genus *Pityrosporum* in 1904 and named it *Pityrosporum malassezii*. This name was altered in 1913 by Aldo Castellani and Albert John Chalmers into *Pityrosporum ovale*. In 1951, Morris Aaron Gordon isolated globose yeasts from patients with pityriasis versicolor and he named this morphological type of the yeast *Pityrosporum orbiculare*.[4,6]

Controversies in the nomenclature of *Malassezia* yeasts continued in the following decades and two taxonomic systems coexisted for many years.[7] In dermatology, the term *Pityrosporum* was used, created by Sabouraud. *Pityrosporum ovale* was used for seborrheic dermatitis and *Pityrosporum orbiculare* for pityriasis versicolor. Botanists and mycologists preferred to use the name *Malassezia*.[7] The main reason for confusion was an inability to culture *Malassezia* yeasts and their dimorphism, recognized only in 1977.

The Microbiology of Skin, Soft Tissue, Bone and Joint Infections
ISSN 2451-9006, http://dx.doi.org/10.1016/B978-0-12-811079-9.00005-7

Table 1 Currently recognized *Malassezia* species, year of first description and main hosts[4,10]

Malassezia spp.	Year of description	Main host
Malassezia furfur	1889	Man Cow, elephant, pig, monkey, ostrich, pelican
Malassezia pachydermatis	1925	Dogs, cats, wild animals, birds
Malassezia sympodialis	1990	Man Horse, pig, sheep
Malassezia globosa	1996	Man Cow, cheetah
Malassezia obtusa	1996	Man
Malassezia slooffiae	1996	Man Pig, goat, sheep
Malassezia restricta	1996	Man
Malassezia dermatis	2002	Man
Malassezia japonica	2003	Man
Malassezia nana	2004	Cat, cow, dog
Malassezia yamatoensis	2004	Man
Malassezia caprae	2007	Goat, horse
Malassezia equina	2007	Horse, cow
Malassezia cuniculi	2011	Rabbit

Molecular techniques implemented major revision of the genus *Malassezia* in 1996. Culture studies and molecular techniques have proved that the genus *Malassezia* and *Pityrosporum* are identical.[8] The name *Malassezia* was given precedence. Gueho revised the *Malassezia* genus using morphology, ultrastructure, physiology and rRNA gene sequencing studies. He classified the genus into seven species: *Malassezia* (M) *furfur*, *M. sympodialis*, *M. globosa*, *M. restricta*, *M. slooffiae*, *M. obtusa*, and *M. pachydermatis*.[9] Within the last 15 years, new diagnostic methods have enabled the identification of new species. Since 2002, seven new species have been identified (*M. dermatis*, *M. japonica*, *M. nana*, *M. yomatoensis*, *M. equina*, *M. caprae*, and *M. cuniculi*), with different host preferences (Table 1).[10–16] Genetic heterogeneity of different isolates has been demonstrated.[17] Various genetic subgroups have been identified within the species *M. furfur*, *M. sympodialis*, and *M. pachydermatis*. Most *Malassezia* species can only be reliably identified using molecular biological techniques.[3] The taxonomy of *Malassezia* yeasts is still in a developing stage.

In this chapter, clinical manifestations of *Malassezia*-related skin and systemic diseases are presented. Ecology and epidemiology of different subtypes of *Malassezia* yeasts and their pathogenic role in diseases are briefly discussed.

2. ECOLOGY OF *MALASSEZIA* YEASTS

Studies have revealed that *Malassezia* yeasts are universally present in the population.[4] Most of the *Malassezia* species are lipid dependent and are mainly found in the infundibulum of the sebaceous glands and in sebum-rich areas of the body, such as the trunk, back, face, scalp and in the external auditory meatus.[18] The highest concentrations of the yeasts are found in areas rich in sebaceous glands.[3,4] Skin involvement in diseases is therefore localized to specific areas. *Malassezia* yeasts are the most common isolated fungi from the head and trunk region and the most common participant of the skin "mycobiome."[3] In the lower parts of the body they are isolated only exceptionally. *Malassezia* species do not attack the hair shaft, nails or mucous membranes.[4]

Human skin in healthy adults can be colonized by various *Malassezia* species (*M. sympodialis, M. globosa, M. restricta, M. slooffiae, M. furfur, M. obtusa, M. dermatis*).[6–8] Different species have been recovered from different body sites and in different skin diseases. Geographical variations in the distribution of *Malassezia* species have been observed.[2] Molecular studies found that *Malassezia* species in healthy persons differ in Asian and Caucasian population.[8] Some *Malassezia* species can probably be transmitted between persons via direct contact. *M. furfur, M. globosa,* and *M. restricta* are the most common species isolated in skin diseases.[10,19]

The highest quantities of *Malassezia* yeasts are reached after puberty and in early adulthood and their density decreases gradually thereafter. They are usually absent in newborns but can be cultured from the nape or shoulders of children by 1 year of age, without any symptoms of skin disease.[20] Transmission usually occurs from direct contact with parents. In hospitals, healthcare workers can be a source of transmission to infants and severely immunosuppressed patients. In intensive neonatal care units, 30%–64% of premature infants are colonized with *M. furfur* within the first weeks of life.[21] Certain strains have been isolated on a neonatal incubator and are capable of long-term colonization of plastic surfaces.[7] They express resistance to superficial cleaning. In nonintensive care settings, fewer than 10% of hospitalized newborns are colonized.[21]

Recently, DNA-based analyses have identified *M. restricta, M. globosa* and *M.* species (not further determined) from the stool samples. *Malassezia* yeasts could be ingested and could constitute a part of the diverse ecosystem of the human gut.[22]

Studies in animals have shown the colonization and pathogenicity of different *Malassezia* yeasts in a wide range of animal hosts (Table 1). Five *Malassezia* species are exclusively zoophilic (*M. pachydermatis, M. equina, M. caprae, M. nana, M. cuniculi*). *M. pachydermatis* is the most common isolated species in pets.[10,23,24]

3. *MALASSEZIA*-ASSOCIATED SKIN DISEASES

Under normal conditions, *Malassezia* yeasts manage to evade local immune response and remain in equilibrium with human skin.[1] Circumstances that disturb the symbiotic status between the yeast and the host and turn *Malassezia* yeast from a commensal to a pathogen remain insufficiently explained.[2] The mechanisms involved in the pathogenesis of skin diseases are diverse. Direct invasion of epidermis by fungal filaments and secretion of enzymes can cause damage of the epidermal barrier function and induce inflammation. *Malassezia* yeasts can be involved in the melanization process. They can upregulate or downregulate immune response and can induce allergic sensitization to their allergens. Keratinocytes can produce both proinflammatory and antiinflammatory cytokines after stimulation with *Malassezia* yeasts.[1,2,25]

The prevalence and clinical presentation of *Malassezia*-associated skin diseases depend on many patient-related factors, such as age, gender and immunological status. Environmental factors also have an important influence, since some of the diseases are more common in regions with hot and humid climates.

3.1 Pityriasis Versicolor

Pityriasis versicolor (PV) is a superficial fungal disease of the skin, localized in the stratum corneum of the epidermis. It is one of the most common skin diseases in adults worldwide. It is generally asymptomatic and some regard it as more of a cosmetic problem. However, if not treated, PV may become widespread and chronic. Recent studies using molecular techniques have shown that *M. globosa* is the most common causative pathogen.[26] Less commonly, *M. furfur*, *M. sympodialis*, *M. restricta*, and *M. obtusa* may also be involved.[3] The frequency of *Malassezia* species varies in reports from different continents. Quantitative analyses have shown that the density of *M. globosa*, *M. restricta*, and *M. sympodialis* is higher in lesional skin compared to nonlesional skin areas.[3]

3.1.1 Predisposing Factors

PV results from proliferation of saprophytic yeast to a pathogenic hyphal form. It can occur in healthy individuals as well as in immunocompromised patients. Endogenous host factors are of the utmost importance for the transformation of the yeast. A lipid-rich moist skin promotes the development of the disease.[20] Young adults around 20 years of age are most commonly affected since sebaceous gland activity is maximum at this age.[18] It only rarely occurs in children and older adults. Increased greasiness of the skin and application of oily products to the skin have also been proposed as favorable factors. Individuals with increased sweating (hyperhidrosis) are more often affected (sportsmen, soldiers). The onset is more common in summer months. Prevalence is increased in tropical climates due to increased temperature and humidity and may reach 40%.[3,26] In colder climates, prevalence is estimated to be less than 1%.[26] Systemic immunosuppression is a well-known predisposing factor. Patients on long-term systemic corticosteroids or treated with other immunosuppressive drugs, transplant patients and HIV-infected individuals are often affected.[3,20,27] Malnutrition and oral contraceptives have also been suggested as predisposing factors.[18]

3.1.2 Pathogenesis

Skin lesions result from tissue invasion of fungal filaments, which occur between and within keratinized cells[26] and lead to damage of the stratum corneum and dyspigmentation, and induce mild inflammation. Histopathological changes are presented in the upper layer of the stratum corneum. Hyperkeratosis, parakeratosis, acanthosis and a mild inflammatory infiltrate with mast cells, T cells, macrophages and Langerhans cells are observed.[3,26] In a hypopigmented variant, melanosomes are smaller, while in a hyperpigmented variant, they appear larger.[26] Alteration in pigmentations probably results from different mechanisms. Azelaic acid, produced by *Malassezia* species, may cause inhibition of the enzyme tyrosinase, needed in the synthesis of melanin. A cytotoxic effect of azelaic acid on melanocytes has also been proposed.[3] *Malassezia* produce bioactive indoles from tryptophan and some of them could be implicated in melanogenesis. Among them, malassezin, induribin, indololcarbazole and formylindololcarbazole are ligands of the aryl hydrocarbon receptor (AhR).[1,3] Malassezin can cause apoptosis of melanocytes.[3] The AhR activity of *Malassezia* indoles also results in downregulation of the inflammatory response. *Malassezia* can downregulate the ability of dendritic cells to present antigens in response to toll-like receptor stimulation.[1] In addition, the increased thickness of the epidermis due to scaling may have an umbrella-like effect and prevent tanning.[20]

3.1.3 Clinical Findings

Skin lesions are distributed on the seborrheic regions of the head, neck, upper trunk, on the shoulders and upper arms. The nape, back and chest are sites of predilection.[20] Lesions may be present along the scalp and in the scalp. Rarely, the face can also be involved. Sharply bordered small round macules present first, usually a few centimeters in diameter, which may be red-brown, dirty yellow, pink, café-au-lait colored, or hypopigmented (Fig. 1). In tanned skin, lesions are hypopigmented. Slight superficial scaling may be seen. The distribution is usually symmetrical. In the course of the disease progression, lesions

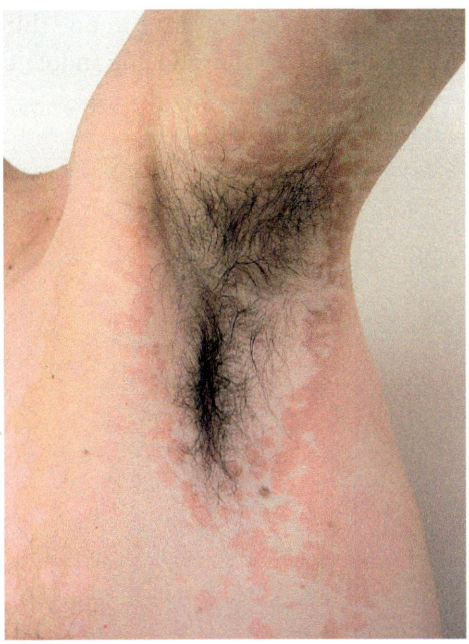

Fig. 1 Pityriasis versicolor.

may coalesce into larger irregularly distributed patches. In a more widespread disease, lesions can reach the umbilicus and thighs and can involve the forearms, back of the hands and popliteal fossae.[3,26] Palmar lesions are unusual in a temperate climate but have been reported in the tropics. Most patients are asymptomatic, although pruritus can occur in some cases.[5,18]

Several distinct forms of PV have been recognised, according to pigmentation and localization of skin lesions: hypopigmented type (PV alba), hyperpigmented type, PV rubra, PV with atrophic lesions and PV localized on the lower abdomen and groin, associated with oval yeasts.[6,19,28]

Differential diagnosis of PV includes several other skin diseases, such as pityriasis rosea, seborrheic dermatitis, tinea, vitiligo, erythrasma, confluent and reticular papillomatosis, pityriasis alba, psoriasis and secondary syphilis.[3]

3.1.4 Laboratory Examination

In clinical practice, diagnosis is usually made on the basis of typical clinical features. It can be confirmed by direct microscopic examination of skin scraping using 10%–20% potassium hydroxide. Characteristic clusters of grape-like round small yeasts are seen, admixed with short thick and fragmented hyphae, in a pattern of "spaghetti and meatballs" appearance (Fig. 2). The presence of mycelium confirms the diagnosis. Microscopic examination does not allow differentiation of *Malassezia* species.[3]

Skin lesions may reveal bright yellow fluorescence with Woods lamp examination. *Malassezia* yeasts produce indole compounds that fluoresce under Wood (UVA) light.[3,18] However, Wood's light examination is only positive in one third of cases.[18] It might be that not all *Malassezia* species are capable of producing indole compounds.

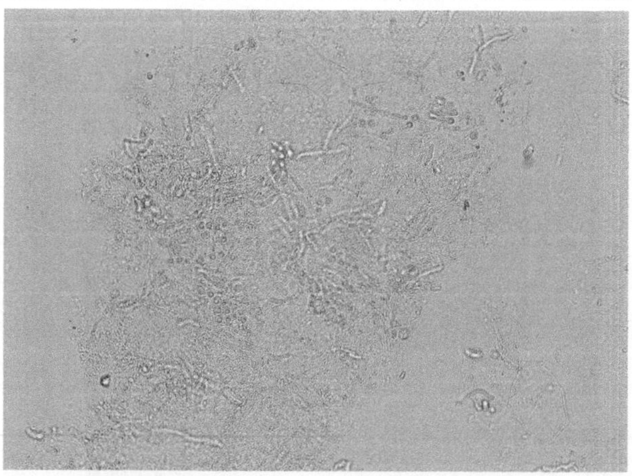

Fig. 2 Microscopic view of *Malassezia* species.

Malassezia yeasts can be cultivated on Sabouraud agar with addition of olive oil and Tween, on modified Dixon agar and Notman's agar.[3,5] Cultivation of *Malassezia* species has no diagnostic value in clinical practice and is not used routinely.

For research purposes, molecular techniques are used. Sequencing of the internal transcribed spacer gen region in ribosomal RNA is the gold standard for identification of *Malassezia* yeasts.[3] Terminal fragment length polymorphism, amplified fragment length polymorphism analysis, pulsed field gel electrophoresis and MALDI-TOF mass spectrometry can also be used.[3,18,26] These new diagnostic techniques can confirm *Malassezia* yeasts at a strain level. PCR-based molecular techniques can detect the yeasts directly from scale samples.[8] Molecular techniques are not used in routine diagnostics since the results do not have any therapeutic consequences.[3]

3.1.5 Treatment

In most patients, treatment with topical antifungal drugs is effective. Topical azoles (ketoconazole, clotrimazole, bifonazole, miconazole) can be applied in creams, solutions, sprays or shampoos.[29] Topical terbinafine is also effective. The recommended period of topical treatment is 2–3 weeks. In some countries, shampoos containing selenium sulfide and zinc pyrithione are used. Oral antifungal treatment is reserved for widespread and recalcitrant cases, resistant to topical treatment. Itraconazole 200 mg daily for 5 days is the drug of choice. Fluconazole can also be used.[3,26]

Hypopigmented lesions usually persist for several months after treatment. Recurrences are very common, since *Malassezia* yeasts are part of normal skin flora and impossible to eradicate. Predisposing factors are also difficult to influence in most patients.

3.2 *Malassezia* Folliculitis

Malassezia folliculitis (MF) was first assumed to be caused by *Pityrosporum ovale* by Graham in 1968.[30] In 1973, Potter et al. confirmed the link between *Pityrosporum* organism and folliculitis.[31] The most common causative pathogens reported are *M. globosa*, *M. pachydermatis* and *M. furfur*.[21] It results from invasion of the hair follicle with large numbers of *Malassezia* yeasts.[18] Follicles become dilated and distended. Secondary inflammatory infiltrate evolves along the follicular epithelium, with lymphocytes, neutrophils and histiocytes and may involve the whole pilosebaceous unit. The inflammatory response results from hydrolization of triglycerides into free fatty acids, induced by *Malassezia* yeasts.[18]

3.2.1 Predisposing Factors

Similar to PV, MF is most often seen in teenagers and young adult males. Triggering factors are occlusion, lack of skin hygiene that can lead to sebum accumulation, and immunosuppression.[6,18,21] Broad spectrum antibacterial therapy can also predispose to this condition. It has been observed in patients with diabetes, leukemia, Hodgkin disease, steroid treatment, bone marrow transplantation, heart and renal transplantation and in AIDS.[21] It has also been observed in severely ill patients admitted to hospital[18,26] and in epidemic forms in intensive care units.[21] Treatment with new biological drugs (infliximab, erlotinib, cetuximab) has also been observed to increase predisposition for this type of folliculitis.[32–34] Colonization of the follicles with *Malassezia* yeasts has also been found in eosinophilic folliculitis in patients with AIDS[18,21] and in steroid-induced folliculitis.[8]

3.2.2 Clinical Findings

MF presents with small papules, 2–3 mm in diameter, dome-shaped, with a central "dell" representing the follicle. Skin lesions are localized around follicular openings. Papules are disseminated on the back, chest and shoulders. They occasionally evolve into pustules. They heal with a brown crust.[20] They are usually very pruritic but may also be asymptomatic. MF is more common during the summer period and in countries with a tropical climate. More severe cases are observed in the tropics, where the face can also be involved and more severe skin lesions with nodules and cysts may appear.[18,21]

A special type of *Malassezia* folliculitis may appear in newborns on the scalp and face.[5,35] Transfer of yeasts occurs from the mother's face during the first week of life. Small pustules with an erythematous border are seen on the cheeks, chin, forehead and scalp. A lipid-rich skin environment is provided by sebaceous hyperplasia due to maternal androgens. The situation is temporary and self-limiting. It can be misdiagnosed as neonatal acne.[35] The causative role of *Malassezia* yeasts in this disease has been a matter of discussion.[5]

MF is probably underdiagnosed in clinical practice since it might be attributed to other causes of folliculitis. It can look similar to bacterial folliculitis, *Candida* folliculitis and acneiform exanthema. Diagnosis can be difficult and is mainly accomplished by microscopic examination of papules and pustules.

3.2.3 Treatment

Topical azole antifungals can be used in creams or ketoconazole and selenium sulfide in shampoo. Topical treatment is usually effective in immunocompetent persons and should be used for one month.[36] Oral itraconazole and fluconazole are recommended for extensive and recalcitrant lesions and in immunocompromised patients.[21]

3.3 Seborrheic Dermatitis and Dandruff

Seborrheic dermatitis (SD) is a chronic inflammatory skin disease of the sebaceous gland-rich regions. The predilection sites are the scalp, skin behind the ears, eyebrows, nasolabial folds (Fig. 3), the chest over the sternum and the back between scapulae. Skin lesions are erythematous and scaly macules that may cause mild itching. The prevalence in the general population is between 1% and 3%.[3,18] It is much higher in immunocompromised patients. It has been observed in 9.5% of renal transplant patients.[37]

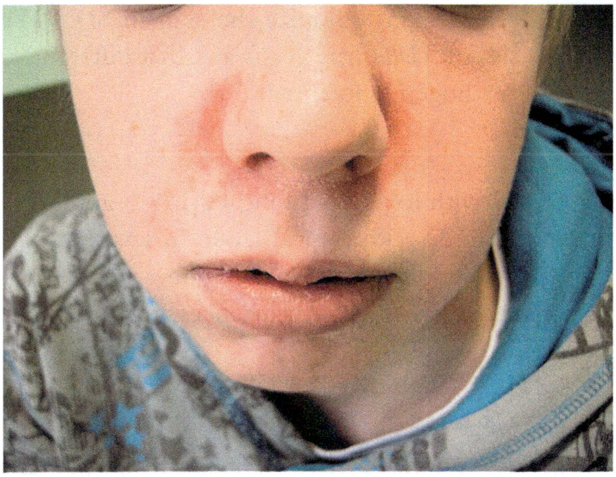

Fig. 3 Seborrheic dermatitis on the face.

In HIV-positive patients, it has been estimated to be between 20% and 40%, and may reach 40%–80% in AIDS patients.[8]

Dandruff (D) is a closely related condition, with oily scaling of the scalp without inflammation. It is very often associated with SD. Adolescents and young adults are most often affected due to the high activity of sebaceous glands in this age. Both SD and D are more common in patients with Parkinson disease, depression and spinal injuries.[18]

3.3.1 Pathogenesis

The exact role of *Malassezia* species in the pathogenesis of SD and D remains insufficiently explained. Various pathogenic mechanisms have been suggested. Lipases produced by *Malassezia* species split triglycerides into free fatty acids, leading to irritation of the skin. Production of lipases and other enzymes also causes damage to the epidermal barrier.[6] *M. globosa* is capable of producing oleic acid, which induces irritation and scaling.[19] *Malassezia* metabolites, such as malassezin and indole-3-carbaldehyde, are involved in immune regulation.[19] *Malassezia* can activate the local immune response and can induce the production of proinflammatory cytokines in antigen-presenting cells and keratinocytes (interleukin (IL)-1, IL-2, IL-4, IL-6, interferon-gamma and tumour necrosis factor alpha).[3,8,38] Some other *Malassezia*-unrelated endogenous host factors are also involved in the complex pathogenesis of SD and D.

The frequency of *Malassaezia* yeasts isolated in SD depends on the method used. Fungal DNA detection results in higher detection rates than cultivation. Different findings have been reported from different countries[3,6] and are hard to compare. The most common species found in Japan are *M. globosa* and *M. restricta*. Some authors from other countries have also reported the detection of *M. furfur*, *M. sympodialis*, *M. yamatoensis*, *M. dermatis*, and *M. slooffiea*.[3,8,18] Genotypic analysis of *M. globosa* and *M. restricta* has found that specific pathogenic strains colonize the skin in SD.[39] These studies have revealed the genetic diversity of *Malassezia* yeasts and that different strains of the same species may have different pathogenic potential.[39]

3.3.2 Treatment

SD and D are usually treated with topical antifungal drugs. Significant clinical improvement is observed in moderate SD with topical azole. In severe cases, oral itraconazole can be used. In some patients, additional treatment regimes are needed, using other antimicrobial and antiinflammatory drugs.[3,29] Recurrences are very common.

3.4 Atopic Dermatitis

Atopic dermatitis (AD) is a chronic allergic skin disease presenting with pruritic eczematous lesions. Genetic predisposition and environmental factors play an important role in the development of the disease. Dry skin and dysfunction in the epidermal barrier leave the skin more vulnerable to microbes and allergens. Cutaneous microorganisms are considered to be an exacerbating factor for AD. IgE-mediated sensitization to food and inhalation allergens can be confirmed in the great majority of patients with AD with skin prick tests and specific serum IgE antibodies. Many patients may have asthma and allergic rhinitis concomitantly. Children are more often affected than adults.

Many reports have indicated that *Malassezia* yeasts may play an important role as an aggravating factor in AD located in the head and neck area in adults.[40] Specific *Malassezia* IgE antibodies can be found in 20% to 100% of patients with AD. The highest prevalence has been found in adult patients with a head and neck distribution and it is about 10 times higher than in children.[3,40] Positive skin prick tests to *Malassezia* yeasts have also been found in adult patients with AD, especially in those with a head and neck distribution.[40]

Molecular studies have found that *M. globosa* and *M. restricta* are the most common representatives of skin microflora in AD lesions, followed by *M. sympodialis*, *M. furfur*, and *M. dermatis*.[23,38] Other species have been isolated rarely.[8] The frequency of *Malassezia* species in AD may vary in different geographic regions.

3.4.1 Pathogenesis

Penetration of *Malassezia* cells through the damaged epidermal barrier could activate sensitization in patients with AD.[6] Scratching could contribute to sensitization. Sensitization of atopic patients to *Malassezia* yeasts is specific to those with skin manifestations and is not present in atopic patients without AD.[6] *Malassezia* can exacerbate AD through a T-cell mediated response. Antigen-presenting cells (dendritic cells) can be stimulated by *Malassezia* yeasts.[8] Studies have shown that keratinocytes and peripheral blood mononuclear cells respond to *Malassezia* species by secreting Th2-type cytokines (IL-4, IL-5, IL-10, IL-13).[38] Some glycoproteins containing mannose chains in the molecules have been identified as specific antigens in *Malassezia* species.[18,40] Three major antigens have been recognized in *M. globosa*, named Mal g46a, Mal g46b, and Mal g46c.[40] Two new antigens (MGp42 and MGL 1304) were recently identified in *M. globosa*.[8] Studies have shown that different *Malassezia* species express different antigens.

3.4.2 Treatment

Malassezia-associated AD can express resistance to conventional treatment with topical corticosteroids, emollients and systemic antihistamines. Treatment with either topical or systemic antifungal agents can have a beneficial effect in such patients. *Malassezia* species are susceptible to ketoconazole and itraconazole.[41] Tacrolimus, a calcineurin inhibitor used for the treatment of AD, has shown antifungal activity against half of the *Malassezia* strains.[41] According to the European guidelines for treatment of atopic eczema, topical ciclopiroxolamin, oral itraconazole and fluconazole are recommended for the "head-neck-form" of AD.[42] However, yeasts are not the primary causative agent for AD.[38] Antifungal treatment may have a beneficial effect in adult patients with head and neck involvement, but other standard treatment regimes are also needed.

3.5 Psoriasis

Psoriasis is a chronic skin disease characterized by hyperproliferation and hyperkeratinization of the epidermis. It is a multifactorial immune mediated disease whose etiology involves a genetic predisposition, as well as environmental and endogenous factors.

Malassezia yeasts have been associated with psoriasis of the scalp and glans penis.[18] Patients with psoriasis may have antibodies to *Malassezia* and T cells reactive to *Malassezia* have been found in psoriatic lesions.[8] *Malassezia* yeasts could invade epidermal cells and induce keratinocytes to produce proinflammatory cytokines. They are able to upregulate keratinocyte expression of growth factor and heat shock protein 70, which are associated with hyperproliferation of epidermis in psoriatic lesions.[23]

The reported prevalence of *Malassezia* species found in psoriasis lesions differs in different studies. Analysis of the skin microflora by PCR assay found that *M. globosa*, *M. restricta* and *M. sympodialis* are the most common detected species.[8] In another report, *M. furfur* was found in the majority of patients.[43] Other species have been found much less frequently.[8] On the other hand, similar species have also been found in nonlesional skin. Treatment with topical antifungal drugs can improve psoriasis on seborrheic parts in some patients. These findings support the view that *Malassezia* could play a role as an aggravating factor in certain types of psoriasis by aggravating the inflammation.[3,6] The exact pathogenic role of the yeasts in psoriasis, however, remains insufficiently explained.

3.6 Other *Malassezia*-Associated Skin Diseases

Confluent and reticular papillomatosis is a rare skin disease, localized on the trunk, with brown scaling noninflammatory macules distributed in a linear network. Clinically, it can resemble pityriasis versicolor. A possible link between *Malassezia* yeasts and this disease has been suggested in some case reports.[5] *Malassezia* yeasts can often be demonstrated in histopathological skin examination. Since the yeasts can also be found in heathy skin, their pathogenic role is therefore difficult to explain. In addition, confluent and reticular papillomatosis is usually treated successfully with systemic tetracyclines.[5] There is insufficient evidence that the lesions respond to antifungal drugs.[4]

Malassezia induced *onychomycosis* in children and in adults has been suggested in some reports.[44,45] *M. globosa* has been isolated from dystrophic nails and antifungal agents have been observed to be effective in the treatment.[45] *Malassezia* yeasts were occasionally observed in fingernail onycholysis.[5] The evidence that *Malassezia* involves the nail plate as a true pathogen is weak, since the nail plate does not enable an appropriate environment for the growth of lipophilic yeasts.[8] More probably, *Malassezia* yeasts can settle the damaged subungual tissue rather as colonizers than etiologic agents.[4,5]

Malassezia yeasts have also been implicated in *blepharitis*, *otitis externa*, *keratitis* and *nonspecific balanitis*.[5] Since they can also colonize the skin in these areas in healthy subjects, their role in these conditions is difficult to estimate. They have also been suspected in *skin carcinogenesis* as a participant in the induction of basal cell carcinoma.[1]

4. SYSTEMIC AND INVASIVE INFECTIONS

Systemic infections induced by *Malassezia* species occur only sporadically. *Malassezia* fungemia and sepsis have been reported in intensive care units, mainly in neonates. Affected patients are premature, low birth weight infants with multiple comorbidities, extended hospitalization, central venous catheters and nutrition with parenteral lipid emulsions.[21] Hospitalized neonates are often colonized with *M. furfur*. In nonneonatal immunocompromised patients, it has been observed as a complication of bowel surgery, bone marrow transplantation and cancer. Exposure to lipid-rich intravenous infusions was the most common risk factor.[21] Numerous reports of catheter-related *Malassezia* fungemia have been published during the last three decades, especially in neonatal intensive care units, but also in immunocompromised children and adults.[46–48]

Three percent of healthy newborn infants are colonized with *Malassezia* spec., whereas 30% to 64% of hospitalized premature infants become colonized by the second week of life.[21] Colonization is associated with premature birth, admission to nenonatal intensive care units and length of hospitalization. Infants may become colonized by skin contact with parents or healthcare workers. Nosocomial outbreak in an intensive care facility has been reported, due to transmission of *Malassezia pachydermatis* from healthcare workers' hands after being colonized from pet dogs at home.[49]

Catheters become infected by the migration of yeasts from the outer wall to the lumen, adherence of the organism to the catheter material, biofilm production and shedding of the organism in the blood.[21] Mycotic thrombi have been found around the tips of catheters. Fungal fungemia may cause metastatic infection of the heart (endocarditis) and lungs (bronchopneumonia, respiratory distress) and may spread to the spleen, liver, kidneys, skin and brain.[21] If untreated, infections are fatal. Laboratory diagnosis includes detection of *Malassezia* species in blood by direct microscopic examination, cultivation and molecular methods.[21]

With appropriate recognition of this rare and severe septic infection and proper and timely management, systemic infection responds well to treatment.[4] Fluconazole and voriconazole are recommended as the first-line treatment option. Amphotericin B should be reserved for refractory infections.[21]

Other invasive *Malassezia* infections, without fungemia, have occasionally been reported, such as *Malassezia* mastitis, thrombophlebitis, sinusitis, severe otitis externa, meningitis, septic arthritis, soft tissue abscesses and catheter-associated peritonitis in peritoneal dialysis patients.[1,2,21,50]

5. CONCLUSION

Malassezia yeasts are the most prevalent fungi of the human skin "mycobiome" and can cause some of the most common skin diseases worldwide. Within the last two decades, new insights into their role in commensalism and in skin diseases have been revealed. The taxonomy of the genus has changed considerably and is still in development. *Malassezia* yeasts express remarkable biodiversity. The genus currently contains 14 species. Molecular studies have revealed their genetic variability and new species and new strains within the same species can therefore be expected in the future.

The complexity of the host-yeast interaction has not been sufficiently explained. *Malassezia* yeasts may be involved in diverse pathogenic processes. They have the ability to elicit both cellular and humoral immune response in the host and to suppress immune response directed against it. The genus *Malassezia* is considered to be an immunological paradox.[2]

Modern molecular techniques have enabled exact identification and will enable further epidemiological and clinical studies that will improve current knowledge about the relationship to human and animal diseases. Geographical variations in the prevalence of specific *Malassezia* yeasts are also open to further research.

Genetic manipulation strategies for *Malassezia* species will aim at investigation of the genes responsible in the pathogenesis of specific skin diseases.[4] The genome sequence will enable better understanding of regulation and expression of the coding genes involved in different pathological processes and may also identify new drug targets.[4] *Malassezia* yeasts remain a challenging field for further research in dermatology and mycology.

REFERENCES

1. Gaitanis G, Magiatis P, Hantschke M, Bassukas ID, Velegraki A. The Malassezia genus in skin and systemic diseases. *Clin Microbiol Rev* 2012;**25**:106–41.

2. Ashbee HR, Evans EGV. Immunology of diseases associated with Malassezia species. *Clin Microbiol Rev* 2002;**15**:21–57.
3. Nenoff P, Krüger C, Mayser P. Cutaneous Malassezia infections and Malassezia associated dermatoses: an update. *Hautarzt* 2015;**66**:465–84.
4. Hay RJ, Midgley G. Introduction: Malassezia yeasts from a historical perspective. In: Boekhout T, Guého-Kellermann E, Mayser P, Velegraki A, editors. *Malassezia and the skin. Science and clinical practice.* Heidelberg: Springer; 2010. p. 1–16.
5. Crespo-Erchiga V, Hay RJ. Pityriasis versicolor and other Malassezia skin diseases. In: Boekhout T, Guého-Kellermann E, Mayser P, Velegraki A, editors. *Malassezia and the skin. Science and Clinical Practice.* Heidelberg: Springer; 2010. p. 175–99.
6. Gaitanis G, Velegraki A, Mayser P, Bassukas ID. Skin diseases associated with Malassezia yeasts: facts and controversies. *Clin Dermatol* 2013;**31**:455–63.
7. Guého E, Boekhout T, Ashbee HR, Guillot J, Van Belkum A, Faergemann J. The role of Malassezia species in the ecology of human skin as pathogens. *Med Mycol* 1998;**36**:220–9.
8. Harada K, Saito M, Sugita T, Tsuboi R. Malassezia species and their associated skin diseases. *J Dermatol* 2015;**42**:250–7.
9. Guého E, Midgley G, Guillot J. The genus Malassezia with description of four new species. *Antonie Van Leeuwenhoek* 1996;**69**:337–55.
10. Cabañes FJ. Malassezia yeasts: how many species infect humans and animals? *PLoS Pathog* 2014;**10**: e1003892.
11. Sugita T, Takashima M, Shinoda T, Suto H, Unno T, Tsuboi R, et al. New yeast species, Malassezia dermatis, isolated from patients with atopic dermatitis. *J Clin Microbiol* 2002;**40**:1363–7.
12. Sugita T, Takashima M, Kodama M, Tsuboi R, Nishikawa A. Description of a new yeast species, Malassezia japonica, and its description in patients with atopic dermatitis and healthy subjects. *J Clin Microbiol* 2003;**41**:4695–9.
13. Sugita T, Tajima M, Takashima M, Amaya M, Saito M, Tsuboi R, et al. A new yeast, Malassezia yamotoensis, isolated from a patient with seborrheic dermatitis, and its distribution in patients and healthy subjects. *Microbiol Immunol* 2004;**48**:79–83.
14. Hirai A, Kano R, Makimura T, Duarte ER, Hamdan JS, Lachance MA, et al. Malassezia nana sp. nov., a novel lipid-dependent yeast species isolated from animals. *Int J Syst Evol Microbiol* 2004;**54**:623–7.
15. Cabañes FJ, Theelen B, Castellá G, Boekhout T. Two new lipid-dependent Malassezia species from domestic animal. *FEMS Yeast Res* 2007;**7**:1064–76.
16. Cabañes FJ, Vega S, Castellá G. Malassezia cuniculi sp. nov., a novel yeast species isolated from rabbit skin. *Med Mycol* 2011;**49**:40–8.
17. Guého-Kellermann E, Boekhout T, Begerow D. Biodiversity, phylogeny and ultrastructure. In: Boekhout T, Guého-Kellermann E, Mayser P, Velegraki A, editors. *Malassezia and the skin. Science and clinical practice.* Heidelberg: Springer; 2010. p. 17–63.
18. Gupta AK, Batra R, Bluhm R, Boekhout T, Dawson TL. Skin diseases associated with Malassezia species. *J Am Acad Dermatol* 2004;**51**:785–98.
19. Hay RJ. Malassezia, dandruff and seborrheic dermatitis: an overview. *Br J Dermatol* 2011;**165**(Suppl. 2):2–8.
20. Korting HC. Fungal infections. In: Burgdorf WHC, Plewig G, Wolff HH, Landthaler M, editors. *Braun Falco's dermatology.* 3rd ed. Heidelberg: Springer; 2009. p. 209–10.
21. Tragiannidis A, Bisping G, Koehler G, Groll AH. Minireview: Malassezia infections in immunocompromised patients. *Mycoses* 2010;**53**:187–95.
22. Gouba N, Raoult D, Drancourt M. Eukaryote culturomics of the gut reveals new species. *PLoS One* 2014;**9**:e106994.
23. Sugita T, Boekhout T, Velegraki A, Guillot J, Hadina S, Cabañes FJ. Epidemiology of Malassezia-related skin diseases. In: Boekhout T, Guého-Kellermann E, Mayser P, Velegraki A, editors. *Malassezia and the skin. Science and clinical practice.* Heidelberg: Springer; 2010. p. 65–119.
24. Sihelská Z, Váczi P, Conková E. Species composition of Malassezia yeasts in dogs in Slovakia. *Berl Munch Tierarztl Wochenschr* 2016;**129**:351–4.
25. Ashbee HR. Update on the genus Malassezia. *Med Mycol* 2007;**45**:287–303.

26. Hay RJ, Ashbee HR. Fungal infections. In: Griffiths CEM, Barker J, Bleiker T, Chlamers R, Creamer D, editors. *Rook's textbook of dermatology*. 8th ed. Oxford: Blackwell; 2016. p. 32.10–4.
27. Ramos-e-Silva M, Lima CMO, Schechtman RC, Trope BM, Carneiro S. Superficial mycoses in immunodepressed patients (AIDS). *Clin Dermatol* 2010;**28**:217–25.
28. Alkhalifah A, Gari-Toussaint M, Cardot-Leccia N, Lacour JP, Passeron T. Centrifugal hypomelanosis: a new clinical phenotype of Malassezia infection. *Eur J Dermatol* 2016; http://dx.doi.org/10.1684/ejd.2016.2896.
29. Faergemann J. Management of seborrheic dermatitis and pityriasis versicolor. *Am J Clin Dermatol* 2000;**1**:75–80.
30. Graham JH. Pityrosporum ovale in the discussion. *Arch Dermatol* 1968;**98**:421.
31. Potter B, Burgoon CF, Johnson C. Pityrosporum folliculitis. Report of seven cases and review of the Pityrosporum organism relative to cutaneous disease. *Arch Dermatol* 1973;**107**:388–91.
32. Nasir A, El Bahesh E, Whitten C, Lawson A, Udall Jr. JN. Pityrosporum folliculitis in a Crohn's disease patient receiving infliximab. *Inflamm Bowel Dis* 2010;**16**:7–8.
33. Cuetara MS, Aguilar A, Martin L, Aspiroz C, del Palacio A. Erlotinib associated with rosacea-like folliculitis and Malassezia sympodialis. *Br J Dermatol* 2006;**155**:477–9.
34. Cholongitas E, Pipili C, Ioannidou D. Malassezia folliculitis presented as acneiform eruption after cetuximab administration. *J Drugs Dermatol* 2009;**8**:274–5.
35. Bieber T. Pustular diseases. In: Burgdorf WHC, Plewig G, Wolff HH, Landthaler M, editors. *Braun Falco's dermatology*. 3rd ed. Heidelberg: Springer; 2009. p. 542.
36. Suzuki C, Hase M, Shimoyama H, Sei Y. Treatment outcomes for Malassezia folliculitis in the dermatology department of a university hospital in Japan. *Med Mycol J* 2016;**57**:E63–6.
37. Lally A, Casabonne D, Network R, Wojnarowska F. Seborrheic dermatitis among Oxford renal transplant recipients. *J Eur Acad Dermatol Venereol* 2010;**24**:561–4.
38. Gaitanis G, Mayser P, Scheynius A, Crameri R. Malassezia yeasts in seborrheic and atopic eczemas. In: Boekhout T, Guého-Kellermann E, Mayser P, Velegraki A, editors. *Malassezia and the skin. Science and clinical practice*. Heidelberg: Springer; 2010. p. 201–28.
39. Tajima M, Sugita T, Nishikawa A, Tsuboi R. Molecular analysis of Malassezia microflora in seborrheic dermatitis patients: comparison with other diseases and healthy subjects. *J Invest Dermatol* 2008;**128**:345–51.
40. Faergemann J. Atopic dermatitis and fungi. *Clin Microbiol Rev* 2002;**15**:545–63.
41. Sugita T, Tajima M, Ito T, Saito M, Tsuboi R, Nishikawa A. Antifungal activities of tacrolimus and azole agents against the eleven currently accepted Malassezia species. *J Clin Microbiol* 2005;**43**:2824–9.
42. Ring J, Alomar A, Bieber T, Deleuran M, Fink-Wagner A, Gelmetti C, et al. Guidelines for treatment of atopic eczema (atopic dermatitis). Part I. *J Eur Acad Dermatol Venereol* 2012;**26**:1045–60.
43. Rudramurthy SM, Honnavar P, Chakrabarti A, Dogra S, Singh P, Handa S. Association of Malassezia species with psoriatic lesions. *Mycoses* 2014;**57**:483–8.
44. Chowdhary A, Randhawa HS, Sharma S, Brandt ME, Kumar S. Malassezia furfur in a case of onychomycosis: colonizer or etiologic agent? *Med Mycol* 2005;**43**:87–90.
45. Zhao Y, Li L, Wang JJ, Kang KF, Zhang QQ. Cutaneous malasseziasis: four case reports of atypical dermatitis and onychomycosis caused by Malassezia. *Int J Dermatol* 2010;**49**:141–5.
46. Arendrup MC, Fisher BT, Zaoutis TE. Invasive fungal infections in the paediatric and neonatal population: diagnostics and management issues. *Clin Microbiol Infect* 2009;**15**:613–24.
47. Sizun J, Karangwa A, Giroux JD, Masure O, Simitzis AM, Alix D, et al. Malassezia furfur-related colonization and infection of central venous catheters. A prospective study in a pediatric intensive care unit. *Intensive Care Med* 1994;**20**:496–9.
48. Devlin RK. Invasive fungal infections caused by Candida and Malassezia species in the neonatal intensive care unit. *Adv Neonatal Care* 2006;**6**:68–77.
49. Chang HJ, Miller HL, Watkins N, Arduino MJ, Ashford DA, Midgley G, et al. An epidemic of Malassezia pachydermatis in an intensive care nursery associated with colonization of health care workers' pet dogs. *N Engl J Med* 1998;**338**:706–11.
50. Rosales CM, Jackson MA, Zwick D. Malassezia furfur meningitis associated with total parenteral nutrition subdural effusion. *Pediatr Dev Pathol* 2004;**7**:86–90.

CHAPTER 6

Skin and Soft Tissue Infections Following Marine Injuries

V. Savini, R. Marrollo, R. Nigro, C. Fusella, P. Fazii
Spirito Santo Hospital, Pescara, Italy

1. INTRODUCTION

Bacterial diseases following aquatic injuries occur frequently worldwide and usually develop on the extremities of fishermen and vacationers, who are exposed to freshwater and saltwater.[1,2]

Though plenty of bacterial species have been isolated from marine lesions, superficial soft tissue and invasive systemic infections after aquatic injuries and exposures are related to a restricted number of microorganisms including, in alphabetical order, *Aeromonas hydrophila*, *Chromobacterium violaceum*, *Edwardsiella tarda*, *Erysipelothrix rhusiopathiae*, *Mycobacterium fortuitum*, *Mycobacterium marinum*, *Shewanella* species, *Streptococcus iniae*, and *Vibrio vulnificus*.[1,2]

In particular, skin disorders represent the third most common cause of morbidity in returning travelers and are usually represented by bacterial infections.[3–12] Bacterial skin and soft tissue infectious conditions in travelers often follow insect bites and can show a wide range of clinical pictures including impetigo, ecthyma, erysipelas, abscesses, necrotizing cellulitis, myonecrosis.[3–12]

In general, even minor abrasions and lacerations sustained in marine waters should be considered potentially contaminated with marine bacteria.[3–12]

Despite variability of the causative agents and outcomes, the initial presentations of skin and soft tissue infections (SSTIs) complicating marine injuries are similar to those occurring after terrestrial exposures and usually include erysipelas, impetigo, cellulitis, and necrotizing infections.[3]

Erysipelas is characterized by fiery red, tender, painful plaques showing well-demarcated edges, and, though *Streptococcus pyogenes* is the major agent of this process, *E. rhusiopathiae* infections typically cause erysipeloid displays.[3]

Impetigo is initially characterized by bullous lesions and is usually due to *Staphylococcus aureus* or *S. pyogenes*. Nevertheless, *S. iniae* infections are also characterized by either impetigo or cellulitis.[3]

Marine necrotizing infections can be either monomicrobial or polymicrobial. In both cases, the process may advance quickly to necrotizing fasciitis and myonecrosis, especially

The Microbiology of Skin, Soft Tissue, Bone and Joint Infections
ISSN 2451-9006, http://dx.doi.org/10.1016/B978-0-12-811079-9.00006-9

in patients with underlying chronic wounds, such as chronic dermatoses and venous stasis ulcers; otherwise this may occur in immunocompromised hosts on corticosteroid treatment or showing liver disease, diabetes mellitus, HIV infection, or cancer.[3]

Grossly contaminated or infected lesions and all puncture wounds should be cultured or biopsied immediately and, although an empirical antibiotic approach is warranted, all definitive antimicrobial therapy should be targeted on precise pathogen identification and in vitro susceptibility testing.[3]

The aim of this chapter is to draw a picture of salient aspects of marine water environment-related SSTIs, by focusing on major bacteria behind the curtain of these clinical conditions and on their habitats, microbiological features, and diseases they may cause.

2. EPIDEMIOLOGY

After acute and chronic management of SSTIs in tsunami survivors with contaminated aquatic injuries, the early predominance of Gram-negative aquatic organisms emerged in early-onset infections, while rapidly growing mycobacteria are mostly agents of sea-related late-onset infectious syndromes.[1,2]

Although Gram-positive bacteria such as *Staphylococcus aureus* and streptococci, as well as *Pseudomonas aeruginosa* (Gram-negative) and several other bacterial species, have been isolated from infected minor marine wounds, some of the more uniquely marine bacterial pathogens observed after more severe aquatic injuries include *Aeromonas* and *Shewanella* species, *Chromobacterium violaceum*, *Edwardsiella tarda*, *Vibrio vulnificus* and marine mycobacteria.[1,2]

Nonetheless, an empiric approach against SSTIs that develop shortly after marine water exposure should still cover *Streptococcus pyogenes* and *S. aureus*, not only marine water-related organisms.[1,2]

Then, in general, treatment could include a first generation cephalosporin or clinda mycin combined with levofloxacin or ciprofloxacin plus doxycycline (if an increased risk for acquiring *V. vulnificus* exists).[1,2]

2.1 *Aeromonas* Species

The genus *Aeromonas* includes Gram-negative bacilli inhabiting warm soil together with fresh and brackish waters worldwide and behaving both as aquatic animal commensals and pathogens.[13]

Most can produce enterotoxins and hemolysins, thus causing acute hemorrhagic diarrheal syndrome and invasive SSTIs in both immunocompetent and immunocompromised hosts exposed to an aquatic environment, including near-drowning.[13]

Aeromonas wound infections may develop after freshwater traumatic lesions, i.e., alligator, fish, snake and leech bites; infected injuries usually involve extremities or other bodily sites where open wounds were immersed in contaminated freshwater in warmer months.[13,14] Within 24 h, erythema, edema and purulent discharge (indistinguishable from streptococcal cellulitis) can be observed.[3] There may be fever and progression to invasive infections, especially in the immunocompromised population, with necrotizing fasciitis, necrotizing myositis and bone involvement.[13,14]

Although most *Aeromonas* SSTIs follow aquatic immersions, an outbreak of *A. hydrophila* wound infections after mud exposure in 26 football players was reported in 2002. None of the players had any immunocompromising underlying condition, and all recovered with antibiotics.[14]

In general, *Aeromonas* strains from human infections are susceptible to aminoglycosides, carbapenems, fluoroquinolones, aztreonam, tetracyclines, as well as third- and fourth-generation cephalosporins, while variable levels of activity have been reported for cotrimoxazole.[15]

In addition to wound drainage and debridement, *Aeromonas* wound infections should be faced initially with either a third-generation cephalosporin or a fluoroquinolone, possibly combined with an aminoglycoside pending antibiotic susceptibility testing and culture results.[15,16]

2.2 *Chromobacterium violaceum*

Chromobacterium violaceum is a Gram-negative, aerobic rod, behaving as a ubiquitous saprophyte in soil and water in tropical and subtropical regions all over the world.[17,18] In spite of its wide geographical distribution, it is a low-grade pathogen that is responsible for few cases of infection in immunocompetent people and is often dismissed as a contaminant at examination of cultures.[17]

The organism is presumptively recognized by the violet color of colonies it forms, with nonpigmented strains being less common.[17–19]

Most cases occur in temperate and tropical regions and are characterized by high fatality rates in immunocompromised patients and the possibility of neutrophil impairment should be taken into consideration when managing patients with fulminant infections of *C. violaceum*.[20] The portal of entry is generally a skin lesion secondary to a laceration or fish bite, followed by contamination with brackish or stagnant water. An ulcerated lesion with a bluish purulent discharge is usually observed at the initial injury site with regional swelling on an extremity. In the following days, bacteremia may occur, more commonly in the immunocompromised hosts, with disseminated macular cutaneous lesions progressing to abscesses; these may also involve bone, liver, and lung as well as spleen. Due to the rarity of this infection, availability of treatment recommendations is still lacking.[18,19]

C. violaceum is usually susceptible to aminoglycosides, carbapenems, chloramphenicol, cotrimoxazole, fluoroquinolones and tetracyclines, but resistant to most penicillins and cephalosporins.[18,19] Because of high mortality rates, treatment of suspected *C. violaceum* infections should start immediately and be based on drainage of purulent abscesses and empiric combination of intravenous antibiotics, which will be narrowed after susceptibility testing results are provided.[19]

2.3 *Edwardsiella tarda*

Edwardsiella tarda is a Gram-negative rod-shaped organism belonging to the family Enterobacteriaceae; it notoriously behaves as an aquaculture pathogen and causes emphysematous putrefactive disease in catfish.[21]

Extraintestinal *E. tarda* infections are uncommon compared with intestinal ones, and are frequently represented by wound abscesses needing surgical incision and drainage in patients with marine exposures or injuries; extensive myonecrosis and fatal septic shock may develop in immunocompromised hosts, especially those with chronic epatopathy.[22]

2.4 *Shewanella* Species

The genus *Shewanella* includes saprophytic Gram-negative microorganisms inhabiting warm and temperate regions worldwide and which are part of the marine environment microflora. More than 50 species of *Shewanella* have been classified thus far, all of which produce yellowish-brown pigmented and mucoid colonies releasing hydrogen sulfide in culture.[23,24]

Several *Shewanella* species have been recently known to be emerging agents of soft tissue and invasive infections following seawater exposure, and these include *Shewanella algae* (which is the most common), *Shewanella haliotis*, *Shewanella putrefaciens* and *Shewanella xiamenensis*.[23,24]

The major clinical manifestations of shewanellosis are deep ulcers associated with hemorrhagic bullae on the lower extremities, otitis externa or media, biliary tract and bloodstream infection.[23–25] Nonhealing ulcers usually evolve to necrotizing fasciitis, compartment syndromes and osteomyelitis.[25] *Shewanella* bacteremia in the past has been complicated by endocarditis and meningitis, while lung infection, cholecystitis and peritonitis may follow aspiration or ingestion, respectively, of *Shewanella*-containing seawater in near-drowning.[23,24] Together with marine exposure and raw seafood ingestion, further common risk factors for shewanellosis are minor trauma or lacerations that occur in a marine environment, preexisting lower extremity ulcers and immune system impairment.[25] The diagnostics of shewanellosis rely on positive culture of blood or lesion aspirate samples, although reliable identification at a species level requires molecular characterization.[23–25] Most species respond in vitro to a broad range of antibiotics, such as third- and fourth-generation cephalosporins, carbapenems, aminoglycosides, and fluoroquinolones. *S. algae*, particularly, is resistant to penicillins and cephalosporins belonging to the first and second generations.[23] Poor data are still available concerning in vitro activity of tigecycline versus *Shewanella* isolates and no species-specific EUCAST breakpoints for antibiotic susceptibility testing are available (Fig. 1).[58] Treatment of

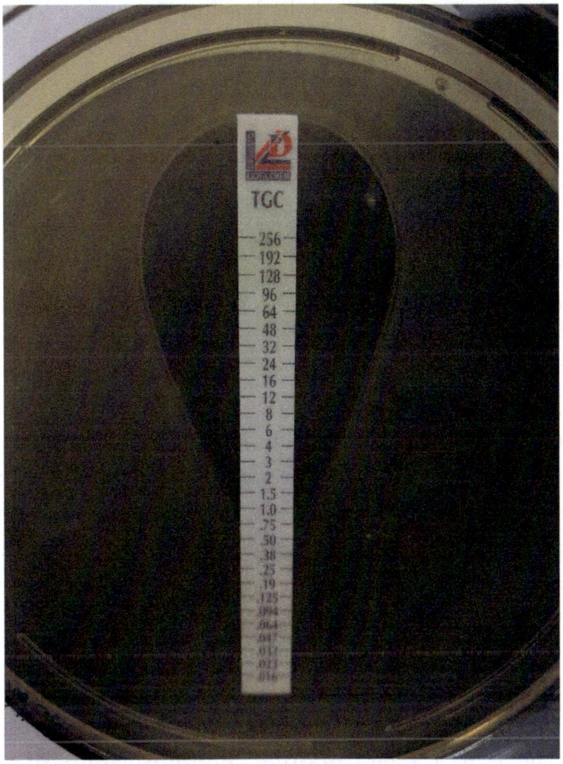

Fig. 1 Tigecycline in vitro activity versus a *Shewanella algae* isolate from the Clinical Microbiology and Virology Unit, Spirito Santo Hospital of Pescara, Italy. MIC has been determined by MIC Test Strip (Liofilchem®, Italy). The isolate is resistant, according to EUCAST 2017 non-species related breakpoints. It is susceptible if breakpoints for Enterobacteriaceae are considered.

invasive infections, especially in immunocompromised hosts, should include, according to most authors, initial intravenous betalactam antibiotics (i.e., cefotaxime, ceftazidime, cefepime) combined with either an aminoglycoside or a fluoroquinolone (ciprofloxacin or levofloxacin), then switch to an antibiotic susceptibility testing-based oral antibiotic therapy.[23–25] An early surgical approach is recommended as well, for drainage of bullous lesions, debridement of ulcers and evaluation of potential compartment syndromes that must be treated with decompressive fasciotomy.[23–25]

2.5 *Vibrio vulnificus*

Vibrio vulnificus has emerged as a highly virulent organism causing three kinds of disease, that is: (1) acute gastroenteritis after consuming raw or undercooked shellfish; (2) invasive septicemia following ingestion of raw or undercooked shellfish, mostly oysters; (3) necrotizing wound infections after marine injuries and exposures.[26,27]

V. vulnificus is a Gram-negative, halophilic bacillus that is free living in marine waters characterized by low to moderate salinities, especially above 18°C.[26–30] Men are uniquely predisposed to infections by *V. vulnificus* due to both recreational and occupational exposures to fish and shellfish, higher blood iron levels, higher rates of alcoholism and chronic epathopathy along with lower levels of protective estrogens.[29,31] Further predisposing factors that are not gender-related include all conditions increasing serum iron levels (such as hemochromatosis and thalassemia major), chronic liver disease (i.e., cirrhosis, hepatoma) and liver transplant, diabetes mellitus, end-stage kidney disease and underlying immunosuppression due to chemotherapy, splenectomy, steroids and HIV infection.[30–32]

V. vulnificus will be cultivated from blood cultures in 30% of wound infections with secondary bacteremia and in 70%–100% of episodes of primary sepsis.[29,32] As delay in starting antibiotics has been related to increasing fatality (until 100% for delays >72 h), treatment should be given in a timely manner by using ceftazidime plus doxycycline as an initial empiric combination for suspected *V. vulnificus* infections.[33] Ceftriaxone and cefotaxime may be used as an alternative, combined with a tetracycline; fluoroquinolones have proven to be effective, too.[29,32,34–36] Early wound debridement and surgical monitoring for compartment syndromes are also recommended and this approach has been shown to decrease mortality rates.[33–35]

2.6 *Erysipelothrix rhusiopathiae*

Erysipelothrix rhusiopathiae is a Gram-positive, nonsporulating rod-shaped bacterium which is frequently misidentified under a microscope as *Lactobacillus* or *Listeria monocytogenes*.[37]

On blood-based culture plates, instead, *E. rhusiopathiae* can be confused with *viridans* streptococci, as it is α-hemolytic.[37]

In fish, the organism persists silently for long periods in the exterior slime.[37] Diseases it causes typically manifest 1–2 days after skin lesions occurred while handling or preparing fish.[37] Most of them, in man, are represented by a localized (erysipeloid) or generalized cutaneous infection; both forms are characterized by pain, throbbing erythema and strong pruritus.[37]

Invasive systemic *E. rhusiopathiae* diseases are uncommon and characterized by bacteremia, and potentially, associated infective endocarditis (IE); the latter, when sustained by *E. rhusiopathiae*, commonly involves the aortic valve.[38]

The organism is susceptible to penicillins, cephalosporins, carbapenems, clindamycin, fluoroquinolones and daptomycin and is typically resistant to aminoglycosides, sulfonamides and vancomycin.[39]

Since this glycopeptide is frequently used as empirical treatment of presumed IE, prompt microbiological differentiation of *E. rhusiopathiae* from other Gram-positive organisms is crucial.[38]

2.7 *Mycobacterium* Species

The aquatic, so-called atypical mycobacteria are the acid-fast causative agents of piscine mycobacteriosis and are responsible for both external and solid organ granulomas in more than 150 fish species.[40,41] Aquatic mycobacteria live at temperatures of 30–33°C in fresh and marine waters and are resistant to chlorine and iodine.[41]

Mycobacterium marinum has been historically associated with exposure to fish, aquariums, swimming pools, or marine fauna other than fish, and is the most frequent cause of external granulomas in those who are professionally exposed as they handle fish or work as aquarium operators.[41–44] Infections it causes typically start as localized zones of red-violet verrucous or crusted plaques that develop at skin inoculation sites 7 or more days after puncture wounds or minor injuries occurred in marine environments on the cooler, distal areas of the extremities.[42,43]

Solitary or multiple granulomatous nodules will be later observed in the inoculation site that may ulcerate with production of a yellowish pus.[42,43] In some occasions, metastatic nodular lesions develop according to a linear or sporotrichoid manner along the proximally draining lymphatic vessels; then they may ulcerate and undergo secondary infection.[43]

Deeper, invasive infectious processes such as bursitis, osteoarthritis, septic arthritis, and tenosynovitis may occur in indolent and untreated cases and, less commonly, in the immunocompromised population.[43] Of particular interest, a case of left wrist tenosynovitis and elbow bursitis following a puncture injury and exposure to coal mine water has been recently reported.[44]

Diagnostics rely on acid-fast staining, of course, along with culture of nodule discharge, aspirate or biopsy and molecular identification of the causative agent.[41–44]

Unlike *M. marinum*, other more rapidly growing marine mycobacterial species (i.e., *Mycobacterium abscessus*, *Mycobacterium fortuitum*, *Mycobacterium perigrinum*, and *Mycobacterium mageritense*) have been clearly related to persistent furunculosis after marine injuries in tsunami survivors.[2]

Among the more unusual outbreaks of mycobacterial furunculosis, rapidly growing *M. fortuitum* infections have occurred after freshwater footbaths and pedicures.[45] Also, razor-shaving of the lower legs has been identified as a relevant risk factor for mycobacterial infections compared with having pedicures without shaving.[45]

Further unusual outbreaks of mycobacterial furunculosis have been caused by rapidly growing *M. fortuitum* and followed ichthyotherapy with freshwater doctor fish (*Garra rufa*) in the United Kingdom.[46] The latter is a freshwater cyprinoid fish that naturally inhabits the river basins of central Eurasia; of interest, it is imported in large numbers to nail and foot spas all over the world for the removal of dead or hyperkeratotic foot skin during ichthyotherapy.[46]

M. marinum is typically susceptible to macrolides (i.e., clarithromycin), sulfonamides, cotrimoxazole, ethambutol and rifampin/rifabutin. A typical therapy regimen includes two of these agents in combination (e.g., clarithromycin plus ethambutol or clarithromycin plus rifampin) for about 3–4 months (until 4–8 weeks post-clinical resolution).[44,47]

M. fortuitum responds instead to fluoroquinolones such as ciprofloxacin and levofloxacin, newer macrolides like azithromycin and clarithromycin, sulfonamides, tetracyclines (minocycline, doxycycline) and amikacin.[44] A 4-month (at least) combined therapy with at least two drugs is recommended.[44]

Similar oral antibiotic combinations may treat nonserious SSTIs by *M. abscessus*. Severe skin diseases caused by this organism are instead initially managed with a combination of intravenous amikacin plus cefoxitin with the possible addition of the carbapenem imipenem.[44]

In general, infections with rapidly growing mycobacteria may benefit from prompt recognition of acquired fluoroquinolone and macrolide resistance, if present.[44]

2.8 *Streptococcus iniae*

Streptococcus iniae is β-hemolytic and has not yet been assigned to a Lancefield group.

It was first identified in 1976 as the etiologic agent of subcutaneous abscesses in Amazon freshwater dolphins living in US aquariums and is nowadays known as a major fish pathogen. Particularly, it may cause epizootic outbreaks of invasive infection in farm-raised fish.[48–50]

The organism initially colonizes the fish surface, thus causing cellulitis, and possible subsequent invasive meningoencephalitis, responsible for 30%–50% mortality in aquaculture ponds.[49]

The first cases of invasive disease described in humans date back to 1996 in Toronto and involved patients who had recently handled live or freshly killed fish.[41] Affected patients developed cellulitis of the hands, or endocarditis (in one case) or arthritis; some of these patients, particularly, had percutaneous injuries while preparing fish and developed cellulitis within 24 h of the injuries' occurrence.[51]

In all such 11 cases, human *S. iniae* isolates matched (by pulse-field gel electrophoresis) to those cultivated from surfaces of infected tilapia from local aquaculture farms.[51] Also, all human isolates showed susceptibility to penicillins, cephalosporins, macrolides, aminoglycosides, and cotrimoxazole.[51]

3. CONTROL AND PREVENTION OF AQUATIC INFECTIONS

People with well-established risk factors for high seriousness of aquatic infections, including subjects with open wounds, immune system deficit, diabetes mellitus, epathopathy, hemochromatosis, alcoholism, hematological disorders, chronic renal diseases, HIV infection and cancer, should be alerted about the risks of infectious conditions due to marine bacteria mediated by exposure to both marine animals and seawater, along with preparation of live or freshly killed seafood and, of course, seawater ingestion or eating of raw or undercooked seafood, especially oysters.

Clinicians, in the meantime, should be aware of potentially catastrophic bacterial infections following marine injuries and exposures.

Initial antibiotic treatment should rely on clinical signs and symptoms (impetigo, erysipelas, pyoderma, cellulitis, necrotizing soft tissue infection), pending culture and antibiotic susceptibility testing results.

Except for minor marine wounds showing localized cellulitis or spreading erysipeloid-type appearance, most other marine-related infectious diseases and all Gram-negative and mycobacterial marine infections will need combined antibiotic approaches, and quite often will benefit from surgical management.

4. CONCLUSION AND FUTURE PERSPECTIVES

Together with consciousness of the risk of insect-related virus infections, travelers should be aware of diseases that may develop following contact with water-inhabiting bacteria. Therefore, it is warranted that travel medicine practitioners maintain a high index of suspicion concerning potentially catastrophic bacterial infections complicating water injuries and exposures, especially those caused by *Vibrio vulnificus* in the Gulf of Mexico, *Chromobacterium violaceum* in the Western Pacific, and *Shewanella* in the Mediterranean and Western Pacific.[3,52–57]

Due to the increasing frequency of travel, both for work and holiday reasons as well as due to migration from war zones, microbiologists should be ready to face unusual diagnostic queries, like those surrounding infections with "foreign," unexpected, bacteria, in a world with fewer and fewer frontiers and barriers.

REFERENCES

1. Hiransuthikul N, Tantisiriwat W, Lertutsahakul K, et al. Skin and soft-tissue infections among tsunami survivors in southern Thailand. *Clin Infect Dis* 2005;**41**:93–6.
2. Finkelstein R, Oren I. Soft tissue infections caused by marine bacterial pathogens: epidemiology, diagnosis, and management. *Curr Infect Dis Rep* 2011;**13**:470–7.
3. Diaz JH. Skin and soft tissue infections following marine injuries and exposures in travelers. *J Travel Med* 2014;**21**:207–13.
4. O'Brien BM. A practical approach to common skin problems in returning travelers. *Travel Med Infect Dis* 2009;**7**:125–46.
5. Hochedez P, Caumes E. Common skin infections in travelers. *J Travel Med* 2008;**15**:252–62.
6. Froelich BA, Phippen B, Fowler P, Noble RT, Oliver JD. Differences in total *Vibrio* spp., *V. vulnificus*, and *V. parahaemolyticus* abundance between clams and oysters in North Carolina. *Appl Environ Microbiol* 2016;**83**:pii: AEM.02265-16 [Epub ahead of print].
7. Cardozo FA, Zimpel CK, Guimaraes AM, Pessoa A, Rivera IN. Draft genome sequence of marine-derived *Aeromonas caviae* CHZ306, a potential chitinase producer strain. *Genome Announc* 2016;**4**:pii: e01293-16.

8. Lodde B, Pougnet R, Roguedas-Contios AM, et al. Skin infection by *Staphylococcus aureus* in a fisherman: difficulty in continuing work on board. *Int Marit Health* 2013;**64**:126–8.
9. Mattioli MC, Sassoubre LM, Russell TL, Boehm AB. Decay of sewage-sourced microbial source tracking markers and fecal indicator bacteria in marine waters. *Water Res* 2017;**108**:106–14. pii: S0043-1354.
10. Thorell K, Collin B, Hernroth B, Sjöling Å. Complete genome sequences of two marine *Vibrio cholerae* strains isolated from the South Coast of Sweden. *Genome Announc* 2016;**4**:pii: e01118-16.
11. Baker-Austin C, Trinanes J, Gonzalez-Escalona N, Martinez-Urtaza J. Non-cholera *Vibrios*: the microbial barometer of climate change. *Trends Microbiol* 2017;**25**:76–84. pii: S0966-842X(16)30140-8.
12. Hmelo LR. Quorum sensing in marine microbial environments. *Annu Rev Mar Sci* 2016;**9**:257–81.
13. Lamy B, Kodjo A, Laurent F, et al. Prospective nationwide study of *Aeromonas* infections in France. *J Clin Microbiol* 2009;**47**:1234–7.
14. Vally H, Whittle A, Cameron S, et al. Outbreak of *Aeromonas hydrophila* wound infections associated with mud football. *Clin Infect Dis* 2004;**38**:1084–9.
15. Janda JM, Abbott SL. The genus *Aeromonas*: taxonomy, pathogenicity, and infection. *Clin Microbiol Rev* 2010;**23**:35–73.
16. Chen PL, Ko WC, Wu CJ. Complexity of β-lactamases among clinical *Aeromonas* isolates and its clinical implications. *J Microbiol Immunol Infect* 2012;**45**:398–403.
17. Midani S, Rathore M. *Chromobacterium violaceum* infection. *South Med J* 1998;**91**:464–6.
18. Brown KL, Stein A, Morrell DS. Ecthyma gangrenosum and septic shock syndrome secondary to *Chromobacterium violaceum*. *J Am Acad Dermatol* 2006;**54**:S224–8.
19. Lee J, Kim JS, Nahm CH, et al. Two cases of *Chromobacterium violaceum* infection after injury in a subtropical region. *J Clin Microbiol* 1999;**37**:2068–70.
20. Macher AM, Casale BT, Fauci AS. Chronic granulomatous disease of childhood and *Chromobacterium violaceum* infections in the South Eastern United States. *Ann Intern Med* 1982;**97**:51–2.
21. Meyer FP, Bullock GL. *Edwardsiella tarda*, a new pathogen of channel catfish (*Ictalurus punctatus*). *Appl Microbiol* 1973;**25**:155–6.
22. Slaven EM, Lopez FA, Hart SM, et al. Myonecrosis caused by *Edwardsiella tarda*: a case report and case series of extraintestinal *E. tarda* infections. *Clin Infect Dis* 2001;**32**:1430–3.
23. Vignier N, Barreau M, Olive C, et al. Human infection with *Shewanella putrefaciens* and *S. algae*: report of 16 cases in Martinique and review of the literature. *Am J Trop Med Hyg* 2013;**89**:151–6.
24. Poovorawan K, Chatsuwan T, Lakananurak N, et al. *Shewanella haliotis* associated with severe soft tissue infection, Thailand, 2012. *Emerg Infect Dis* 2013;**19**:1019–21.
25. Chen YS, Liu YC, Yen MY, et al. Skin and soft-tissue manifestations of *Shewanella putrefaciens* infection. *Clin Infect Dis* 1997;**25**:225–9.
26. Oliver JD. Wound infections caused by *Vibrio vulnificus* and other marine bacteria. *Epidemiol Infect* 2005;**133**:383–91.
27. Lillis RA, Dugan V, Mills T, et al. A fish hook and liver disease: revisiting an old enemy. *J La State Med Soc* 2002;**154**:20–5.
28. Froelich B, Oliver JD. The interactions of *Vibrio vulnificus* and the oyster *Crassotrea virginica*. *Microb Ecol* 2013;**65**:807–16.
29. Horseman MA, Surani S. A comprehensive review of *Vibrio vulnificus*: an important cause of severe sepsis and skin and soft tissue infection. *Int J Infect Dis* 2011;**15**:157–66.
30. Shapiro RL, Altekruse S, Hutwagner L, et al. The role of Gulf Coast oysters harvested in the winter months in *Vibrio vulnificus* infections in the United States, 1988–1996. *J Infect Dis* 1998;**178**:752–9.
31. Menon MP, Yu PA, Iwamato M, et al. Pre-existing medical conditions associated with *Vibrio vulnificus* septicaemia. *Epidemiol Infect* 2013;**10**:1–4.
32. Slaven EM, Lopez FA. *Vibrio vulnificus*. *Infect Dis Clin Pract* 2000;**24**:77–80.
33. Jones JL, Kinsey TP, Johnson LW, et al. Effects of intertidal harvest practices on *Vibrio parahaemolyticus* and *Vibrio vulnificus* levels in oysters. *Appl Environ Microbiol* 2016;**82**:4517–22. pii: AEM.00721-16 [Epub ahead of print].
34. Chen SC, Lee YT, Tsai SJ, et al. Antibiotic therapy for necrotizing fasciitis caused by *Vibrio vulnificus*: retrospective analysis of an 8 year period. *J Antimicrob Chemother* 2012;**67**:488–93.

35. Hsueh P-R, Lin C-Y, Tang H-J, et al. *Vibrio vulnificus* in Taiwan. *Emerg Infect Dis* 2004;**10**:1363–8.
36. Piper KE, Steckelberg JM, Patel R. In vitro activity of daptomycin against clinical isolates of gram-positive bacteria. *J Infect Chemother* 2005;**11**:207–9.
37. Romney M, Cheung S, Montessori V. *Erysipelothrix rhusiopathiae* endocarditis and presumed osteomyelitis. *Can J Infect Dis* 2001;**12**:254–6.
38. Hill DC, Ghassemian JN. *Erysipelothrix rhusiopathiae* endocarditis: clinical features of an occupational disease. *South Med J* 1997;**90**:1147–8.
39. Venditti M, Gelfusa V, Tarasi A, et al. Antimicrobial susceptibilities of *E. rhusiopathiae*. *Antimicrob Agents Chemother* 1990;**34**:2038–40.
40. Savini V, Fazii P, Favaro M, et al. Tuberculosis-like pneumonias by the aerobic actinomycetes *Rhodococcus*, *Tsukamurella* and *Gordonia*. *Microbes Infect* 2012;**14**:401–10.
41. Decostere A, Hermans K, Haesebrouck F. Piscine mycobacteriosis: a literature review covering the agent and the disease it causes in fish and humans. *Vet Microbiol* 2004;**99**:159–66.
42. Piersimoni C, Scarparo C. Extrapulmonary infections associated with nontuberculous mycobacteria in immunocompetent persons. *Emerg Infect Dis* 2009;**15**:1351–8.
43. Lahey T. Invasive *Mycobacterium marinum* infections. *Emerg Infect Dis* 2003;**9**:1496–8.
44. Huaman MA, Ribes JA, Lohr KM, Evans ME. *Mycobacterium marinum* infection after exposure to coal mine water. *Open Forum Infect Dis* 2015;**3**(1):ofv205.
45. Winthrop KL, Abrams M, Yakrus M, et al. An outbreak of mycobacterial furunculosis associated with a footbath at a nail salon. *N Engl J Med* 2002;**346**:1366–71.
46. Verner-Jeffreys DW, Baker-Austin C, Pond MJ, et al. Zoonotic disease pathogens in fish used for pedicure. *Emerg Infect Dis* 2012;**18**:1006–8.
47. Safdar N, Abad CL, Kaul DR, et al. Clinical problem solving. Skin deep. *N Engl J Med* 2012;**366**:1336–40.
48. Pier GB, Madin SH. *Streptococcus iniae* sp. nov., a β-hemolytic *Streptococcus* isolated from an Amazon freshwater dolphin, *Inia geoffrensis*. *Int J Syst Bacteriol* 1976;**26**:545–53.
49. Kitao T, Aoki T, Sakoh R. Epizootic caused by b-hemolytic *Streptococcus* species in cultured freshwater fish. *Fish Pathol* 1981;**15**:301–7.
50. Baiano JCF, Barnes AC. Towards control of *Streptococcus iniae*. *Emerg Infect Dis* 2009;**15**:1891–6.
51. Weinstein MR, Litt M, Kertesz DA, et al. Invasive infections due to a fish pathogen, *Streptococcus iniae*. *N Engl J Med* 1997;**337**:589–94.
52. Lau SK, Chan SY, Curreem SO, Hui SW, Lau CC, Lee P, et al. *Burkholderia pseudomallei* in soil samples from an oceanarium in Hong Kong detected using a sensitive PCR assay. *Emerg Microbes Infect* 2014;**3**:e69.
53. Eronen-Rasimus E, Lyra C, Rintala JM, Jürgens K, Ikonen V, Kaartokallio H. Ice formation and growth shape bacterial community structure in Baltic Sea drift ice. *FEMS Microbiol Ecol* 2015;**91**:1–13.
54. Dong HP, Hong YG, Lu S, Xie LY. Metaproteomics reveals the major microbial players and their biogeochemical functions in a productive coastal system in the northern South China Sea. *Environ Microbiol Rep* 2014;**6**:683–95.
55. Oliver AE, Newbold LK, Whiteley AS, van der Gast CJ. Marine bacterial communities are resistant to elevated carbon dioxide levels. *Environ Microbiol Rep* 2014;**6**:574–82.
56. Szeinbaum N, Burns JL, DiChristina TJ. Electron transport and protein secretion pathways involved in Mn(III) reduction by *Shewanella oneidensis*. *Environ Microbiol Rep* 2014;**6**:490–500.
57. Hao J, Liu J, Sun M. Identification of a marine *Bacillus* strain C5 and parathion-methyl degradation characteristics of the extracellular esterase B1. *Biomed Res Int* 2014;**2014**:863094.
58. Marisa N Almuzara, Maria Isabel Encalada Barzallo, Angela MR Famiglietti, Carlos A Vay. Comparative Activities of Tigecycline and Other Tetracyclines against Nonfermenting Gram-Negative Bacilli. *Excluding Acinetobacter spp Antimicrob Agents Chemother* 2011;**55**:3961–3.

CHAPTER 7

Gordonia and Related Genera as Opportunistic Human Pathogens Causing Infections of Skin, Soft Tissues, and Bones

H. Sowani*, M. Kulkarni*, S. Zinjarde*, V. Javdekar†
*Savitribai Phule Pune University, Pune, India
†MES Abasaheb Garware College, Pune, India

1. INTRODUCTION

Actinomycetes are filamentous, Gram-positive bacteria that form branching filaments at some stage of their development. Actinomycetes were initially described by Ferdinand Cohn in 1875 when he observed a filamentous organism in a concretion from a human lacrimal duct and named it *Streptothrix foersteri* and later C.O. Harz in 1878 named an organism that he isolated from a case of bovine lumpy jaw, as *Actinomyces bovis*.[1] The term actinomycete means "ray fungus" in Greek and actinomycetes are still referred to as "ray fungi."[2]

Actinomycetes are found to occur either as spores or vegetative forms in varied habitats such as soil, aquatic environments, plant litter, composts, and food products. They are also associated with plants, animals, and man. Actinomycetes mediate degradation of complex polymers such as chitin, lignin, and cellulose, help in the biocontrol of other bacteria and fungi by the production of antimicrobial agents and cause deterioration of stored food products.[3] They are an important group of prokaryotes that are able to synthesize metabolic products such as antibiotics, pigments, enzyme inhibitors, and biotechnologically relevant enzymes. They are also pathogenic to plants, animals, and man. Plant diseases like rot of sweet potato and human diseases such as tuberculosis, paratuberculosis, mycetomas, actinomycosis, allergic pneumonias, and various types of abscesses are caused by actinomycetes.

Actinomycetes have been classified into more than 150 genera. This classification is based on morphological and chemical criteria.[4–6] Nine different cell wall chemotypes have been distinguished on the basis of the type of the diaminopimelic acid

The Microbiology of Skin, Soft Tissue, Bone and Joint Infections
ISSN 2451-9006, http://dx.doi.org/10.1016/B978-0-12-811079-9.00007-0

(DAP) isomer and the presence of characteristic amino acids and sugars in their cell walls.

The order Actinomycetales includes phylogenetically diverse but morphologically similar aerobic and anaerobic bacteria that exhibit filamentous branching structures which fragment into bacillary or coccoid forms. The anaerobic genera of medical importance include *Actinomyces*, *Arachnia*, *Rothia*, and *Bifidobacterium*. Aerobic actinomycetes are a large, diverse group of Gram-positive bacteria that includes *Gordonia*, *Nocardia*, and *Rhodococcus* besides various other genera.

Actinomycetes belonging to the genera *Gordonia*, *Nocardia*, and *Rhodococcus* are widely distributed in natural environments. However, they are known to cause infections of the skin and soft tissues in immunocompromised human patients. These organisms are opportunistic human pathogens and infections may commonly be postoperational and hospital acquired.

The genera *Gordonia*, *Nocardia*, and *Rhodococcus* all belong to the cell wall chemotype IV and whole cell sugar pattern Type A (WCSP Type A) characterized by the presence of meso-DAP, arabinose and galactose in their whole cell hydrolysates. Actinomycetes belonging to this cell wall type are further classified on the basis of presence or absence of mycolic acids and further on the basis of molecular weights of the mycolates.[7,8] All these three genera contain medium nocardomycolates (34–60 carbon atoms), as compared to the small corynomycolates (22–38 carbon atoms) or the large mycolates of mycobacteria (60–90 carbon atoms).

The genera *Gordonia* and *Rhodococcus* differ with respect to the molecular weights of their mycolates and menaquinone patterns. Gordonias contain mycolic acids with 44–66 carbon atoms and dehydrogenated menaquinones with nine isoprene units [MK-9 (H_2)] as the predominant menaquinone. Rhodococci, on the other hand, have mycolic acids with 34–52 carbon atoms and dehydrogenated menaquinones with eight isoprene units [MK-8 (H_2)] as the predominant menaquinone.[9]

The present chapter details the role of actinomycetes belonging to the genus *Gordonia* and the related genera *Nocardia* and *Rhodococcus* as opportunistic human pathogens causing infections of skin, soft tissues, bones, and joints.

2. *GORDONIA*

2.1 Microbiology and Epidemiology

The members of genus *Gordonia* belong to the family Gordoniaceae in the suborder Corynebacteriales.[9,10] *Gordonia* species are Gram-positive, partially acid-alcohol fast, catalase-positive, aerobic, chemoorganotrophic mycolic acid–containing coccobacillary actinomycetes with high G + C contents in their DNA (63–69%). They can be differentiated from the related genus *Nocardia* on the basis of their ability to grow in the presence of lysozyme, to reduce nitrate and by the lack of mycelial forms. *Gordonia* can be distinguished from *Rhodococcus*, the other related genus, on the basis of 16S rRNA.[11–14]

Some *Gordonia* species are opportunistic human pathogens.[15] They have been isolated from clinical samples and are often associated with implants. *Gordonia* species can cause systemic infections in immunocompromised patients and local infections in immunocompetent patients.[16] *G. amicalis*, *G. araii*, *G. bronchialis*, *G. otitidis*, *G. polyisoprenivorans*, *G. rubripertincta*, *G. sputi*, and *G. terrae* are reported to cause human infections. Clinical isolates have been primarily identified on the basis of morphology, biochemical profiles, and high-performance liquid chromatography. The identity of these species has generally been confirmed by molecular typing methods such as 16S rRNA gene sequencing, random amplified polymorphic DNA (RAPD) or restriction fragment length polymorphism (RFLP) analysis.[11]

Gordonia species are widely distributed in aquatic and terrestrial environments. As stated earlier, they cause human infections in immunocompromised patients with different underlying conditions and some cases are also reported in immunocompetent patients.[11,12] *Gordonia* infections are reported to be local (confined to small areas) or systemic. In most cases, they have been contained with antimicrobial therapy although individual strains differ in susceptibility patterns. In some instances, surgical procedures were essential to contain *Gordonia* infections.

2.2 Skin and Soft Tissue Infections

From the anatomical point of view, soft tissues are tissues connecting, supporting or surrounding body structures and organs. These include connective tissues (tendons, ligaments, fascia, skin, fibrous tissues, fat, and synovial membranes) and nonconnective tissue (muscles, nerves, and blood vessels). *Gordonia* species can cause the following soft tissue infections.

Strains of *G. terrae* are reported to cause a variety of infections in immunocompetent as well as immunocompromised hosts. There is a report on *G. terrae* being cultured from a patient having cellulitis, an acute disseminating infection of dermis, epidermis, and subcutaneous tissues.[11] The infection could be effectively controlled by treatment with amoxicillin or clavulanate. *G. terrae* strains also cause eye-related (palpebral) abscesses in immunocompetent hosts. Granulomatous exudates obtained by surgical excision yielded this bacterium. The abscess was eliminated within short periods of time by surgical drainage without the administration of antibiotics.[17] In a patient with chronic hepatitis B virus infection, a traumatic wound infection was caused by *G. terrae*.[11] The wound infection could be effectively managed by repeated surgical debridement and treatment with amoxicillin or clavulanate. *G. terrae* has also been implicated in subcutaneous mycetoma developing after minor injuries. The initially developed subcutaneous nodules were painless. These eventually spread and discharged grains that contained clusters of bacteria. The resultant effect was the generation of purulent debris and a painful condition. To avoid recurrence of disease after surgical removal of debris, antibiotic therapy (oral doxycycline) was effective. In addition to the aforementioned soft tissue and skin infections, *G. terrae* also causes central venous catheter (CVC) related bacteremia.[18] This condition was observed in immunocompromised hosts. The organism could be isolated from blood samples through CVC and skin. The infection could be overcome by vancomycin, imipenem, ciprofloxacin, amoxicillin-clavulanate, or ceftriaxone.

G. bronchialis is reported to cause recurrent breast infection and associated granulomatous mastitis in immunocompetent hosts. The bacterial culture isolated from the abscess was confirmed to be a strain of *G. bronchialis* on the basis of 16S rRNA sequencing. With prolonged antibiotic therapy, the infection could be cured.[19] *G. bronchialis* is also known to cause bacteremia. Surgical drainage from a paravertebral abscess led to the isolation of this bacterium. After intravenous administration of vancomycin and ceftriaxone and oral amoxicillin-clavulanate for a longer duration, recovery occurred. There are more cases wherein *G. bronchialis* associated bacteremia was detected in an immunocompromised patient with nonketotic hyperosmolar coma and a patient with autoimmune thyroiditis. With antibiotic therapy, the infection could be cured.

G. amicalis and *G. araii* are other species associated with cutaneous infections. *G. amicalis* has been implicated in a case of cutaneous infection developed after a traumatic injury. Subcutaneous edema and chronic granulomatous inflammation with yellowish discharge were clinical features of this condition. The bacterium could be isolated from the debrided tissue and amoxicillin-clavulanate therapy could control the infection successfully. In another case, *G. araii* was responsible for the pain, swelling, and fever associated with an implant introduced after knee injury. The patient recovered after intravenous administration of trimethoprim-sulfamethoxazole (TMP-SMX) antibiotic treatment.

In a case of CVC-related bacteremia that had spread to lungs (miliary nodule), the causative agent was identified as *G. otitidis*.[14] With the removal of the CVC and initial use of vancomycin and trimethoprim-sulfamethoxazole followed by treatment with imipenem and clarithromycin the infection could be controlled. In another case of *G. otitidis* associated bacteremia,[13] the patient had an indwelling Hickman catheter with underlying condition of AML s/p-MUD allogeneic PBSCT (peripheral arterial disease; MUD, matched unrelated donor; PBSCT, peripheral blood stem cell transplant). The organism could be contained by using vancomycin.

In addition, two more species of *Gordonia*, namely *G. sputi* and *G. polyisoprenivorans*, have been reported in cases of bacteremia. *G. sputi* was isolated from blood samples of an immunocompromised patient infected with HIV. To cure the *Gordonia* infection, vancomycin, imipenem, linezolid, levofloxacin antibiotics were effective. *G. polyisoprenivorans* was responsible for bacteremia in immunocompromised patients with indwelling central catheters and in a patient with chronic myelogenous leukemia who had received a bone marrow transplant.[20] The condition was cured after the indwelling catheter was removed and antimicrobial therapy initially of piperacillin-tazobactam followed by treatment with amoxicillin and ciprofloxacin was given.

2.3 Bone and Joint Infections

The main bone-related infection that *Gordonia* species cause is osteomyelitis. *G. bronchialis* was found to be involved in a nosocomial sternal wound infection. Cellulitis and focal osteomyelitis around the sternotomy site were the major effects. *G. bronchialis* has been isolated from a healthy immunocompetent host who developed tibial osteomyelitis after arthoscopic knee surgery. The condition was characterized by erythema, swelling, and pain around the harvest incision site. A small abscess draining purulent fluid was observed. With excisional debridement, removal of the inserted bioresorbable polymer screw and administration of oral ciprofloxacin the condition was completely controlled.[21] In yet another case, *G. terrae* was associated with meta tarsal osteomyelitis. The infection was characterized by swelling and pus formation in cloaca on the foot sole. Antibiotic therapy with ciprofloxacin and azithromycin improved the case. A summary of the infections caused by *Gordonia* species have been detailed in Table 1. From this table it is evident that *G. terrae*, *G. bronchialis*, *G. amicalis*, *G. sputa*, *G. polyisoprenivorans*, *G. araii*, and *G. otitidis* are the important species causing infections of skin, soft tissues, bones, and joints. Fig. 1 shows representative pictures of *G. amicalis*.

Table 1 Summary of the skin, soft tissue, and bone infections caused by *Gordonia* species

***Gordonia* species**	**Skin and soft tissue infections**	**Bone**
G. terrae	Cellulitis, wound infection, palpebral abscess, mycetoma of hand, bacteremia	Metatarsal osteomyelitis
G. bronchialis	Breast abscess, granulomatous abscess, bacteremia	Sternal wound infections, tibial osteomyelitis
G. amicalis	Cutaneous infection	–
G. sputi	Localized breast abscess in expander implant, bacteremia	–
G. polyisoprenivorans	Bacteremia	–
G. araii	Cutaneous infection	
G. otitidis	Bacteremia	–

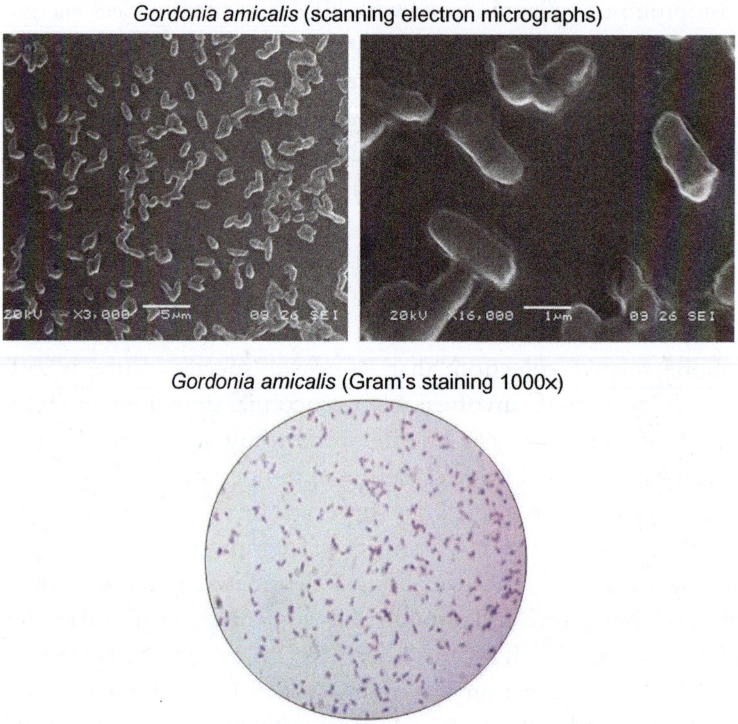

Fig. 1 *Gordonia amicalis*, scanning electron micrograph and light micrograph.

3. *NOCARDIA*

3.1 Microbiology and Epidemiology

Members of the genus *Nocardia* are soil saprophytes that show a weak acid-fast nature due to the presence of mycolic acids in their cell walls. After Gram staining, they display an irregular beaded appearance. They can grow slowly on Löwenstein-Jensen medium, blood agar, and Sabouraud's glucose agar medium.[22,23] Biochemical profiles and molecular techniques such as 16S rRNA gene sequence analysis, DNA-DNA hybridization, RAPD, and a combination polymerase chain reaction-restriction fragment length polymorphism (PCR-RFLP) analysis allow related species to be differentiated.[24–26]

More than 85 species of *Nocardia* have been identified to date. Among these, 25 are associated with human infections. These are mainly included in the *N. asteroides* complex (accounting for more than 50% of human cases). *N. brasiliensis, N. abscessus, N. cyriacigeorgica, N. farcinica, N. nova, N. pseudobrasiliensis* and members of the *N. transvalensis* and *N. nova* complex are human pathogens. *N. veteran* and *N. cerradoensis* are other recently reported human pathogenic species. *Nocardia* infections are generally initiated via inhalation or through cuts in the skin (cutaneous nocardiosis/cellulitis). From these sites, infections can spread to the brain, kidneys, joints, heart, eyes, and bones. *Nocardia* species are opportunistic pathogens and these infections are often hospital-acquired.

Nocardia species cause a condition referred to as nocardiosis. This is most commonly noticed in India, Central and South America, and North Africa. Cell-mediated immunity is crucial to contain the infection. Organ transplantation, cancer, diabetes mellitus, alcoholism, AIDS and long-term use of corticosteroids are reported to increase the risk of nocardiosis.[27] Nocardiosis can be classified in two forms: (i) primary cutaneous nocardiosis and (ii) disseminated nocardiosis. Primary cutaneous nocardiosis is characterized by three distinct clinical presentations: (i) superficial infection with cellulitis, abscess formation, or severe ulceration; (ii) development of the lymphocutaneous form; (iii) generation of a chronic granulomatous mass with sinus formation referred to as actinomycetoma or madurafoot.[23] *N. brasiliensis* is usually associated with subcutaneous infections. Disseminated nocardiosis occurs most frequently in immunocompromised patients with clinical history of diabetes mellitus, alcoholism, malignancy, chronic granulomatous diseased condition or in patients who are on long-term corticosteroids.[28,29] Immunosuppressant conditions associated with organ transplantation, HIV infection, leukemia or lymphoma and preexisting chronic pulmonary disease are also predisposing to disseminated nocardiosis.[23,30] Members of the *N. asteroides* group cause serious invasive lung infections in such immunocompromised patients.[31] From pulmonary or cutaneous sites, the infection can be disseminated to virtually any organ.[27] Sulfonamides and a combination of trimethoprim and sulfamethoxazole antibiotics have been found to be effective against species such as *N. brasiliensis*, *N. asteroides*, and *N. transvalensis*. Amikacin, minocycline, amoxicillin-clavulanate, and cephalosporins are useful for curing the cutaneous form of the disease.[32]

3.2 Skin and Soft Tissue Infections

N. brasiliensis is the main causative agent of primary cutaneous nocardiosis. There is a report on the isolation of this species from an immunocompetent patient with cellulitis.[30] After entry via a superficial abrasion on the forearm, a severely painful condition was observed. There was swelling and blazing at the site of injury. Surgical debridement of forearm pustules, application of silvadene to the wound and an oral dose of trimethoprim and sulfamethoxazole cured the condition. *N. brasiliensis* has also been implicated in a rare case oflymphocutaneous infection instigated in the eyelid.[32] The patient suffered from preseptal cellulitis from an abrasion of the eyelid that subsequently developed into submandibular lymph node suppuration. The patient was cured with antibiotic therapy using amoxicillin and clavulanate.

As stated earlier, actinomycetoma is a condition that involves the formation of a chronic granulomatous mass. This is usually initiated as a localized cutaneous infection following a traumatic injury related to contaminated soil.[27,33,34] *N. brasiliensis* is the second most frequent isolate responsible for actinomycetoma. This can be effectively treated with a combination of 4,4′-diaminodiphenylsulfone and trimethoprim-sulfamethoxazole (TMP-SMZ)-amikacin.[35] *N. brasiliensis* along with *N. asteroids* as causative agents of nocardial mycetoma involving characteristic pain, swelling, and pus have also been reported. Successful recovery has been observed when treated with cotrimoxazole, amikacin, and rifampicin. In addition, *N. asteroides* is also implicated in cutaneous infections. There has been a report on the development of leg-associated cellulitis characterized by swelling, erythema, and multiple pus-filled bullae.[23] The condition could be cured by draining the abscesses and by treatment with cotrimoxazole. In another case, *N. asteroides* formed abscesses in the thigh and brain. After surgical draining of the abscesses, treatment with ceftriaxone and amikacin and oral administration of minocycline cured the infection.[36] In addition to the previously mentioned species, there are reports on actinomycetoma being caused by *N. aobensis*. This infection could be cured by treatment with antibiotics such as amikacin and trimethoprim-sulfamethoxazole.[37] *N. transvalensis* is another species reported to cause actinomycotic mycetoma, wherein papules and nodules were observed. The abrasions were erythematous, nodular, and slightly tender on palpation. The patient recovered after surgical removal of the lesion and treatment with cotrimoxazole.[38] *N. nova* was found to be the causative agent of mycetoma in another case[39] and *N. yamanashiensis* was identified to be the causative agent associated with a case of primary cutaneous nocardiosis.[40]

Endocarditis associated with disseminated skin and soft tissue infection has been caused by *N. cyriacigeorgica*.[41] *N. abscessus* is another species allied with soft tissue abscesses or empyemas, pericarditis, endoprosthesis infection and fibular abscesses. The soft tissue infection could be cured by surgical cleaning and repetitive application of an antiseptic. Oral antibiotic therapy of trimethoprim-sulfamethoxazole cured the infection.[42] A psoas muscle abscess was reported due to the infection caused by *N. beijingensis* in a renal transplant recipient.[43]

In another case, *N. neocaldoniensis* caused skin and soft tissue infection in a patient with rheumatoid arthritis. The patient displayed the formation of a large abscess, with the area being erythematous and tender to palpation. The patient's condition improved upon surgical debridement and treatment with trimethoprim-sulfamethoxazole and intravenous injection of imipenem. Later on, doxycycline in combination with moxifloxacin was prescribed.[44]

Bacteremia caused by *Nocardia* is generally rare. This can lead to infection in the eyes (keratitis), heart valves, liver, spleen, adrenal glands, thyroid gland, and other organ tissues. This is usually prevalent as postoperative infections associated with catheterized patients.[45,46] Isolates belonging to the *N. nova* complex have been reported to cause catheter-related bacteremia. These rarely assault remote anatomical structures and form deep-seated infections. Disseminated *Nocardia* bacteremia is mainly associated with patients from intensive care units and cancer patients. The severity of bacteremia mainly depends on the underlying conditions. *Nocardia* species have the capability of adhering to a polyurethane and silicon central venous catheter (CVC) and forming a biofilm on the catheter surfaces.

Primary CVC-associated bacteremia is reported to be caused by *N. asteroides*.[47] Catheter-associated *N. higoensis* bacteremia was detected in a patient with acute lymphocytic leukemia.[45] *N. caviae* causing bacteremia was reported in patients with thalassemia and postbone marrow transplant.[46] Members of the *N. nova* complex caused bacteremia in central venous catheter-dwelling patients with cancer and acute lymphocytic leukemia. The *Nocardia* isolates adhered to and formed extensive biofilms on the surface of polyurethane and silicone CVCs. In all cases, discontinuation of the use of the catheter and antibiotic treatment with trimethoprim sulfamethoxazole and minocycline imipenem and amikacin cured the condition.[48]

3.3 Bone and Joint Infections

N. asteroides has been associated with a case of osteomyelitis.[49] In addition, *N. brasiliensis* was also allied with a case of femorotibial osteomyelitis. This condition was characterized by tissue edema, multiple radiolucent areas alternating with healthy bone, and thickening of the cortical bone of the distal femur. Surgical debridements and combinations of amikacin and imipenem or TMP-SMX drugs were used to cure the disease.[34] *N. cyriacigeorgica* and *N. nova* have been associated with knee arthritis. The infections were controlled by surgical debridement and by treatment with antibiotics such as amoxicillin, cefotaxime, imipenem, meropenem, and cotrimoxazole. Knee arthritis has also been caused by *N. farcinica*. Despite multiple-antibiotic therapy, a condition of osteomyelitis was developed. With surgical synovectomy, administration of cefuroxime, cotrimoxazole and later doxycycline, the osteolytic lesions were controlled. However, due to secondary infection with *Klebsiella pneumoniae*, septic shock ensued which resulted in the death of the patient.[50] Skin, soft tissue, bone, and joint infections associated with *Nocardia* species have been summarized in Fig. 2.

4. *RHODOCOCCUS*

4.1 Microbiology and Epidemiology

Members of genus *Rhodococcus* are aerobic, Gram-positive, partially acid-fast rod-cocci. *Rhodococcus* species are well-recognized pathogens in veterinary medicine that can also affect human subjects. Most cases have been reported from immunocompromised patients with organ transplants or those with human immunodeficiency virus infection. The most common species reported is *R. equi* and the most common clinical manifestation is a slowly progressive pneumonia that may cavitate. *R. equi* is a common

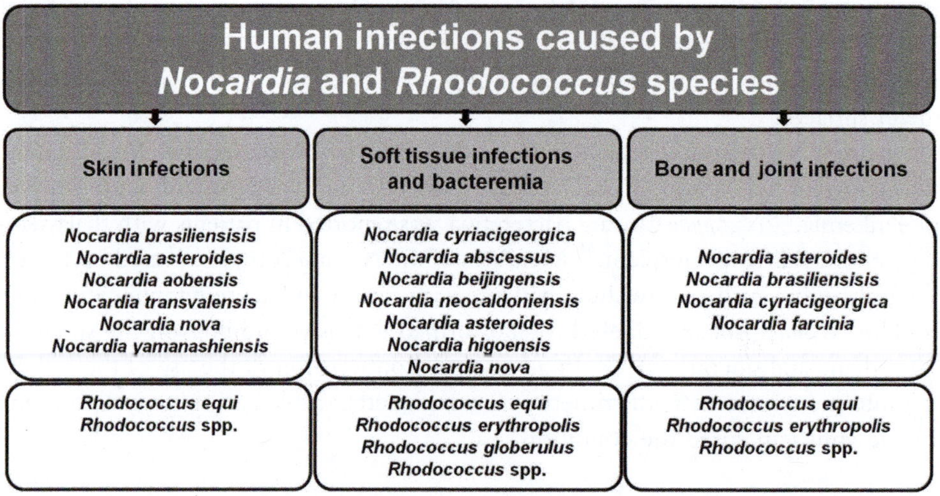

Fig. 2 A summary of skin, soft tissue, bone, and joint infections caused by different species of *Nocardia* and *Rhodococcus.*

inhabitant of the soil and inhalation of dust is supposed to be the primary route of exposure for both animals and man. Infection often takes place through lungs, alimentary tract or via injured skin. From here it may spread to other tissues and organs. Infected patients often have a history of contact with animals, manure, or soil. *R. equi* infection is also called rhodococcosis. This is regarded as an opportunistic disease which is often chronic and recurrent. The host's cell-mediated immunity seems to be of major importance in conferring resistance to the infection. The ability of *R. equi* to persist in macrophages and destroy them is important in the pathogenesis. Rhodococcal infections are difficult to treat, usually requiring prolonged administration of parenteral antibiotics and surgical drainage. A combination of rifampicin and erythromycin is generally used to cure such infections.

4.2 Skin and Soft Tissue Infections

Most of the skin infections caused by *R. equi* occur in immunocompromised hosts, whereas immuno-competent individuals are rarely affected and usually show localized disease symptoms. Microbiological analysis is very important in the identification of the organism since it may be mistaken for common skin flora. There is a report on *R. equi* infection after reduction mammaplasty in an immunocompetent patient without medical problems who showed right breast cellulitis and purulent drainage from the surgical wound.[51] The patient underwent incision and drainage, and cultures of the wound yielded *R. equi*. Antimicrobial therapy with moxifloxacin and rifampin resulted in complete resolution. A kidney transplant recipient who developed multiple disseminated abscesses caused by *R. equi* while on adequate antimicrobial therapy had subcutaneous abscesses in the hip region and mamma and had two intracerebral abscesses. *R. equi* was cultured from all abscesses. The patient died of progressive neurologic complications.[52]

In the case of two HIV patients with disseminated *R. equi*, infection occurred despite highly active antiretroviral therapy (HAART) and antirhodococcal treatment. In one patient, pulmonary rhodococcosis was followed by an extracerebral, intracranial and a cerebral rhodococcal abscess. In the other patient, the infection spread to various organs, including subcutaneous tissue, skin, colon and other intraabdominal tissues and central nervous system.[53] Submandibular lymphadenitis was observed in an HIV-infected patient and *R. equi* was isolated from the lymph node.[54]

In a case of actinomycetoma caused by *Rhodococcus* spp. in a patient with multiple draining sinuses of the left foot, microabcesses showed granules containing rod and coccoid Gram-positive and partially acid-fast elements, typical of *Rhodococcus* spp. on culture.[55] *Rhodococcus* spp. has also been isolated from a granulomatous skin lesion from a patient, who developed a nodule with axillary lymphadenopathy.[56]

R. equi is the most common species associated with human soft tissue infections and bacteremia. *R. equi* infection characterized by multiple abscesses of soft tissues and muscles was observed in a kidney transplant patient. Combining specific antibiotic therapy and surgical management of the abscesses without immune-suppression discontinuation led to a complete recovery of both patient and graft.[57] A patient with diabetes and liver cirrhosis on long-term corticosteroid therapy after skin-grafting surgery was diagnosed with *R. equi* bacteremia. The patient recovered soon after he was treated with vancomycin.[58]

Patients with leukemia are also susceptible to *Rhodococcus* bacteremia.[59–61] In such cases, *R. equi* was almost always isolated from blood cultures during the course of disease. Response to broad spectrum antimicrobials was poor. Patients failed to improve even after combined antibiotic therapy, leading to death in serious cases. These authors have discussed the importance of immune reconstitution or immune supportive treatment in addition to combined antimicrobial treatment in such cases.

Catheter-related bacteremia due to *R. equi* infections is also very common.[62,63] *Rhodococcus* bacteremia in cancer patients is mostly catheter related and associated with biofilm formation. Through a detailed study on cancer patients with *Rhodococcus* bacteremia and indwelling central venous catheters, these authors found that *Rhodococcus* bacteremia was, for the most part, central line–associated bloodstream infection (CLABSI). Most of the bacteremia isolates were efficient biofilm formers and thus resisted antimicrobial therapy. In a rare case, hemodialysis catheter–related *Rhodococcus* bacteremia in an immunocompetent host has also been reported.[64]

The incidence of *R. equi* infection in humans has increased with the advent of HIV and use of immune suppressants.[65–67] Bacteremia is observed in most cases along with other clinical manifestations like diarrhea, granulomatous soft tissue masses, pericardial effusion, and pericardial cysts. Infections are generally treated with drainage and combined antimicrobial therapy with vancomycin, azithromycin, and levofloxacin.

R. equi bacteremia has also been reported in infants and immunocompetent children and is therefore an important pediatric pathogen.[68–71] *R. equi* was isolated from blood cultures in all cases along with other manifestations like recurrent swellings on different parts of the body, fever, and respiratory distress. Treatment with multiple antibiotics such as vancomycin, aminoglycosides, chloramphenicol, and erythromycin was successful in most cases.

R. erythropolis has been reported to cause septicemia.[72,73] Bacteremia caused by *R. globerulus* was diagnosed in an allogeneic hematopoietic stem cell transplant recipient, who had subacute systemic illness accompanied by severe hepatitis.[74] Although these *Rhodococcus* species are rarely associated with human infections, they should be considered as potential causative agents of bacteremia.

Infection of the central nervous system, an uncommon infection site, was seen in a patient with pneumonia, brain abscess, and recurrent meningitis.[75] The patient initially had pneumonia with necrosis, which progressed to brain abscess and recurrent meningitis. *Rhodococcus* spp. was identified from the cerebrospinal fluid (CSF). The patient was cured with treatment comprising vancomycin and amikacin administered directly into the CSF in conjunction with intravenous and oral antibiotics.

4.3 Bone and Joint Infections

Rhodococcus species have been reported to cause bone and joint infections in immunocompromised patients. A case of acute osteomyelitis of the mandible caused by *R. equi* was observed in a patient diagnosed with a swelling of the left submental and submandibular spaces.[76] The patient was immunocompromised owing to medication against myasthenia gravis and type II diabetes mellitus. The patient underwent surgical debridement under local anesthesia. Histologic examination showed acute osteomyelitis and *R. equi* was isolated from both blood and pus cultures. The patient was cured after prolonged treatment with oral linezolid.

Mastoiditis is a bacterial infection of the mastoid bone, which lies behind the ear. A patient with AIDS was diagnosed with otomastoiditis with systemic dissemination due to *R. equi*.[77] A similar case of severe otitis media (infection of the middle ear) and mastoiditis due to *R. equi* was seen in a patient with AIDS.[78] The patient had cough and right facial palsy. Imaging revealed right otitis media and severe temporal bone destruction, associated with pneumonia. *R. equi* was isolated from ear secretions, blood, and sputum.

A rare case of acute osteomyelitis caused by *R. equi* in an immunocompetent child has been reported.[79] There are reports on osteomyelitis due to *R. equi* in a renal transplant recipient[80] and vertebral osteomyelitis in a liver transplant recipient.[81] In the latter case, the liver transplant recipient presented with recurrent pneumonia and a pleura-based lung abscess and subsequently developed osteomyelitis of the lower thoracic spine. The clinical treatment was comprised of surgical debridement and prolonged treatment with rifabutin and clarithromycin.

A rare case of *R. erythropolis* osteomyelitis in the toe was seen in a woman as a musculoskeletal infection that developed after a metatarsophalangeal joint fusion.[82] The patient continued to have pain and swelling with nonunion even a year after surgery. Tissue samples after repeated surgery revealed *R. erythropolis* infection. The patient's symptoms improved with oral antibiotics and the fusion had united 1 year after the revision surgery.

A case of septic arthritis and osteomyelitis was observed in a previously healthy, immunologically normal patient who presented painless swelling of fingers, a toe, and one knee. Rice bodies (corpora oryzoidea) were identified in viscous fluid obtained from the knee during arthroscopy. Culture of this fluid resulted in growth of an organism identified as *Rhodococcus* spp. Treatment with a combination of erythromycin and amoxicillin proved successful.[83]

Invasive *R. equi* infections in Taiwan have been reported with an emphasis on the emergence of multidrug-resistant strains.[84] The molecular mechanisms responsible for antimicrobial resistance in *R. equi* include alterations in the cell envelope composition and structure, activity of the efflux pumps, enzymatic destruction or inactivation of antibiotics and changes in the target site.[85] A summary of skin, soft tissue, bone, and joint infections caused by *Rhodococcus* species is given in Fig. 2.

5. CONCLUSIONS

Actinomycetes of genera *Gordonia*, *Nocardia*, and *Rhodococcus* are opportunistic pathogens causing infections of skin, soft tissues, bones, and joints in human subjects. Infections in humans may commonly be postoperative and hospital acquired and most cases have been reported from immunocompromised individuals such as organ transplant patients, those with HIV infection or patients with acute leukemia under chemotherapy. Although such infections are rare in immunocompetent individuals, premature infants and children, the probability of such infections needs serious consideration. Infections caused by these organisms may be more common than expected and, if neglected, may lead to serious clinical manifestations in patients and even death in extreme cases. Therefore, an early diagnosis of such infections among immunosuppressed patients, infants, and children is very important due to the high virulence associated with these organisms. Emergence of multidrug resistant strains among these genera poses additional risks. Treatment comprising combinations of effective antibiotics and immune reconstitution or immune supportive treatment in appropriate situations would prove useful in most cases. More recently, treatment procedures like vacuum-assisted wound closure (VAC) therapy have been reported wherein chronic tibial osteomyelitis caused by mixed infection comprising *Nocardia* spp. could be treated by postoperative VAC therapy directly to the medullary cavity, followed by residual wound coverage with a gastrocnemius myocutaneous flap.[86] A very recent report[87] describes a multicenter European case-control study carried out between 2000 and 2014 in 36 European centers on *Nocardia* infection after solid organ transplant (SOT) in recipients. The study resulted in identification of risk factors for nocardiosis after SOT and also showed that it could not be prevented by low-dose cotrimoxazole. Such findings and surgical procedures would help to improve management of transplant recipients and patients suffering from infections due to opportunistic pathogens.

REFERENCES

1. Waksman SA. *The actinomycetes—nature, occurrence and activities*, vol. I. Baltimore, MD: Williams and Wilkins Company; 1959.
2. Lechevalier HA, Lechevalier MP. *The prokaryotes: a handbook of habitat, isolation and identification of bacteria*, vol. II. Berlin: Springer-Verlag; 1981.
3. Goodfellow M, Williams E. Ecology of actinomycetes. *Annu Rev Microbiol* 1983;**37**:189–216.
4. Lechevalier MP. Identification of aerobic actinomycetes of clinical importance. *J Lab Clin Med* 1968;**71**:934–44.
5. Lechevalier MP, Lechevalier HA. Chemical composition in the classification of aerobic actinomycetes. *Int J Syst Bacteriol* 1970;**20**:435–43.
6. Lechevalier HA, Lechevalier MP, Gerber NN. Chemical composition as a criterion in the classification of actinomycetes. *Adv Appl Microbiol* 1971;**14**:47–72.
7. Alshamaony L, Goodfellow M, Minnikin DE. Free mycolic acids as criteria in the classification of *Nocardia* and the 'rhodochrous' complex. *J Gen Microbiol* 1976;**92**:188–99.
8. Collins MD, Goodfellow M, Minnikin DE. A survey of the structures of mycolic acids in *Corynebacterium* and related taxa. *J Gen Microbiol* 1982;**128**:129–49.
9. Arenskötter M, Broker D, Steinbuchel A. Biology of the metabolically diverse genus *Gordonia*. *Appl Environ Microbiol* 2004;**70**:3195–204.
10. Goodfellow M, Kämpfer P, Busse H, Trujillo ME, Suzuki K, Ludwig W, et al. *Bergey's manual of systematic bacteriology*. 2nd ed. *The* actinobacteria, *part A*, vol. 5; 2012. http://dx.doi.org/10.1007/978-0-387-68233-4.
11. Lai CC, Wang CY, Liu CY, Tan CK, Lin SH, Liao CH, et al. Infections caused by *Gordonia* species at a medical centre in Taiwan, 1997 to 2008. *Clin Microbiol Infect* 2010;**16**:1448–53.
12. Verma P, Brown JM, Nunez VH, Morey RE, Steigerwalt AG, Pellegrini GJ, et al. Native valve endocarditis due to *Gordonia polyisoprenivorans*: case report and review of literature of bloodstream infections caused by *Gordonia* species. *J Clin Microbiol* 2006;**44**:1905–8.
13. Ramanan P, Deziel PJ, Wengenack NL. *Gordonia* bacteremia. *J Clin Microbiol* 2013;**51**:3443–7.
14. Blaschke AJ, Bender J, Byington CL, Korgenski K, Daly J, Petti CA, et al. *Gordonia* species: emerging pathogens in pediatric patients that are identified by 16S ribosomal RNA gene sequencing. *Clin Infect Dis* 2007;**45**:483–6.
15. Tsukamura M. Proposal of a new genus, *Gordona*, for slightly acid-fast organisms occurring in sputa of patients with pulmonary disease and in soil. *J Gen Microbiol* 1971;**68**:15–26.
16. Chang JH, Misuk J, Hong HL, Choi SH, Kim YH, Chung CH, et al. Sternal osteomyelitis caused by *Gordonia bronchialis* after open-heart surgery. *Infect Chemother* 2014;**46**:110–4.
17. Blanc V, Markarian A, Debunne MV. *Gordonia terrae*: a difficult to diagnose emerging pathogen? *J Clin Microbiol* 2007;**45**:1076–7.

18. Pham AS, De I, Rolston KV, Tarrand JJ, Han XY. Catheter-related bacteremia caused by the nocardioform actinomycete *Gordonia terrae*. *Clin Infect Dis* 2003;**36**:524–7.
19. Werno AM, Anderson TP, Chambers ST, Laird HM, Murdoch DR. Recurrent breast abscess caused by *Gordonia bronchialis* in an immunocompetent patient. *J Clin Microbiol* 2005;**43**:3009–10.
20. Kempf VAJ, Schmalzing M, Yassin AF, Schaal KP, Baumeister D, Kempf VAJ, et al. *Gordonia polyisoprenivorans* septicemia in a bone marrow transplant patient. *Eur J Clin Microbiol Infect Dis* 2004;**3**:226–8.
21. Siddiqui N, Toumeh A, Georgescu C. Tibial osteomyelitis caused by *Gordonia bronchialis* in an immunocompetent patient. *J Clin Microbiol* 2012;**50**:3119–21.
22. Corti ME, Fioti MFV. Nocardiosis: a review. *Int J Infect Dis* 2003;**7**:243–50.
23. Wong YW, Waghorn D, Ratnavel RC. Antibiotic-resistant cellulitis acquired in Turkey. *J R Soc Med* 2003;**96**:32–3.
24. Kampfer P, Dott W, Kroppenstedt RM. Numerical classification and identification of some nocardioform bacteria. *J Gen Appl Microbiol* 1990;**36**:309–31.
25. Chun J, Goodfellow M. A phylogenetic analysis of the genus *Nocardia* with 16S rRNA gene sequences. *Int J Syst Bacteriol* 1995;**45**:240–5.
26. Goodfellow M, Isik K, Yates E. Actinomycete systematics: an unfinished synthesis. *Nova Acta Leopold NF* 1999;**80**:47–82.
27. Johnson P, Ammar H. *Nocardia brasiliensis* vertebral osteomyelitis and epidural abscess. *BMJ Case Rep* 2013. http://dx.doi.org/10.1136/bcr-2012-008400.
28. Farina C, Boiron P, Goglio A, Provost F. Human nocardiosis in northern Italy from 1982 to 1992 Northern Italy collaborative group on Nocardiosis. *Scand J Infect Dis* 1995;**27**:23–7.
29. Dorman SE, Guide SV, Conville PS. *Nocardia* infection in chronic granulomatous disease. *Clin Infect Dis* 2002;**35**:390–4.
30. Hearne CB, Eckford J, Forjuoh SN. The gardener's cellulitis. *Am J Med* 2009;**122**:27–8.
31. Sharma NL, Mahajan VK, Agarwal S, Katoch VM, Das R, Kashyap M. Nocardial mycetoma: diverse clinical presentations. *Indian J Dermatol Venereol Leprol* 2008;**74**:635–40.
32. Brannan PA, Kersten RC, Hudak DT, Anderson HK, Kulwin DR. Primary *Nocardia brasiliensis* of the eyelid. *Am J Ophthalmol* 2004;**138**:498–9.
33. Wilson JW. Nocardiosis: updates and clinical overview. *Mayo Clin Proc* 2012;**87**:403–7.
34. Vanegas S, Franco-Cendejas R, Cicero A, López-Jácome E, Colín C, Hernández M. *Nocardia brasiliensis*-associated femorotibial osteomyelitis. *Int J Infect Dis* 2014;**20**:63–5.
35. Sousa JMP. Mycetoma caused by *Nocardia brasiliensis* in an immunocompetent patient. *J Dtsch Dermatol Ges* 2014;**12**:903–5.
36. Ozaras R, Mert A, Hakko E, Tabak F, Ozturk R. Pulmonary, cerebral, and soft tissue nocardiosis. *Lancet Infect Dis* 2004;**4**:222.
37. Vongphoumy I, Dance DAB, Dittrich S, Logan J, Davong V, Rattanavong S. Case report: actinomycetoma caused by *Nocardia aobensis* from Lao PDR with favourable outcome after short-term antibiotic treatment. *PLoS Negl Trop Dis* 2015;**9**. http://dx.doi.org/10.1371/journal.pntd.0003729.
38. Mirza SH, Campbell C. Mycetoma caused by *Nocardia transvalensis*. *J Clin Pathol* 1994;**47**:85–6.
39. Dhingra M, Kaistha N, Bansal N, Solanki LS, Chander J, Thami GP, et al. *Nocardia nova* mycetoma over forehead in a lepromatous leprosy patient. *Dermatol Online J* 2012;**18**:3.
40. Anzalone CL, Cohen PR, Tarrand JJ, Diwan AH, Prieto VG. *Nocardia yamanashiensis* in an immunocompromised patient presenting as an indurated nodule on the dorsal hand. *Tumori* 2013;**99**(4): e156–8.
41. Cargill JS, Boyd GJ, Weightman NC. *Nocardia cyriacigeorgica*: a case of endocarditis with disseminated soft-tissue infection. *J Med Microbiol* 2010;**59**:224–30.
42. Daeschlein G, Fetouh Yassin AA, Franke A, Kramer A, Schaal KP. Unusual infections: femoral abscess due to *Nocardia abscessus* in a patient suffering from metastatic peripheral bronchial carcinoma and hygienic consequences. *GMS Krankenhhyg Interdiszip* 2011;**6**. http://dx.doi.org/10.3205/dgkh000160.
43. Palavutitotai N, Chongtrakoo P, Ngamskulrungroj P, Chayakulkeeree M. *Nocardia beijingensis* psoas abscess and subcutaneous phaeohyphomycosis caused by *Phaeoacremonium parasiticum* in a renal transplant recipient: the first case report in Thailand. *Southeast Asian J Trop Med Public Health* 2015;**46**:1049–54.

44. McGhie T, Fader R, Carpenter J, Brown-Elliott BA, Vasireddy R, Wallace RJ. *Nocardia neocaledoniensis* as a cause of skin and soft tissue infection. *J Clin Microbiol* 2012;**50**:3139–40.
45. Kontoyiannis DP, Jacobson KL, Whimbey EE, Rolston KVI, Raad II. Central venous catheter-associated *Nocardia* bacteremia: an unusual manifestation of Nocardiosis. *Clin Infect Dis* 2000;**31**:617–8.
46. Kandi V. Human *Nocardia* infections: a review of pulmonary nocardiosis. *Cureus* 2015;**7**(8):e304. http://dx.doi.org/10.7759/cureus.304.
47. Feng YH, Huang WT, Tsao CJ. Venous access port-related *Nocardia* bacteremia with successful short-term antibiotics treatment. *J Chin Med Assoc* 2004;**67**:416–8.
48. Akhrass FA, Hachem R, Mohamed JA, Tarrand J, Kontoyiannis DP, Chandra J, et al. Central venous catheter-associated *Nocardia* bacteremia in cancer patients. *Emerg Infect Dis* 2011;**17**:1651–8.
49. Shin KH, Lee WS, Son YK, Lee K, Chong Y. *Nocardia* oteomyelitis in a pachymeningitis patient: an example of a difficult case to treat with antimicrobial agents. *Yonsei Med J* 1998;**39**:604–10.
50. Chaussade H, Lebeaux D, Gras G, Catherinot E, Rammaert B, Poiree S, et al. *Nocardia* arthritis: 3 cases and literature review. *Medicine (Baltimore)* 2015;**94**(12):e1671.
51. Sandkovsky U, Sandkovsky G, Sordillo EM, Polsky B. *Rhodococcus equi* infection after reduction mammaplasty in an immunocompetent patient. *Rev Inst Med Trop Sao Paulo* 2011;**53**:291–4.
52. Rahamat-Langendoen JC, van Meurs M, Zijlstra JG, Lo-Ten-Foe JR. Disseminated *Rhodococcus equi* infection in a kidney transplant patient without initial pulmonary involvement. *Diagn Microbiol Infect Dis* 2009;**65**:427–30.
53. Ferretti F, Boschini A, Iabichino C, Gerevini S, De Nardi P, Guffanti M, et al. Disseminated *Rhodococcus equi* infection in HIV infection despite highly active antiretroviral therapy. *BMC Infect Dis* 2011;**11**:343.
54. Chiewchanvit S, Mahanupab P, Baosoung V, Khamwan C. Uncommon manifestations of opportunistic infections in an HIV infected patient. *J Med Assoc Thai* 1998;**81**:923–6.
55. Severo LC, Petrillo VF, Coutinho LM. Actinomycetoma caused by *Rhodococcus* spp. *Mycopathologia* 1987;**98**:129–31.
56. Martin T, Hogan DJ, Murphy F, Natyshak I, Ewan EP. *Rhodococcus* infection of the skin with lymphadenitis in a non-immunocompromised girl. *J Am Acad Dermatol* 1991;**24**:328–32.
57. Varotti G, Barabani C, Dodi F, Bertocchi M, Mondello R, Cupo P, et al. Unusual extrapulmonary *Rhodococcus equi* infection in a kidney transplant patient. *Exp Clin Transplant* 2016;**14**:676–8. http://dx.doi.org/10.6002/ect.2014.0176.
58. Chen XY, Xu F, Xia JY, Cheng YS, Yang Y. Bacteremia due to *Rhodococcus equi*: a case report and review of the literature. *J Zhejiang Univ Sci B* 2009;**10**:933–6.
59. Shahani L. *Rhodococcus equi* pneumonia and sepsis in an allogeneic haematopoietic stem cell transplant recipient. *BMJ Case Rep* 2014. http://dx.doi.org/10.1136/bcr-2014-204721.
60. Zhang J, Xu J, Du X, Yu Y, Lv F. Rare and persistent *Rhodococcus equi* infection in a diffuse large B cell lymphoma patient: case report and review of the literature. *J Thorac Dis* 2014;**6**:E281–8.
61. Sladek GG, Frame JN. *Rhodococcus equi* causing bacteremia in an adult with acute leukemia. *South Med J* 1993;**86**:244–6.
62. Guerrero R, Bhargava A, Nahleh Z. *Rhodococcus equi* venous catheter infection: a case report and review of the literature. *J Med Case Rep* 2011;**5**:358.
63. Akhrass F, Wohoush I, Chaftari AM, Reitzel R, Jiang Y, Ghannoum M, et al. *Rhodococcus* bacteremia in cancer patients is mostly catheter related and associated with biofilm formation. *PLoS One* 2012;**7**: e32945.
64. Dalal P, Dalal T, Shah G. Hemodialysis catheter related *Rhodococcus* bacteremia in immunocompetent host. *Saudi J Kidney Dis Transpl* 2011;**22**:761–3.
65. Fierer J, Wolf P, Seed L, Gay T, Noonan K, Haghighi P. Non-pulmonary *Rhodococcus equi* infections in patients with acquired immune deficiency syndrome (AIDS). *J Clin Pathol* 1987;**40**:556–8.
66. Gundelly P, Thornton A, Greenberg RN, McCormick M, Myint T. *Rhodococcus equi* pericarditis in a patient living with HIV/AIDS. *J Int Assoc Provid AIDS Care* 2014;**13**:309–12.
67. Emmons W, Reichwein B, Winslow DL. *Rhodococcus equi* infection in the patient with AIDS: literature review and report of an unusual case. *Rev Infect Dis* 1991;**13**:91–6.
68. Bildik HN, Takcı S, Yurdakök M, Kara A. Neonatal sepsis due to *Rhodococcus equi* in two preterm infants. *Turk J Pediatr* 2013;**55**:229–31.

69. Devi P, Malhotra S, Chadha A. Bacteremia due to *Rhodococcus equi* in an immunocompetent infant. *Indian J Med Microbiol* 2011;**29**:65–8.
70. Sarangi G, Chayani N, Mahapatra A, Mahapatra D, Paty BP, Parida B. Bacteremia due to *Rhodococcus equi*—a case report. *Indian J Pathol Microbiol* 2004;**47**:553–5.
71. McGowan KL, Mangano MF. Infections with *Rhodococcus equi* in children. *Diagn Microbiol Infect Dis* 1991;**14**:347–52.
72. Park SD, Uh Y, Jang IH, Yoon KJ, Kim HM, Bae YJ. *Rhodococcus erythropolis* septicaemia in a patient with acute lymphocytic leukaemia. *J Med Microbiol* 2011;**60**:252–5.
73. Baba H, Nada T, Ohkusu K, Ezaki T, Hasegawa Y, Paterson DL. First case of bloodstream infection caused by *Rhodococcus erythropolis*. *J Clin Microbiol* 2009;**47**:2667–9.
74. Ramanan P, Deziel PJ, Razonable RR. *Rhodococcus globerulus* bacteremia in an allogeneic hematopoietic stem cell transplant recipient: report of the first transplant case and review of the literature. *Transpl Infect Dis* 2014;**16**:484–9.
75. Lee K, Rho M, Yu M, Kwak J, Hong S, Kim J, et al. A case of recurrent meningitis caused by *Rhodococcus* species successfully treated with antibiotic treatment and intrathecal injection of Vancomycin through an Ommaya reservoir. *Infect Chemother* 2015;**47**:183–9.
76. Rallis G, Dais P, Gkinis G, Mourouzis C, Papaioannou V, Mezitis M. Acute osteomyelitis of the mandible caused by *Rhodococcus equi* in an immunocompromised patient: a case report and literature review. *Oral Surg Oral Med Oral Pathol Oral Radiol* 2012;**114**:e1–5.
77. Kim SC, Jorgensen J, Graybill JR, Smith J. Otomastoiditis caused by *Rhodococcus equi* in a patient with AIDS. *P R Health Sci J* 1999;**18**:285–8.
78. Ibarra R, Jinkins JR. Severe otitis and mastoiditis due to *Rhodococcus equi* in a patient with AIDS. Case report. *Neuroradiology* 1999;**41**:699–701.
79. Sistla S, Karthikeyan S, Biswas R, Parija SC, Patro DK. Acute osteomyelitis caused by *Rhodococcus equi* in an immunocompetent child. *Indian J Pathol Microbiol* 2009;**52**:263–4.
80. Novak RM, Polisky EL, Janda WM, Libertin CR. Osteomyelitis caused by *Rhodococcus equi* in a renal transplant recipient. *Infection* 1988;**16**:186–8.
81. Fischer L, Sterneck M, Albrecht H, Krupski G, Polywka S, Rogiers X, et al. Vertebral osteomyelitis due to *Rhodococcus equi* in a liver transplant recipient. *Clin Infect Dis* 1998;**26**:749–52.
82. Roy M, Sidhom S, Kerr KG, Conroy JL. Case report: *Rhodococcus erythropolis* osteomyelitis in the toe. *Clin Orthop Relat Res* 2009;**467**:3029–31.
83. Broughton RA, Wilson HD, Goodman NL, Hedrick JA. Septic arthritis and osteomyelitis caused by an organism of the genus *Rhodococcus*. *Clin Microbiol* 1981;**13**:209–13.
84. Hsueh PR, Hung CC, Teng LJ, Yu MC, Chen YC, Wang HK, et al. Report of invasive *Rhodococcus equi* infections in Taiwan, with an emphasis on the emergence of multidrug-resistant strains. *Clin Infect Dis* 1998;**27**:370–5.
85. Cisek AA, Rzewuska M, Witkowski L, Binek M. Antimicrobial resistance in *Rhodococcus equi*. *Acta Biochim Pol* 2014;**61**:633–8.
86. Miyamura S, Tsuji S, Iwai T, Hamada M. VAC therapy direct to the medullary cavity for chronic tibial osteomyelitis. *Wounds* 2016;**28**:E22–6.
87. Coussement J, Lebeaux D, van Delden C, Guillot H, Freund R, Marbus S, et al. *Nocardia* infection in solid organ transplant recipients: a multicenter European case-control study. *Clin Infect Dis* 2016;**63**:338–45.

SECTION II

Bone and Joint Infections: Diagnosis and Treatment

CHAPTER 8

The Molecular Mechanics of Inflammatory Bone and Joint Disease Caused by Microbial Infection

N. Kavanagh, F.J. O'Brien, S.W. Kerrigan
Royal College of Surgeons in Ireland and Trinity College Dublin, Dublin, Ireland

1. INTRODUCTION

Bone infection has been documented as far back as Hippocrates (460–370 BC) after a bone fracture became infected. It has had various terms throughout history, including "Abcessus in medulla" as described by Broomfield in 1773, and later by the first professor of surgery at Yale University, Smith, who described the dead bone due to infection as "disease necrosis." It is, however, Nelaton who is credited with introducing the term "osteomyelitis" in 1844.[1] French pharmacist and microbiologist Louis Pasteur first suggested microbes were present in osteomyelitis clinical pathology samples, and described it as "boil of the bone marrow." Pioneering work by Ogston identified two types of micrococci, which were later termed staphylococci and streptococci.[2,3] Modern-day understanding of osteomyelitis indicates that it is an inflammatory bone disease caused by an infecting microorganism leading to progressive bone loss and bone destruction.[4] In osteomyelitis, infection is established when a microorganism gains entry to the sterile bone, accompanied by edema in the bone marrow and/or surrounding soft tissue. This subsequently leads to restricted vascularization of the infected area and small vessel thrombosis. Due to the inflammation and reduced vascularization of the area, the infected section of bone forms an agglomerate of necrotic bone cells, also known as sequestrum, a phenomenon considered the hallmark sign of osteomyelitis.[4–7] Likewise, septic arthritis is a joint disease whereby a microorganism gains entry to the joint, leading to rapid articular deterioration. This occurs due to the proliferation of bacteria within the synovium generating an immune response. Infiltration of the white blood cells results in increased pressure, restricting the blood flow and nutrients to the area. This further exacerbates the joint damage, leaving permanent destruction of the articular cartilage and subchondral bone within a few days of the offset of infection.[8–11]

1.1 Classification

There are two main systems used to classify osteomyelitis suggested by Waldvogel et al.[12] and Cierny et al.[7] The Waldvogel system classifies osteomyelitis based on the etiology of

The Microbiology of Skin, Soft Tissue, Bone and Joint Infections
ISSN 2451-9006, http://dx.doi.org/10.1016/B978-0-12-811079-9.00008-2

the infection. The infecting microorganism is capable of gaining entry to bone via the bloodstream (hematogenous infection), via invasion from a contiguous focus of infection, or cause infection due to vascular insufficiency. The Cierny system takes into account the anatomy of the infection and the physiology of the bone in the host.[13] It classifies long bone osteomyelitis and provides guidelines for treatment and management of infection over 12 stages. These stages have been outlined based on disease progress, regardless of where the infection is located, how the area became infected or chronicity.[14] Classification for septic arthritis has been defined by Choi and Newman based on late stage diagnosis of children under 1 year of age with septic arthritis.[15] Notably, Newman et al. splintered their classification based on the presence of the microorganism isolated from the infected joint, followed then by the radiological, histological and clinical presentation.[16] Like that seen in osteomyelitis, infection of the joints can occur through hematogenous spread of bacteria from the synovial capillaries into the joint, contiguous focus of infection or direct inoculation from foreign-body penetration, such as a contaminated prosthesis.[17]

1.2 Acute and Chronic Infection

Regardless of the classification, osteomyelitis and septic arthritis can be described as either acute, subacute or chronic based on histopathological findings of the infected area.[13] Acute osteomyelitis is defined as an infection diagnosed within 2 weeks of the onset of symptoms, whereas chronic osteomyelitis is typically diagnosed months later.[18,19] Characteristics of acute osteomyelitis include pus-induced inflammation at the site of infection, bone matrix damage, and vascular compression and destruction due to the migration of the infection into the surrounding tissues. Dissemination of the infection into these surrounding tissues can further exacerbate bone necrosis. However, if treated promptly and aggressively, development to chronic osteomyelitis may be prevented.[20,21] Due to the inflammation and vascular congestion of the area, the infection can detach from the viable bone and form its own separate infectious island or sequestrum. This sequestrum is protected from antibiotics and immune cells and can therefore lead to chronic persistence of the infection.[10,20] The pathological characteristics of chronic infection are low-grade inflammation, sequestra and possibly the presence of the involucrum (an area of new bone formation surrounding the affected bone).[19] Through the involucrum, pus from the infection can leak out into the surrounding tissue and eventually drain onto the skin surfaces, forming a chronic sinus.[3,20,22,23]

Acute septic arthritis results in an increase in synovial fluid and exudate. In an effort to eradicate the infection, polymorphonuclear leukocytes enter the joint to phagocytose the bacteria and release proteolytic enzymes that break down the hyaline cartilage. Irreversible damage leads to what is considered the subacute stage. As the subacute phase progresses, a pannus of synovial tissue grows over intraarticular structures, releasing lytic enzymes, which erode the articular cartilage. Purulent material in the joint leads to subluxation. Once damage extends to the extraarticular cartilage, the disease has reached chronicity. Chronic infection has one or more of the following symptoms: osteomyelitis, destruction of the joint, fracture, joint dislocation, avascular necrosis and ankylosis.[24–26]

1.3 Microorganisms Involved in Bone Infection

Ogston's report in 1881 identified the presence of various microorganisms in osteomyelitis patients. Since then bacteria, fungi and viruses have been discovered to cause both bone and joint infection.[9,27,28] The most common microorganism isolated from osteomyelitis and septic arthritis patients is *Staphylococcus aureus*.[5,25,29] Additionally, *Staphylococcus epidermidis*, the second most causative agent of bone infection, is commonly implicated with implant infections, such as those used as prosthetic joint implants.[30–32] A report by Miyahara et al. in 2014 demonstrated that men were most commonly affected while the most commonly isolated microorganism was *S. aureus* (73.3%), a result supported by previous epidemiological findings.[33] A long-term study (1969–2009) by Kremers et al. using a population size of 760 males and females found that *S. aureus* was most prominent in 44% of patients, followed by *S. epidermidis* (17%), *Streptococci* species (14%), *Pseudomonas* species (8%), *Escherichia coli* (3%) and anaerobes (5%).[34]

Although bacteria are the main cause of osteomyelitis and septic arthritis, fungal and viral infections are also seen but are mostly associated with immunocompromised individuals. In osteomyelitis, the regions regularly affected by fungal infection include the skull and vertebrae in both adults and children. Fungal osteomyelitis typically occurs as a secondary infection to otogenic, odontogenic or sinus infection though it is uncommon.[35] Fungal arthritis is usually associated with chronic infection due to its indolent nature.[36] The most common fungal pathogen involved in osteomyelitis and septic arthritis is the *Candida* species, with *C. albicans* the most common offender. Other species responsible include *Aspergillus*, *Coccidioides*, *Histoplasma*, *Blastomyces dermatitidis* and *Cryptococcus*.[35,37,38] Additionally, viral bone infection is extremely rare, with one study by Tabrizi et al. implicating *Herpes zoster* as the instigator of mandibular osteomyelitis.[28] Septic arthritis due to viral infection is also considered rare; however, cases of septic arthritis caused by non–type b *Haemophilus influenzae* in children have been described.[39] This chapter will focus mainly on bacterial-induced bone infection.

2. MECHANISM OF INFECTION

Bone and joints are typically sterile environments; however, they can be infected in a number of ways including hematogenous seeding of bacteria, contiguous focus of infection due to vascular insufficiency, and contiguous focus of infection without vascular insufficiency.[13]

2.1 Hematogenous Infection

Hematogenous infection is a result of bacteremia or spread of bacteria from the blood supply. In children less than 1 year of age, *S. aureus*, *S. agalactiae* and *E. coli* are most commonly isolated from the blood and bone.[20] In infants above 1 year of age, the most prominent bone infection is acute hematogenous osteomyelitis (AHO), usually caused by *S. aureus*, *Streptococcus pyogenes*, *Streptococcus pneumoniae* and *Haemophilus influenza* type B.[18,24,40] These microorganisms are also isolated from septic arthritic children with the addition of *Neisseria gonorrhoeae*.[8,17,24] AHO usually arises in the metaphysis of long bones in children or the vertebral bodies in adults. *S. aureus* causes 70%–90% of AHO in children. Chronic osteomyelitis is not as frequent in children, though can occur as a result of failed antimicrobial therapy or the presence of an orthopedic implant.[18]

2.2 Secondary to Contiguous Focus of Infection

Contiguous infection is transmitted from local tissue. Bone infection caused by secondary to contiguous focus of infection accounts for at least 50% of osteomyelitis cases. It is biphasic in regards to the age it affects, with infection occurring in younger patients due to bone trauma and older individuals as a result of decubitus ulcers and surgical procedures.[22] Interestingly, the incidence of joint infection stemming from prosthetic implants is a rising phenomenon perhaps linked to an aging population and augmented orthopedic-related infection.[41] Infection secondary to contiguous focus of infection is infection of the bone and/or joint as a result of direct introduction of the microorganism (without vascular insufficiency) or by spread from an adjacent infected tissue (vascular insufficiency).[5]

2.2.1 Secondary to Contiguous Focus of Infection Without Vascular Insufficiency

Contiguous focus osteomyelitis without vascular insufficiency results from direct inoculation of bacteria to the bone. This may occur through trauma of the bone, for example, from an open fracture or through contaminated prosthetic implants.[6,42–44] In the latter, approximately 65% of prosthetic hip and knee infections are caused by *S. aureus* and coagulase negative staphylococci. Interestingly, *E. faecalis*, another type of Gram-positive cocci, is responsible for 2.5% of prosthetic joint infections (PJI), making it the most commonly identified enterococcal species.[45] Open fractures lead to infection rates of up to 50%, which can cause not only osteomyelitis but also sepsis, delayed union of the bone or possible amputation of the limb.[44] Fracture of the tibia is the most frequent fracture in adults, meaning extensive research has been conducted into managing and treating the deep resulting infection. The microorganism causing infection is dependent on the type of fracture and how the trauma occurred. Type I and type II fractures are usually infected by Gram-positive microorganisms, whereas Gram-negative microorganisms are more often isolated from type III fractures. Additionally, if it is a soil contaminated injury, treatment would require coverage for anaerobic bacteria. Surgeons must also take into account nosocomial infections such as *S. aureus* and *Pseudomonas* species.[46] Osteomyelitis triggered by a contaminated implant device can also occur, with *S. epidermidis* reported to be the pathogen isolated in more than 90% of cases.[47]

2.2.2 Secondary to Contiguous Focus of Infection with Vascular Insufficiency

Osteomyelitis initiated by vascular insufficiency most commonly arises from a foot ulcer and is usually polymicrobial. *S. aureus* is the most common isolate (31%–47% of cases) followed by streptococci (27–61% of cases) and Gram-negative enteric bacteria (20%–50% of cases).[48] Unfortunately, neuropathy seen in patients with diabetes often means soft tissue infections can go unnoticed. Such diabetic foot infections can lead to osteomyelitis, with 15% of diabetic foot ulcers resulting in underlying foot osteomyelitis.[49] Infection was equally due to Gram-negative (*E. coli*, *Klebsiella* and *Pseudomonas aeruginosa)* and Gram-positive bacteria (*Streptococci* and *S. aureus*).[50]

3. SITE OF INFECTION

3.1 Bone and Cartilage as a Target

Bone is 95% type I collagen and hydroxyapatite, with the remaining 5% consisting of proteoglycans and various noncollagenous proteins. Bone is continually remodeling through the action of three bone cell types: osteoblasts, osteoclasts and osteocytes. Osteoblasts are the bone-forming cells and are responsible for the production of the osteoid, a structure consisting of type I collagen, proteoglycans and glycoproteins, such as bone salioprotein and osteocalcin. Once osteoblasts have terminally differentiated, they become osteocytes, which have been implicated in initiating and directing the bone remodeling process through their ability to respond to mechanical load and to detect mechanical strain and associated bone microdamage.[51] Osteoclasts are responsible for bone resorption, the dissolution of hydroxyapatite and degradation of organic matrix components of bone matrix. This is achieved through the generation of an acidic environment by ATPase, located in the osteoclast ruffled border, thereby decalcifying local hydroxyapatite in the bone matrix. Additionally, the lysosomal enzyme cathespin K, which is produced extracellularly by osteoclasts, can degrade collagen fibers in an acidic environment, thus working in synergy with the ATPase towards bone resorption.[52] The balance of activity between osteoblasts and osteoclasts is vital to maintain bone homeostasis, with a shift in favor of either cell type resulting in altered bone pathology. In order to keep the balance, there is an intimate relationship between osteoblasts and osteoclasts. Osteoprotegerin (OPG), receptor activated for factor nuclear factor kappa B (RANK) and receptor activated for factor nuclear factor kappa B ligand (RANK-L) are a carefully orchestrated triad that are crucial components for osteoclastogenesis.[53,54] RANK-L is expressed by osteoblasts and binds to its receptor, RANK, found on preosteoclastic cells. This results in activation and differentiation of preosteoclasts into mature osteoclasts, which begin bone resorption. To counteract this resorption, OPG is secreted by osteoblasts and acts as a decoy for RANK, causing cessation of osteoclast differentiation and inducing their apoptosis.[51,54] Infection by a pathogen, such as *S. aureus*, can disturb this balance between osteoblasts and osteoclasts, with inflammation also responsible for imbalance due to infection.[10]

Articular cartilage is made of hyaline cartilage and lines where bones meet to form a joint. It has a dense extracellular matrix (ECM) and, like bone, collagen accounts for two-thirds its dry weight in adults. The cells responsible for generating cartilage are chondrocytes and account for 2% of the total articular cartilage volume.[55] Types II, IX and XI collagen are the primary collagen components in mammals, with collagen type II representing 90%–95% of the collagen present. Cartilage turnover is primarily dependent on metalloproteinases, namely collagenases, gelatinase, stromelysin, and cathepsins B and D. Collagenase 3 (MMP13) degrades collagen fibrils, specifically type II collagen. Gelatinase breaks down the denatured type II collagen and is also active against other ECM components such as fibronectin and elastin. Stromelysin degrades the protein core of aggrecan.[56,57] As in the case of osteomyelitis, an imbalance of this regulation of cartilage turnover can lead to the development of disease such as osteoarthritis. This cartilage damage is seen upon infection, with *S. aureus* isolated from the majority of cases.[58]

4. CURRENT TREATMENTS FOR BONE INFECTION

Current treatment strategies for osteomyelitis patients are proving to be inadequate, with recurrent infection occurring in up to 50% of patients with chronic osteomyelitis caused by *P. aeruginosa*, for example.[5] Generally, for bone and joint infections, the treatment is patient dependent, with the course of antibiotics selected based on the infecting microorganism and with surgical intervention required in the majority of cases. Notably, most treatments require a combination of antimicrobial therapy and surgery to ensure an acceptable level of cure.[6,59]

4.1 Antibiotic Treatment

In the absence of results indicating a microbiological cause for infection, a broad-spectrum antibiotic is usually administered parenterally though it is recommended to delay antibiotic treatment if possible until microbial culture results are available.[5,6,59] Depending on the microorganism isolated, there are a variety of antibiotics available that are delivered intravenously in osteomyelitis cases. However, delivery of antibiotics systemically may not penetrate the bone effectively and can result in suboptimal levels of antibiotics reaching the area. This is partially due to the poor penetration of antimicrobials into bone, which is not as well vascularized as other tissues.[60] Additionally, in chronic osteomyelitis, accessibility of the antibiotics to the infection may be prohibited due to the avascular nature of the sequestra.[47] Therefore, in an effort to facilitate the off-loading of antibiotics directly into the infected area of bone as opposed to systemically, various treatment strategies have been developed. One such strategy is the use of polymethylmethacrylate (PMMA) beads containing antibiotics, in particular gentamicin, that is surgically embedded into the infected bone.[61] From this, however, other problems have arisen, including the emergence of highly resistant strains of *S. aureus* allowing them to withstand the antimicrobial arsenal.[62] Furthermore, PMMA beads require a second surgery in order to be removed, which can provide another opportunity for reinfection.[63] In children, it is recommended that the duration of antibiotic therapy be 3–6 weeks. Treatment for less than 3 weeks increases the risk of reinfection. In hematogenous infection, surgical intervention is usually only required in infection of the long bones.[40,48] In the case of PJI, there is a grading system whereby the patient is evaluated based on certain characteristics, following which the choice of treatment is then decided. The use of antibiotics is important for infection eradication. For Gram-positive infections, vancomycin is the drug of choice. For Gram-negative infections, ceftriaxone and azithromycin are recommended.[64,65] Unfortunately, infections caused by antibiotic-resistant *S. aureus* have reached epidemic proportions worldwide. Methicillin-resistant *Staphylococcus aureus* (MRSA) is at present the most acknowledged antibiotic-resistant pathogen globally, in areas including Europe, the Americas, North Africa, the Middle East, and East Asia.[66] Due to the increasing resistance to antibiotics, treatment of osteomyelitis and septic arthritis is of growing concern in the medical community.[47,65]

4.2 Surgical Management

Surgical management of osteomyelitis is often required in conjunction with antibiotic treatment. Bone debridement addresses three issues: decreasing the bacterial load, removal of the necrotic tissue and providing an opportunity for the host and antibiotics to arrest the infection.[67,68] Extensive debridement may lead to "dead space." Management of dead space is important as it may predispose the area to recurrent infection and may compromise bone integrity.[20,21] In the case of chronic osteomyelitis, patients may undergo several surgeries to remove necrotic bone and be subjected to weeks of oral antibiotic adherence. In order to fill the dead space, bone autografts and allografts can be used. Autogenous grafting is the removal of bone from the patient's own skeleton and implanting into the required site. However, this method has limitations due to donor site pain, loss of blood and potential for infection during surgery. Implanting bone from a sterilized cadaver provides several advantages over autogenous grafting; however, there are issues in regards to viral transmission and extended immune rejection.[69] Depending on the size of the defect, different approaches can be taken. Defects less than 6 cm can be bridged using conventional autogenous bone grafts. Defects larger than 6 cm would require an external fixator, such as the ilizarov intercalary bone transport method.[70] Additionally, PMMA bone cements impregnated with antibiotics are utilized for local delivery of the antibiotic. However, this treatment has limitations such as toxicity, stability of the antibiotic and PMMA cement, and conferring antimicrobial resistance. Notably, systemic administration of antibiotics is recommended when using this method.[71,72] The grading system for PJI can be crudely divided into three stages: debridement and retention of a well-fixed prosthesis; removal of implant followed by immediate implantation (first revision) or removal and delayed reimplantation (second revision); and permanent removal of the prosthesis in conjunction with extended antibiotic administration with possibility of amputation if unsuccessful. For septic arthritis with no implant involvement, removal of any purulent material is necessary to preserve the joint. Usually treatment using the relevant antibiotics administered for 2–3 weeks alone can kill the infection.[59,64,73]

5. BACTERIAL INTERACTION

5.1 Immune Response Involvement

Adherence of staphylococci to surfaces and cells is mediated by the presence of cell-surface proteins present on the bacterial cell wall.[30] These proteins are known as microbial surface components recognizing adhesive matrix molecules (MSCRAMMs) and are used to establish a foothold for infection.[74] This interaction often induces host cell production of chemokines and cytokines, which recruit and activate innate and adaptive immune cells.[75,76] Sakiniene et al. demonstrated that the host immune response was partially responsible for the progression of septic arthritis whereby if the infection is not fully cleared, a sustained immune response can lead to joint damage.[58] In an effort to clear the infection, cytokines interleukin 6 (IL-6) and interleukin 1β (IL-1β) are secreted into the synovial fluid by synovial cells. Activation of these cytokines results in the production of acute phase proteins that bind the bacteria and promote opsonization and activation of the complement system. Phagocytosis by macrophages and polymorphonuclear leukocytes encourages the production of additional cytokines such as tumor necrosis factor α (TNF-α) and interleukin 8 (IL-8). IL-1, IL-6 and TNFα are important in bone physiology and pathology.[53,77] It has been well documented that these cytokines have a potent effect on bone remodeling by driving progenitor osteoclast proliferation and differentiation in the presence of osteoblastic/stromal cells.[77] In osteomyelitis murine models, increased levels of IL-1β and IL-6 have been shown locally and systemically in the early postinfection period. TNFα was documented to rise later in the infection.[78]

5.2 Direct Interaction of Microorganisms with ECM and Bone/Cartilage Cells

Due to the prevalence of *S. aureus* in bone and joint infections, interactions between this pathogen and bone and cartilage are well documented.[75] A well-known virulence factor found both on the surface and secreted by *S. aureus* is protein A (SpA). SpA is a surface and secreted protein that is ubiquitous among *S. aureus* strains.[30,79] Structurally it consists of five immunoglobulin binding domains (IgBDs) that bind the Fc portion of immunoglobulins G (IgG), leading to reduced antibody mediated opsonization.[80] This immune evasion mechanism has been proposed to be the basic contribution of SpA to the virulence of *S. aureus*.[81] Recent publications have revealed that SpA can bind to tumor necrosis factor receptor 1 (TNFR1) in osteoblasts, allowing increased bone destruction and bone loss. Binding of SpA to TNFR-1 activates nuclear factor kappa B (NFκB), resulting in translocation of NFκB to the nucleus and causing the release of IL-6, an important osteotropic factor. IL-6 release from osteoblasts leads to osteoclast activation, which, as mentioned previously, is responsible for bone resorption.[76,82,83] The presence of bone specific surface proteins on *S. aureus* allows colonization and infection of the tissue. Collagen adhesion protein (Cna) in *S. aureus* is an MSCRAMM shown to be a virulence factor in osteomyelitis. This protein was shown to have a dock-and-lock mechanism to collagen fibers, providing the bacteria with an attachment point to colonize. Cna is part of the collagen binding family of MSCRAMMs in Gram-positive bacteria; however, there are several species that have close relatives of can, such as ACE from *E. faecalis*, ACM from *Enterococcus faecium*, CNE from *Streptococcus equi* and CMN from *Streptococcus mutans*.[30,75,84,85] *S. aureus* also expresses bone sialoprotein binding protein (Bbp)[86] which binds to human fibrinogen, thus providing *S. aureus* with yet another attachment point to establish infection.[75,87] Interestingly, Jasin et al. found that lipopolysaccharide (LPS), found on Gram-negative bacteria, interacts directly with chondrocytes resulting in cartilage degradation via protease and collagenase activation *in vitro*.[88,89]

5.3 Invasion of Nonprofessional Phagocytic Cells and Small Colony Variants

Invasion and persistence of *S. aureus* in naturally nonphagocytic cells has been described in a variety of cell types, including endothelial and keratinocytes.[90,91] One family of surface proteins found to mediate this effect are the fibronectin binding proteins (Fnbps) expressed on a number of bacterial species. Fnbp-A and Fnbp-B can bind to the extracellular matrix protein fibronectin and use this to crosslink to $\alpha5\beta1$ integrins to internalize themselves into bone cells.[92] This internalization has two possible outcomes: the bacteria activates production of the tumor necrosis factor–related apoptosis inducing ligand (TRAIL), which in turn causes osteoblast apoptosis, or it can persist intracellularly as a small colony variant (SCV) and cause recurrent infection months or years later.[62,93] Apoptosis induced by TRAIL occurs due to it binding to its receptor on the osteoblast membrane. There are five known receptors for TRAIL: death receptors DR5 and DR4, decoy receptors DcR1 and DcR2, and soluble receptor OPG. Binding of TRAIL to these receptors leads to the activation of caspases 8 and 10. When activated, these caspases cleave caspase 3, causing cellular apoptosis via mitochondrial dysregulation.[94] Strains lacking in both FnBPs have been shown to invade host cells very poorly.[30] Additionally, *S. epidermidis* does not internalize using the fibronectin–integrin bridge, unlike *S. aureus*.[10] If *S. aureus* persists intracellularly, it will not activate this pathway. A variant subpopulation of *S. aureus* was described over 100 years ago, but the connections between this subpopulation with persistent and recurrent infections has only been appreciated in recent years.[95,96] A phenotypical trait of this subpopulation includes slow growth, which leads to the development of microcolonies or SCVs.[97] It has been suggested that these mutations have been acquired to evade the immune response and promote intracellular survival.[95] Methicillin-resistant SCVs of *S. aureus* have been reported in an intensive care unit, with *S. epidermidis* SCVs linked to implanted medical devices.[97] Moreover, vancomycin intermediate resistant (VISA) SCVs were isolated from a male patient with septic arthritis undergoing long-term treatment with daptomycin, which is the first documentation of VISA SCVs to date.[98] SCVs have been described in osteomyelitis cases and have been deemed responsible for the recurrent infection associated with the disease.[97,99] The mechanism underlying how *S. aureus* decides which path to take is still unknown.

5.4 Non-cell Surface-Associated Proteins

Other non–cell surface–associated proteins such as coagulase, Panton–Valentine leucocidin (PVL) and alpha toxins play a role in bone and joint infection. Jin et al. demonstrated that coagulase and PVL inhibit osteoblast proliferation, induced apoptosis, inhibited mineral deposition and upregulated RANK-L, leading to an overall significant contribution to bone destruction and bone loss.[100] In addition to non–cell surface–associated proteins, bacteria can form biofilms, which can cause chronic and persistent infection in diseases such as cystic fibrosis, endocarditis and osteomyelitis. A biofilm is comprised of organized communities of microorganisms embedded in an extracellular matrix attached to a surface.[101] It was found that bacteria embedded in biofilms are more resistant to antimicrobial drugs and host immune defenses than planktonic bacteria and both *S. aureus* and *S. epidermidis* are capable biofilm formers.[102] In chronic osteomyelitis, the involucrum has decreased vascularity and oxygen tension, so therefore access to the area by the host immune system is diminished. These conditions provide a perfect attachment point for bacteria, allowing their biofilm formation. The presence of biofilms has been suggested as the main cause of clinical quiescence of chronic osteomyelitis.[68,103] Although *S. aureus* is the main biofilm former found in chronic wounds, other microorganisms have been isolated, such as *P. aeruginosa* and *E. faecalis*.[45] Prosthetic joint infections owing to prosthetic implants are predicted to rise due to improved detection techniques of biofilms on the implant, an aging population and the increased use of prosthetic devices.[104]

6. CONCLUSION

Osteomyelitis and septic arthritis still remain global issues that affect both adults and children. Although *S. aureus* is a well-investigated pathogen involved in several infectious diseases, effective treatment strategies for osteomyelitis and septic arthritis caused by *S. aureus* are a challenge. Notably, there are many microbes involved in causing infection, so diagnosis and treatments can be difficult. Moreover, not only are bone and joints available to infection in their native state, they also propagate infection due to the availability of bonding sites in the ECM and also the cells themselves. Additionally, the host's response to the invading microbes can facilitate the progression of the infection and disease in an attempt to combat infection. Currently, treatment strategies involve patient adherence to long courses of oral antibiotics and possible surgical debridement of dead and infected bone. Vancomycin is the last line of antibiotic available to treat difficult infections. However, the first vancomycin-resistant *S. aureus* (VRSA) strain was identified in 1997, with eight further strains isolated from patients in 2002.[105–107] In light of the antibiotic resistance emerging, research into the use of nonantibiotic compounds such as metals is being investigated. Silver and copper as antimicrobial compounds are well documented throughout the literature, with a broad variety of microorganisms affected by their natural properties. Species affected include *P. aeruginosa*, *S. aureus*, *B. subtilis* and *E.coli* with biofilm formation targeted in addition to planktonic bacterial eradication.[108–112] Other nonantibiotic antimicrobials described are antimicrobial peptides (AMPs). AMPs are a diverse group of peptides that are produced by a variety of invertebrate, plant and animal species.[113,114] Cationic AMPs have been shown to be active against both Gram-positive and Gram-negative bacteria, fungi, certain viruses and protozoans.[115] Additionally, *S. aureus* is also susceptible to AMPs, with a study conducted by Midorikawa demonstrating the effectiveness of defensin produced by keratinocytes against *S. aureus* and MRSA isolates.[116] The ability of these AMPs to be manipulated, designed and synthesized in the laboratory exposes a large potential for their use as a therapeutic for infection. These promising antimicrobial studies, in conjunction with knowledge of the disease state, have potential for generating new opportunities for treatment of osteomyelitis and septic arthritis in the future.

REFERENCES

1. Klenerman L. A history of osteomyelitis from the Journal of Bone and Joint Surgery: 1948 to 2006. *J Bone Joint Surg* 2007;**89-B**(5):667–70.
2. Ogston A. Report upon micro-organisms in surgical diseases. *Br Med J* 1881;**1**(1054):369.b2-75.

3. Radcliffe G. (iii) Osteomyelitis—a historical and basic sciences review. *Orthopaed Trauma* 2015;**29**(4):243–52.
4. Ikpeme IA, Ngim NE, Ikpeme AA. Diagnosis and treatment of pyogenic bone infections. *Afr Health Sci* 2010;**10**(1):82–8.
5. Hatzenbuehler J, Pulling TJ. Diagnosis and management of osteomyelitis. *Am Fam Physician* 2011;**84**(9):1027–33.
6. Lew DP, Waldvogel FA. Osteomyelitis. *N Engl J Med* 1997;**336**(14):999–1007.
7. Cierny 3rd G, Mader JT, Penninck JJ. A clinical staging system for adult osteomyelitis. *Clin Orthop Relat Res* 2003;**414**:7–24.
8. Goldenberg DL. Septic arthritis. *Lancet* 1998;**351**(9097):197–202.
9. Munoz-Egea MC, Blanco A, Fernandez-Roblas R, Gadea I, Garcia-Canete J, Sandoval E, et al. Clinical and microbiological characteristics of patients with septic arthritis: a hospital-based study. *J Ortho* 2014;**11**(2):87–90.
10. Wright JA, Nair SP. Interaction of staphylococci with bone. *Int J Med Microbiol* 2010;**300**(2–3):193–204.
11. Gupta MN, Sturrock RD, Field M. A prospective 2-year study of 75 patients with adult-onset septic arthritis. *Rheumatology (Oxford)* 2001;**40**(1):24–30.
12. Waldvogel FA, Medoff G, Swartz MN. Osteomyelitis: a review of clinical features, therapeutic considerations and unusual aspects. 3. Osteomyelitis associated with vascular insufficiency. *N Engl J Med* 1970;**282**(6):316–22.
13. Mader JT, Shirtliff M, Calhoun JH. Staging and staging application in osteomyelitis. *Clin Infect Dis* 1997;**25**(6):1303–9.
14. Lima ALL, Oliveira PR, Carvalho VC, Cimerman S, Savio E. Recommendations for the treatment of osteomyelitis. *Braz J Infect Dis* 2014;**18**(5):526–34.
15. Choi IH, Pizzutillo PD, Bowen JR, Dragann R, Malhis T. Sequelae and reconstruction after septic arthritis of the hip in infants. *J Bone Joint Surg* 1990;**72**(8):1150–65.
16. Newman JH. Review of septic arthritis throughout the antibiotic era. *Ann Rheum Dis* 1976;**35**(3):198–205.
17. Nade S. Septic arthritis. *Best Pract Res Clin Rheumatol* 2003;**17**(2):183–200.
18. Harik NS, Smeltzer MS. Management of acute hematogenous osteomyelitis in children. *Expert Rev Anti Infect Ther* 2010;**8**(2):175–81.
19. Davis JS. Management of bone and joint infections due to *Staphylococcus aureus*. *Intern Med J* 2005;**35**(Suppl. 2):79–96.
20. Lazzarini L, Mader JT, Calhoun JH. Osteomyelitis in long bones. *J Bone Joint Surg* 2004;**86**(10):2305–18.
21. Lew DP, Waldvogel FA. Osteomyelitis. *Lancet* 2004;**364**(9431):369–79.
22. Brady RA, Leid JG, Costerton JW, Shirtliff ME. Osteomyelitis: clinical overview and mechanisms of infection persistence. *Clin Microbiol Newsl* 2006;**28**(9):65–72.
23. Surendra G, Shetty U. Chronic recurrent multifocal osteomyelitis: a rare entity. *J Med Imaging Radiat Oncol* 2015;**59**(4):436–44.
24. Shirtliff ME, Mader JT. Acute septic arthritis. *Clin Microbiol Rev* 2002;**15**(4):527–44.
25. Horowitz DL, Katzap E, Horowitz S, Barilla-LaBarca ML. Approach to septic arthritis. *Am Fam Physician* 2011;**84**(6):653–60.
26. Perry CR. *Bone and joint infections.* London: Taylor & Francis; 1996. p. 116–24.
27. Gamaletsou MN, Walsh TJ, Sipsas NV. Epidemiology of fungal osteomyelitis. *Curr Fungal Infect Rep* 2014;**8**(4):262–70.
28. Tabrizi R, Dehghani Nazhvani A, Vahedi A, Gholami M, Zare R, Etemadi Parsa R. *Herpes zoster* induced osteomyelitis in the immunocompromised patients: A 10-year Multicenter Study. *J Dent (Shiraz, Iran)* 2014;**15**(3):112–6.
29. Dubost JJ, Soubrier M, De Champs C, Ristori JM, Bussiere JL, Sauvezie B. No changes in the distribution of organisms responsible for septic arthritis over a 20 year period. *Ann Rheum Dis* 2002;**61**(3):267–9.
30. Foster TJ, Geoghegan JA, Ganesh VK, Hook M. Adhesion, invasion and evasion: Many functions of the surface proteins of *Staphylococcus aureus*. *Nat Rev Microbiol* 2014;**12**(1):49–62.

31. Montanaro L, Testoni F, Poggi A, Visai L, Speziale P, Arciola CR. Emerging pathogenetic mechanisms of the implant-related osteomyelitis by *Staphylococcus aureus*. *Int J Artif Organs* 2011;**34**(9):781–8.
32. Montanaro L, Speziale P, Campoccia D, Ravaioli S, Cangini I, Pietrocola G, et al. Scenery of *Staphylococcus* implant infections in orthopedics. *Future Microbiol* 2011;**6**(11):1329–49.
33. de Souza Miyahara H, Helito CP, Oliva GB, Aita PC, Croci AT, Vicente JRN. Clinical and epidemiological characteristics of septic arthritis of the hip, 2006 to 2012, A seven-year review. *Clinics* 2014;**69**:464–8.
34. Kremers HM, Nwojo ME, Ransom JE, Wood-Wentz CM, Melton LJ, Huddleston PM. Trends in the epidemiology of osteomyelitis. A population-based study, 1969 to 2009. *J Bone Joint Surg Am* 2015;**97**(10):837–45.
35. Blyth CC, Gomes L, Sorrell TC, da Cruz M, Sud A, Chen SC. Skull-base osteomyelitis: Fungal vs. bacterial infection. *Clin Microbiol Infect* 2011;**17**(2):306–11.
36. Cuéllar ML, Silveira LH, Espinoza LR. Fungal arthritis. *Ann Rheum Dis* 1992;**51**(5):690–7.
37. Bariteau JT, Waryasz GR, McDonnell M, Fischer SA, Hayda RA, Born CT. Fungal osteomyelitis and septic arthritis. *J Am Acad Orthop Surg* 2014;**22**(6):390–401.
38. Kumar M, Thilak J, Zahoor A, Jyothi A. Septic arthritis due to tubercular and aspergillus co-infection. *Indian J Orthop* 2016;**50**(3):327–30.
39. Thorgrimson J, Ulanova M. *Haemophilus influenzae* type A as a cause of paediatric septic arthritis. *JMM Case Reports* 2016;**3**(5):1–4.
40. Peltola H, Pääkkönen M. Acute osteomyelitis in children. *N Engl J Med* 2014;**370**(4):352–60.
41. Mathews CJ, Weston VC, Jones A, Field M, Coakley G. Bacterial septic arthritis in adults. *Lancet* 2010;**375**(9717):846–55.
42. Penn-Barwell JG, Bennett PM, Fries CA, Kendrew JM, Midwinter MJ, Rickard RF. Severe open tibial fractures in combat trauma: management and preliminary outcomes. *Bone Joint J* 2013;**95-B**(1):101–5.
43. Berbari EF, Hanssen AD, Duffy MC, Steckelberg JM, Ilstrup DM, Harmsen WS, et al. Risk factors for prosthetic joint infection: case-control study. *Clin Infect Dis* 1998;**27**(5):1247–54.
44. Otchwemah R, Grams V, Tjardes T, Shafizadeh S, Bäthis H, Maegele M, et al. Bacterial contamination of open fractures—pathogens, antibiotic resistances and therapeutic regimes in four hospitals of the trauma network Cologne. *Germany Injury* 2015;**46**(Suppl 4):104–8.
45. Frank KL, Vergidis P, Brinkman CL, Greenwood Quaintance KE, Barnes AMT, Mandrekar JN, et al. Evaluation of *the Enterococcus faecalis* biofilm-associated virulence factors AhrC and Eep in rat foreign body osteomyelitis *in vitro* biofilm-associated antimicrobial resistance. *PLoS One* 2015;**10**(6):e0130187.
46. Cross WW, Swiontkowski MF. Treatment principles in the management of open fractures. *Indian J Orthop* 2008;**42**(4):377–86.
47. Gogia JS, Meehan JP, Di Cesare PE, Jamali AA. Local antibiotic therapy in osteomyelitis. *Semin Plast Surg* 2009;**23**(2):100–7.
48. Fritz JM, McDonald JR. Osteomyelitis: Approach to diagnosis and treatment. *Phys Sportsmed* 2008;**36**(1):50–4.
49. Malhotra R, Chan CS, Nather A. Osteomyelitis in the diabetic foot. *Diabetic Foot & Ankle* 2014;**5**:1–8.
50. Lavery LA, Sariaya M, Ashry H, Harkless LB. Microbiology of osteomyelitis in diabetic foot infections. *J Foot Ankle Surg* 1995;**34**(1):61–4.
51. Raggatt LJ, Partridge NC. Cellular and molecular mechanisms of bone remodeling. *J Biol Chem* 2010;**285**(33):25103–8.
52. Bilezikian JP, Raisz LG, Martin TJ. 3rd ed. *Principles of bone biology*, vol. 1. California, USA: Elsevier Inc; 2008. p. 221–4.
53. Steeve KT, Marc P, Sandrine T, Dominique H, Yannick F. IL-6, RANKL, TNF-alpha/IL-1: Interrelations in bone resorption pathophysiology. *Cytokine Growth Factor Rev* 2004;**15**(1):49–60.
54. Hadjidakis DJ, Androulakis II. Bone remodeling. *Ann N Y Acad Sci* 2006;**1092**:385–96.
55. Eyre D. Articular cartilage and changes in arthritis: Collagen of articular cartilage. *Arthritis Res* 2002;**4**(1):30–5.
56. Sophia Fox AJ, Bedi A, Rodeo SA. The basic science of articular cartilage: Structure, composition, and function. *Sport Health* 2009;**1**(6):461–8.
57. Yasuda T. Cartilage destruction by matrix degradation products. *Mod Rheumatol* 2006;**16**(4):197–205.

58. Sakiniene E, Bremell T, Tarkowski A. complement depletion aggravates *Staphylococcus aureus* septicaemia and septic arthritis. *Clin Exp Immunol* 1999;**115**(1):95–102.
59. Matthews PC, Berendt AR, McNally MA, Byren I. Diagnosis and management of prosthetic joint infection. *BMJ* 2009;**338**:2569–78.
60. Landersdorfer CB, Kinzig M, Bulitta JB, Hennig FF, Holzgrabe U, Sörgel F, et al. Bone penetration of amoxicillin and clavulanic acid evaluated by population pharmacokinetics and Monte Carlo simulation. *Antimicrob Agents Chemother* 2009;**53**(6):2569–78.
61. Faber C, Hoogendoorn RJ, Stallmann HP, Lyaruu DM, van Nieuw Amerongen A, Wuisman PI. In Vivo Comparison of Dhvar-5 and Gentamicin in an MRSA osteomyelitis prevention model. *J Antimicrob Chemother* 2004;**54**(6):1078–84.
62. Garcia LG, Lemaire S, Kahl BC, Becker K, Proctor RA, Denis O, et al. Antibiotic activity against small-colony variants of *Staphylococcus aureus*: review of in vitro animal and clinical data. *J Antimicrob Chemother* 2013;**68**:1455–64.
63. Klemm K. Gentamicin-PMMA-beads in treating bone and soft tissue infection. *Zentralbl Chir* 1979;**101**(14):934–42.
64. Sharff KA, Richards EP, Townes JM. Clinical management of septic arthritis. *Curr Rheumatol Rep* 2013;**15**(6):1–9.
65. Moran E, Masters S, Berendt AR, McLardy-Smith P, Byren I, Atkins BL. Guiding empirical antibiotic therapy in orthopaedics: The microbiology of prosthetic joint infection managed by debridement, irrigation and prosthesis retention. *J Infect* 2007;**55**(1):1–7.
66. Laxminarayan R, Duse A, Wattal C, Zaidi AKM, Wertheim HFL, Sumpradit N, et al. Antibiotic resistance—The need for global solutions. *Lancet Infect Dis* 2013;**13**(12):1057–98.
67. Wagner JM, Zollner H, Wallner C, Ismer B, Schira J, Abraham S, et al. Surgical debridement is superior to sole antibiotic therapy in a novel murine posttraumatic osteomyelitis model. *PLoS One* 2016;**11**(2):e0149389.
68. Calhoun JH, Manring MM, Shirtliff M. Osteomyelitis of the long bones. *Semin Plast Surg* 2009;**23**(2):59–72.
69. Roberts TT, Rosenbaum AJ. Bone grafts, bone substitutes and orthobiologics: The bridge between basic science and clinical advancements in fracture healing. *Organogenesis* 2012;**8**(4):114–24.
70. Green SA, Jackson JM, Wall DM, Marinow H, Ishkanian J. Management of segmental defects by the Ilizarov intercalary bone transport method. *Clin Orthop Relat Res* 1992;**280**:136–42.
71. Bistolfi A, Massazza G, Vern E, Mass A, Deledda D, Ferraris S, et al. Antibiotic-loaded cement in orthopedic surgery: a review. *ISRN Orthoped* 2011;**2011**:1–8.
72. Hinarejos P, Guirro P, Puig-Verdie L, Torres-Claramunt R, Leal-Blanquet J, Sanchez-Soler J, et al. Use of antibiotic-loaded cement in total knee arthroplasty. *World J Ortho* 2015;**6**(11):877–85.
73. Berendt T, Byren I. Bone and joint infection. *Clin Med (Lond)* 2004;**4**(6):510–8.
74. Jenkins A, Diep BA, Mai TT, Vo NH, Warrener P, Suzich J, et al. Differential expression and roles of *Staphylococcus aureus* virulence determinants during colonization and disease. *MBio* 2015;**6**(1):02272-02214.
75. Josse J, Velard F, Gangloff SC. *Staphylococcus aureus* vs. osteoblast: Relationship and consequences in osteomyelitis. *Front Cell Infect Microbiol* 2015;**5**:1–10.
76. Claro T, Widaa A, McDonnell C, Foster TJ, O'Brien FJ, Kerrigan SW. *Staphylococcus aureus* protein A binding to osteoblast tumour necrosis factor receptor 1 results in activation of Nuclear Factor kappa B and release of Interleukin-6 in bone infection. *Microbiology* 2013;**159**(Pt 1):147–54.
77. Manolagas SC. Role of cytokines in bone resorption. *Bone* 1995;**17**(2, Suppl. 1):63–7.
78. Yoshii T, Magara S, Miyai D, Nishimura H, Kuroki E, Furudoi S, et al. Local levels of interleukin-1β, -4, -6, and Tumor Necrosis Factor α in an experimental model of murine osteomyelitis due to *Staphylococcus aureus*. *Cytokine* 2002;**19**(2):59–65.
79. Becker S, Frankel MB, Schneewind O, Dominique M. Release of protein A from the cell wall of *Staphylococcus aureus*. *Proc Natl Acad Sci U S A* 2014;**111**(4):1574–9.
80. Falugi F, Kim HK, Missiakas D, Schneewind O. Role of protein A in the evasion of host adaptive immune responses by *Staphylococcus aureus*. *mBio* 2013;**4**(5):e00575-13.
81. Heilmann C. Adhesion mechanisms of staphylococci. *Adv Exp Med Biol* 2011;**715**:105–23.

82. Widaa A, Claro T, Foster TJ, O'Brien FJ, Kerrigan SW. *Staphylococcus aureus* protein A plays a critical role in mediating bone destruction and bone loss in osteomyelitis. *PLoS One* 2012;**7**(7):e40586.
83. Claro T, Widaa A, O'Seaghdha M, Miajlovic H, Foster TJ, O'Brien FJ, et al. *Staphylococcus aureus* protein A binds to osteoblasts and triggers signals that weaken bone in osteomyelitis. *PLoS One* 2011;**6**(4): e18748.
84. Zong Y, Xu Y, Liang X, Keene DR, Höök A, Gurusiddappa S, et al. A 'collagen hug' model for *Staphylococcus aureus* CNA binding to collagen. *EMBO J* 2005;**24**(24):4224–36.
85. Peacock SJ, Moore CE, Justice A, Kantzanou M, Story L, Mackie K, et al. Virulent combinations of adhesin and toxin genes in natural populations of *Staphylococcus aureus*. *Infect Immun* 2002;**70**(9):4987–96.
86. Rydén C, Tung HS, Nikolaev V, Engström A, Oldberg A. *Staphylococcus aureus* causing osteomyelitis binds to a nonapeptide sequence in bone sialoprotein. *Biochem J* 1997;**327**(Pt 3):825–9.
87. Vazquez V, Liang X, Horndahl JK, Ganesh VK, Smeds E, Foster TJ, et al. Fibrinogen is a ligand for the *Staphylococcus aureus* microbial surface components recognizing adhesive matrix molecules (MSCRAMM) bone sialoprotein-binding protein (Bbp). *J Biol Chem* 2011;**286**(34):29797–805.
88. Abdelnour A, Arvidson S, Bremell T, Rydén C, Tarkowski A. The Accessory Gene Regulator (agr) controls *Staphylococcus aureus* virulence in a murine arthritis model. *Infect Immun* 1993;**61**(9):3879–85.
89. Jasin HE. Bacterial lipopolysaccharides induce i*n vitro* degradation of cartilage matrix through chondrocyte activation. *J Clin Invest* 1983;**72**(6):2014–9.
90. Mempel M, Schnopp C, Hojka M, Fesq H, Weidinger S, Schaller M, et al. Invasion of human keratinocytes by *Staphylococcus aureus* and intracellular bacterial persistence represent haemolysin-independent virulence mechanisms that are followed by features of necrotic and apoptotic keratinocyte cell death. *Br J Dermatol* 2002;**146**(6):943–51.
91. Vesga O, Groeschel MC, Otten MF, Brar DW, Vann JM, Proctor RA. *Staphylococcus aureus* small colony variants are induced by the endothelial cell intracellular milieu. *J Infect Dis* 1996;**173**(3):739–42.
92. Brouillette E, Talbot BG, Malouin F. The fibronectin-binding proteins of *Staphylococcus aureus* may promote mammary gland colonization in a lactating mouse model of mastitis. *Infect Immun* 2003;**71**(4):2292–5.
93. Alexander EH, Bento JL, Hughes Jr FM, Marriott I, Hudson MC, Bost KL. *Staphylococcus aureus* and *Salmonella enterica* serovar Dublin induce tumor necrosis factor-related apoptosis-inducing ligand expression by normal mouse and human osteoblasts. *Infect Immun* 2001;**69**(3):1581–6.
94. Wang S, El-Deiry WS. TRAIL and apoptosis induction by TNF-family death receptors. *Oncogene* 2003;**22**(53):8628–33.
95. Kahl BC. Small colony variants (SCVs) of *Staphylococcus aureus*—A bacterial survival strategy. *Infect Genet Evol* 2014;**21**:515–22.
96. Melter O, Radojevic B. Small colony variants of *Staphylococcus aureus*—Review. *Folia Microbiol (Praha)* 2010;**55**(6):548–58.
97. Proctor RA, von Eiff C, Kahl BC, Becker K, McNamara P, Herrmann M, et al. Small colony variants: a pathogenic form of bacteria that facilitates persistent and recurrent infections. *Nat Rev Micro* 2006;**4**(4):295–305.
98. Lin YT, Tsai JC, Yamamoto T, Chen HJ, Hung WC, Hsueh PR, et al. Emergence of a small colony variant of vancomycin-intermediate S*taphylococcus aureus* in a patient with septic arthritis during long-term treatment with daptomycin. *J Antimicrob Chemother* 2016;**71**(7):1807–14.
99. Seebach E, Holschbach J, Buchta N, Bitsch RG, Kleinschmidt K, Richter W. Mesenchymal stromal cell implantation for stimulation of long bone healing aggravates *Staphylococcus aureus* induced osteomyelitis. *Acta Biomater* 2015;**21**:165–77.
100. Jin T, Zhu YL, Li J, Shi J, He XQ, Ding J, et al. Staphylococcal protein a, Panton–Valentine leukocidin and coagulase aggravate the bone loss and bone destruction in osteomyelitis. *Cell Physiol Biochem* 2013;**32**(2):322–33.
101. Gupta P, Sarkar S, Das B, Bhattacharjee S, Biofilm Tribedi P. pathogenesis and prevention—a journey to break the wall: a review. *Arch Microbiol* 2015;**198**:1–15.
102. Valour F, Trouillet-Assant S, Rasigade JP, Lustig S, Chanard E, Meugnier H, et al. *Staphylococcus epidermidis* in orthopedic device infections: the role of bacterial internalization in human osteoblasts and biofilm formation. *PLoS One* 2013;**8**(6):e67240.

103. Brady RA, Leid JG, Calhoun JH, Costerton JW, Shirtliff ME. Osteomyelitis and the role of biofilms in chronic infection. *FEMS Immunol Med Microbiol* 2008;**52**(1):13–22.
104. Ribeiro M, Monteiro FJ, Ferraz MP. Infection of orthopedic implants with emphasis on bacterial adhesion process and techniques used in studying bacterial-material interactions. *Biomatter* 2012;**2**(4):176–94.
105. Chang S, Sievert DM, Hageman JC, Boulton ML, Tenover FC, Downes FP, et al. Infection with vancomycin-resistant *Staphylococcus aureus* containing the Vana resistance gene. *N Engl J Med* 2003;**348**(14):1342–7.
106. Hiramatsu K. Vancomycin-resistant *Staphylococcus aureus*: a new model of antibiotic resistance. *Lancet Infect Dis* 2001;**1**(3):147–55.
107. Hiramatsu K. Vancomycin resistance in staphylococci. *Drug Resist Updat* 1998;**1**(2):135–50.
108. Shahverdi AR, Fakhimi A, Shahverdi HR, Minaian S. Synthesis and effect of silver nanoparticles on the antibacterial activity of different antibiotics against *Staphylococcus aureus* and *Escherichia coli*. *Nanomed Nanotech Biol Med* 2007;**3**(2):168–71.
109. Mikolay A, Huggett S, Tikana L, Grass G, Braun J, Nies DH. Survival of bacteria on metallic copper surfaces in a hospital trial. *Appl Microbiol Biotechnol* 2010;**87**(5):1875–9.
110. Harrison JJ, Turner RJ, Joo DA, Stan MA, Chan CS, Allan ND, et al. Copper and quaternary ammonium cations exert synergistic bactericidal and antibiofilm activity against *Pseudomonas aeruginosa*. *Antimicrob Agents Chemother* 2008;**52**(8):2870–81.
111. Lemire JA, Harrison JJ, Turner RJ. Antimicrobial activity of metals: mechanisms, molecular targets and applications. *Nat Rev Micro* 2013;**11**(6):371–84.
112. Ruparelia JP, Chatterjee AK, Duttagupta SP, Mukherji S. Strain specificity in antimicrobial activity of silver and copper nanoparticles. *Acta Biomater* 2008;**4**(3):707–16.
113. Brogden KA. Antimicrobial peptides: pore formers or metabolic inhibitors in bacteria? *Nat Rev Micro* 2005;**3**(3):238–50.
114. Zasloff M. Antimicrobial peptides of multicellular organisms. *Nature* 2002;**415**(6870):389–95.
115. Hancock REW, Diamond G. The role of cationic antimicrobial peptides in innate host defences. *Trends Microbiol* 2000;**8**(9):402–10.
116. Midorikawa K, Ouhara K, Komatsuzawa H, Kawai T, Yamada S, Fujiwara T, et al. *Staphylococcus aureus* susceptibility to innate antimicrobial peptides, β-Defensins and CAP18. Expressed by Human keratinocytes. *Infect Immun* 2003;**71**(7):3730–9.

FURTHER READING

1. Boyce BF, Rosenberg E, de Papp AE, Duong LT. The osteoclast, bone remodelling and treatment of metabolic bone disease. *Eur J Clin Invest* 2012;**42**(12):1332–41.

CHAPTER 9

Microbiological Diagnosis of Prosthetic Joint Infections

L. Prieto-Borja, C. Pérez-Jorge, J. Esteban
IIS-Fundación Jiménez Díaz, UAM, Madrid, Spain

1. INTRODUCTION

Prosthetic joint infections (PJI) are a rare but devastating complication of orthopedic surgery. The incidence of this infection is low (between 1% and 3%),[1] but the number of patients who undergo prosthetic surgery is growing almost exponentially,[2] so the number of patients with this disease is also expected to increase in the future. It is now recognized that the complex management of these cases implies multidisciplinary teams.[3] One key part of the team is the microbiology, especially due to the importance of diagnosis.

Diagnosis of PJI is difficult in many cases. It starts with the detection of a clinical syndrome that can range from detection of clear inflammatory events in the affected joints (as happens in many acute PJIs) to a subtle set of unspecific symptoms and signs (as happens in some delayed PJIs).[1,4,5] Different serological markers (like C-reactive protein or erythrocyte sedimentation rate) together with other newly developed markers (such as alpha-defensin)[6] can be used with image tests in order to confirm the clinical diagnosis. However, proper management of these patients requires the etiological diagnosis, because the essential antibiotic therapy must be selected according to the microorganism that causes the disease.[7] Etiology of these infections includes a broad spectrum of microorganisms, with different antibiotic susceptibilities[8] (Table 1), so this knowledge must be a cornerstone of all management.

Etiological diagnosis is performed with different microbiological tools, and the results must be combined with all other data in order to be correctly interpreted.[7,9] The microbiological diagnosis for PJIs should include the analysis of different samples and technologies. Preoperative and intraoperative samples, such as blood cultures, synovial fluid, periprosthetic tissue and the explanted prosthesis culture, are necessary to diagnose PJIs.[10]

2. CULTURE-BASED DIAGNOSIS OF PJI

2.1 Preoperative Samples

Blood and synovial fluid culture are considered the available preoperative samples[10] to be processed for a diagnosis of PJIs. Blood cultures are carried out in both aerobic and anaerobic bottles, usually employing automated systems. However, their use is limited by the lack of bacteremia in most infections, especially delayed ones.

The Microbiology of Skin, Soft Tissue, Bone and Joint Infections
ISSN 2451-9006, http://dx.doi.org/10.1016/B978-0-12-811079-9.00009-4

Table 1 Microorganisms frequently involved in PJI

Gram-positive aerobic/facultative
Staphylococcus aureus *Staphylococcus epidermidis* and other coagulase-negative staphylococci *Enterococcus sp.* *Streptococcus sp.* *Corynebacterium sp* and other Gram-positive rods.
Gram-negative
Enterobacteriaceae *Pseudomonas aeruginosa* Other nonfermenting Gram-negative rods
Anaerobic bacteria
Propionibacterium acnes
Fungi
Candida sp.
Uncommon organisms (*Mycobacterium, Capnocytophaga, Legionella,* etc.) **Polymicrobial infections** **Culture-negative PJI**

Synovial fluid culture for aerobic and anaerobic organisms is carried out by means of conventional microbiological techniques such as agar media and broth culture.[10] However, the lack of sensitivity for conventional culture due to the low quantity of microorganisms and the presence of inhibitors such as antibiotics makes it necessary to study alternative culture techniques. The inoculation into blood culture bottles[11–13] represents an alternative to conventional cultures, increasing the sensitivity of the synovial fluid and the quantity of bacteria, while decreasing the effect of inhibitors by means of special resins.[11] This methodology can also have limitations, because potential contaminants of blood cultures (like coagulase-negative staphylococci) can also be the cause of infection, so interpretation of the results must be performed carefully. Moreover, these cultures are especially useful in acute and hematogenous infections, where planktonic bacteria can be found in high amounts, but not in delayed infections, since the bacteria in these latter infections are embedded in the biofilm attached to the implant surface.[11,14–17]

Synovial fluid can also be processed for rapid diagnosis by using Gram stain. This is the commonest technique for rapid diagnosis in most microbiology laboratories, and it has showed high specificity, but low sensitivity, for synovial fluids.[18] For this reason, the use of synovial fluid Gram stain cannot be used to discard a PJI.

Another potential sample to be considered is the exudate from superficial wounds or fistulae. The study of Cune et al.[19] showed a good correlation between culture wound exudates and intraoperative samples in acute infections when *Staphylococcus aureus* or

Enterobacteriaceae were isolated. However, except for this circumstance, cultures from exudates are not recommended for the diagnosis of PJI, because many skin colonizers can be isolated without any value. The fact that many of these organisms can be the actual cause of infection may lead to a misdiagnosis and erroneous treatments, so the use of these cultures cannot be recommended and are considered useless.[1,4,20]

2.2 Intraoperative Samples

Periprosthetic tissue cultures are usually considered the gold standard for determining the presence or absence of periprosthetic infection.[21] The conventional microbiological diagnosis for intraoperative periprosthetic tissue samples includes classical culture into agar plates or broth medium for aerobic and anaerobic microorganisms.[10] However, similarly to synovial fluid, the low sensitivity of these techniques (between 37% and 61%), has led to the necessity to study new systems and methodologies to perform the microbiological diagnosis.[22] The low sensitivity could be related to the fact that some organisms causing PJI may be difficult to isolate in cultures because of their phenotypic state (i.e., anaerobic microorganism, biofilm).[23] The results obtained when periprosthetic tissue samples are inoculated into blood culture bottles show an improvement in the sensitivity for diagnosis of PJIs.[24,25] On the one hand, the time decreases for microorganism detection, obtaining microorganism isolation within the first 24 h of surgery.[25] However, the approach allows recovery of slow-growing microorganisms, such as *Propionibacterium* spp. without prolonged culture periods, given the fact that the time estimated for anaerobic culture bottles is a maximum of 14 days.[24,25] In this sense, an approach claimed to improve the results of conventional cultures is to prolong the incubation time of cultures. Shafer et al. report a study where incubation time was extended from 7 to 14 days, allowing an increase of 26.4% in the number of positive results .[26] Nevertheless, these impressive results have not been confirmed in other reports, either with periprosthetic tissue samples[27] or sonicated implants.[28] Taking into account these studies and the logistic implications of the increased culture time for clinical microbiology laboratories, a more conservative approach could be to prolong the incubation time only in cases of highly suspected infection and negative cultures after 7 days, as well as in potential shoulder prosthesis infections, where *Propionibacterium acnes* is a relevant slowly growing pathogen.[29]

Regarding the methodology used for sample processing, the article of Roux et al.[30] showed an increasing sensitivity (83.7%) for detection of organisms if the samples were processed using glass beads together with mechanical agitation with a beadmill. Moreover, reports have shown the usefulness of dithiothreitol (DTT) for this processing. DTT dislodges bacteria from the biofilm, thus increasing sensitivity of cultures (88%) too for detection of microorganisms in tissue samples,[31,32] maintaining a high specificity.

Another important topic that merits discussion is the interpretation of positive cultures of periprosthetic tissue samples. Highly aggressive microorganisms, like *S. aureus*, *Pseudomonas aeruginosa* or *Enterobacteriaceae*, are usually the cause of acute PJI.

However, chronic/delayed infections are mostly caused by less pathogenic organisms, mainly coagulase-negative staphylococci, *Corynebacterium* or *P. acnes*.[8] These latter organisms are common contaminants of samples and culture media, so the actual significance of their scanty growth could be difficult to obtain. Aiming to increase the accuracy of cultures, Atkins et al.[33] establish the criteria currently used for this purpose: several samples of periprosthetic tissue need to be obtained (3–6 samples are most often recommended), and the bacteria is considered to be the cause of infection when it grows in 3 or more samples (Fig. 1A). This criterion is now considered the gold standard for the microbiological diagnosis of PJI in most reviews and guidelines.[1,10,34–37]

Together with periprosthetic tissue samples, the other available intraoperative sample is, obviously, the retrieved implant. According to the "race for the surface" theory, the implant surface is the site where bacteria attach and develop the biofilm.[38] Taking this theory into account, the detection of implant-attached bacteria would seem to be a good approach for microbiological diagnosis. However, it is not so simple, because these implants are difficult to manage in sterile conditions because of their size, and it is not possible to know where biofilm is in the implant surface. Different strategies have been used in the past, aiming to isolate organisms from the implant.[39] But it was not until 2007 that a useful methodology was described by Trampuz et al.,[40] followed by the description of this technique in other laboratories,[1,41,42] showing a high sensitivity with a good specificity as well: sonication of the retrieved implant.

Sonication uses low-intensity ultrasound to release the adhered biofilm from the implant. In this technology, a buffer covers the implant and then ultrasound is applied using a water bath sonicator. Ultrasound generates microbubbles that subsequently explode, releasing the biofilm from the surface. An *in vitro* study showed that ultrasound can also affect bacterial viability,[43] so selection of the time and intensity are factors of extreme importance in the obtained results.

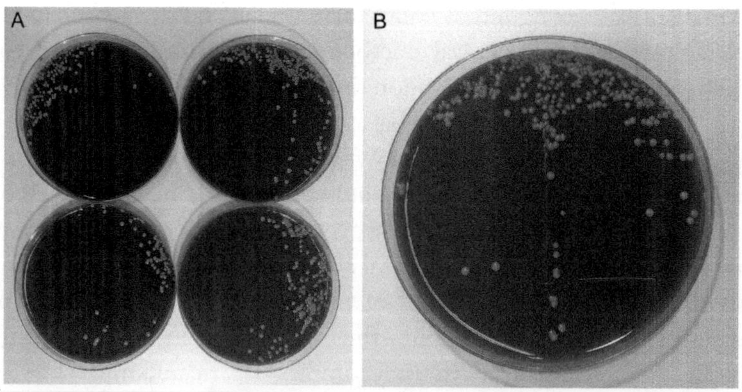

Fig. 1 PJI caused by *Corynebacterium striatum*. (A) Culture from four periprosthetic tissue samples. (B) Culture from sonicated implant.

During the years since the first description of this technique was made, many different reports have shown its usefulness, and a meta-analysis recovered most of these reports in 2014,[44] showing high sensitivity (80%) and specificity (95%). A new meta-analysis performed in 2016[45] confirms these results.

In addition, a confluence of all described methods has led to a more common protocol, which includes vortexing or agitation of the implant inside the buffer (which can vary between authors, using Ringer lactate,[40] phosphate buffer,[28] etc.), followed by sonication of the implant inside rigid plastic containers.[1,20,46] After sonication, some authors inoculate the different culture media,[20,47,48] while others concentrate the sample by centrifugation before culture media inoculation.[1,28] Cultures were then incubated and a quantification of the isolated organisms was performed. Variations of this procedure were also performed, using differential sonication of the prosthesis components,[49,50] inoculation of sonicate fluid in blood culture bottles,[51] or different incubation times.[28] The results (Fig. 1B) have been used to establish a breakpoint of the number of recovered colony-forming units (CFUs) in order to help in the interpretation of the results.[1] However, these results can be affected by different variables, especially the use of antibiotics prior to surgery, and some authors have found an extremely broad spectrum of results that makes it difficult to establish such a breakpoint.[41]

Sonication is extremely useful for those patients who have taken antibiotics prior to surgery and in chronic/delayed infections, but it can be useful in other circumstances as well.[1,46,47] Sonication has also been used for diagnosis of osteosynthesis implant-related infection[52,53] or PMMA spacer-related infection.[54] It has also been shown that preoperative antibiotic prophylaxis does not affect the results of these cultures, as well as those of other intraoperative samples.[55] Sonication methodology is easy to perform, and due to its high sensitivity and specificity can be incorporated into the laboratory routine. Nevertheless, it cannot replace conventional diagnosis, but should be added to it in order to improve the possibilities of obtaining a microbiological diagnosis of PJI.

3. MOLECULAR BIOLOGY-BASED DIAGNOSIS OF PJI

Despite the improvements in the last few years in culture-based diagnosis of PJI, there is still a not-insignificant number of patients who remain undiagnosed using conventional techniques.[46] Molecular biology methods have been used in an attempt to obtain better results, apart from the several advantages they provide.

3.1 Homemade PCR

In 1995, the first case of polymerase chain reaction (PCR) used for the diagnosis of PJI was published,[56] concluding that PCR had better results than culture of implants in terms of sensitivity and specificity. However, since that time, several studies have been carried out in periprosthetic tissue, obtaining good results of both parameters in general,[57–67] but they were not conclusive when compared with those values obtained from culture. In some cases, sensitivity of PCR was higher than that of culture, but specificity was lower[61]; in other cases, sensitivity obtained with PCR assays was lower, whereas specificity was higher in comparison with culture,[62,63,65] or equal[68]; and in some cases, both parameters were higher with PCR[66].

Following the use of periprosthetic tissue for the development of broad range homemade PCR, studies were carried out from sonicate fluid, obtaining very good statistical parameters.[67] Some authors concluded that PCR and culture of sonicate fluid showed equivalent values of sensitivity and specificity.[69] However, others determined that PCR was not able to detect microorganisms that culture did detect.[70]

Apart from the drawbacks inherent to homemade PCR, 16S rRNA gene homemade PCR has the main limitation of ease of contamination, as it detects any bacterial DNA. In an attempt to solve this problem, specific homemade PCR has been developed, which specifically detects the desired bacteria. For example, in 2006 a study was published in which a specific homemade PCR detected *P. acnes*.[71] The aim was to diagnose prosthetic shoulder infection, due to this slow-growth anaerobic microorganism, from intraoperative tissue samples and sonicate fluid. It showed good values of sensitivity and specificity. Nevertheless, PJI can be polymicrobial and when this is the case, specific PCR is not enough, since it only can detect one microorganism. In fact, that is the main limitation of specific PCR and the reason why multiplex PCR was developed.

In 2013, a homemade multiplex real-time PCR designed for the diagnosis of PJI[72] could detect most of the etiological agents from sonicate fluid in hip and knee infection: *Staphylococcus* spp., *Propionibacterium* spp., Enterobacteriaceae, *Streptococcus* spp., *Bacteroides fragilis* group, *Actinomyces* spp., *Pseudomonas aeruginosa*, *Enterococcus* spp., *Finegoldia magna*, *Corynebacterium* spp., and other less-common bacteria. However, there were two undetected cases, because the microorganisms presented in the implants were not targeted by the PCR panel (*Candida albicans* and *Capnocytophaga canimorsus*). This reflects how difficult is to achieve a good microbiological diagnosis of PJI. PCR even detected 11 polymicrobial infections from 144 patients with PJI, which would not have been diagnosed otherwise. In 11 other cases, the only positive result was that from PCR, which explains the better sensitivity result of the PCR assay with regard to culture of sonicate fluid and culture of periprosthetic tissue (77.1% vs. 72.9% and 70.1%); and PCR obtained almost the same high level of specificity as both culture-based techniques (all of them around 98%).

3.2 Commercial PCR

Commercial PCR for the diagnosis of PJI was developed to overcome the inconvenience of the lack of reproducibility between different centers. Initially, there were no specific kits for the diagnosis of PJI, so kits for blood cultures, which included most organisms involved in PJI, were used for this purpose.

3.2.1 Commercial PCR Designed for Blood Cultures

The first approach in this regard was SeptiFast multiplex PCR. It was tested from sonicate fluid of retrieved implants, once DNA was extracted, and it could identify 19 species of microorganisms (not only bacteria, but also fungi). Two studies tested it with good results. The first one[73] included a smaller group of patients with PJI (n=37) and PCR from sonicate fluid had a higher sensitivity than that obtained by culture of implants and periprosthetic tissues (78% vs. 62% and 65%). However, specificity data was not available, as no patients without infection were included. Two years later, another team used the same kit for the same purpose but with a larger number of patients (n=86) and including a negative control of patients (they presented aseptic loosening instead of PJI).[74] In this case, both parameters of sensitivity and specificity of PCR were excellent (96% and 100%, respectively).

Another commercial kit, GenoType BC, was tested from sonicate fluid of prostheses previously retrieved,[75] once PCR was extracted. Based on 16S rRNA gene PCR, it detected 32 species of bacteria, but not fungi, from a total of 126 patients. It showed a high number of false positives. On the other hand, the main benefit of this kit was the increment of patients diagnosed when it was performed in addition to the culture.

Cartridge technology was used for Xpert MRSA/SA SSTI real-time PCR assay,[76] designed for the detection of *Staphylococcus* spp. in clinical samples, as well as gene *mecA* related to the presence of methicillin resistance. Briefly, cartridge assays are performed as follows: osteoarticular samples, once cultured, are inoculated directly into a closed cartridge in which all reactives are included and where DNA extraction, amplification and detection are carried out. This way, the turnaround time (just 72 min) is much shorter than that of culture and other PCR. And the other main advantage is the easy management, the process being fully automated, thus avoiding contaminations due to manual manipulation. However, the obvious limitation of this assay in particular is that only *Staphylococcus* spp. were detected.

Also based on the cartridge technology, FilmArray multiplex PCR is the latest commercial assay designed for blood cultures, used for the diagnosis of PJI from sonicates fluid.[77] Although specificity was 99%, it had worse sensitivity than culture (53% vs. 69%), even when only specimens included in the panel were taken into account (58%).

All these PCR assays have the inconvenient aspect of detecting a low number of organisms, and do not include *Propionibacterium* spp. and *Corynebacterium* spp., which are considered etiological agents of PJI.

3.2.2 Commercial PCR Specifically Designed for the Diagnosis of PJI

To date, two commercial kits haven been specifically developed for the diagnosis of PJI, but without a common conclusion as to their statistical parameters.

The first one, the novel Prove-it Bone and Joint assay,[78] was tested in periprosthetic tissue samples in a total of 61 patients with and without the criteria of PJI. It is a

BR-PCR and microarray-based assay targeting over 60 bacterial species. It gave results in 6 h and showed a good sensitivity (82%), but a lower specificity (74%), resulting in a high number of false positives. Furthermore, some important pathogens involved in PJI were not included in the panel, such as *P.acnes*, *Streptococcus viridans* and *Mycoplasma* spp.

The last one is Unyvero i60 ITI multiplex PCR, also based on the cartridge system (as Xpert MRSA/SA SSTI realtime PCR and FilmArray multiplex PCR). It was initially tested as a pilot study[79] in tissue samples from patients with and without the criteria of PJI. This assay was compared not only with culture but also with another PCR, in particular with a homemade 16S rRNA PCR. The Unyvero system was used only in those patients with positive culture and showed a very high specificity (95.2%), but a low sensitivity (42.9%). Therefore, there was no evidence of superiority of this kit with regard to conventional culture. And again, the number of pathogens included in the panel was limited. Very recently, the same kit, Unyvero i60 ITI multiplex PCR, has been tested in a larger group of patients ($n=68$) which includes a noninfected group, and compared with conventional culture for the diagnosis of PJI.[80] Data of both sensitivity and specificity were better in this work than those obtained in the pilot study (60.5% and 98% vs. 42.9% and 95.2%, respectively). This assay allowed the detection of resistance genes or mutations, with a good correlation in general with the conventional antimicrobial susceptibility testing performed. The panel detected 24 specimens of bacteria (including *Propionibacterium* spp. and *Corynebacterium* spp.), and 3 yeasts. However, some organisms were not included, so if there had been a case of PJI due to those organisms, it would not be diagnosed by this test.

4. CONCLUSIONS

PJI is still a devastating complication of orthopedic surgery and it needs a full multidisciplinary approach for an adequate management of patients. Etiological diagnosis is a key part of this management, and many different approaches have been used for this goal. Conventional culture-based techniques include cultures of synovial fluid, periprosthetic tissue biopsies and sonicated implants. All these techniques can identify about 80%–90% of all patients. Aiming to improve this diagnosis, several molecular biology-based methodologies have been used, maintaining good specificity with variable sensitivity. Further improvements are needed in order to perform such diagnosis in all patients, because the knowledge of the etiology is essential to select the proper antimicrobial therapy and surgical management, and this even has implications in the design of future biomaterials aimed at minimizing the risk of infection. The increasing number of patients at risk makes this issue of even greater importance in the interest of all present and future patients.

REFERENCES

1. Tande AJ, Patel R. Prosthetic joint infection. *Clin Microbiol Rev* 2014;**27**:302–45.
2. Kurtz SM, Lau E, Schmier J, Ong KL, Zhao K, Parvizi J. Infection burden for hip and knee arthroplasty in the United States. *J Arthroplasty* 2008;**23**:984–91.
3. Perez-Jorge C, Gomez-Barrena E, Horcajada JP, Puig-Verdie L, Esteban J. Drug treatments for prosthetic joint infections in the era of multidrug resistance. *Expert Opin Pharmacother* 2016;**17**:1233–46.
4. Del Pozo JL, Patel R. Clinical practice. Infection associated with prosthetic joints. *N Engl J Med* 2009;**361**:787–94.
5. Trampuz A, Widmer AF. Infections associated with orthopedic implants. *Curr Opin Infect Dis* 2006;**19**:349–56.
6. Ahmad SS, Becker R, Chen AF, Kohl S. EKA survey: diagnosis of prosthetic knee joint infection. *Knee Surg Sports Traumatol Arthrosc* 2016; http://dx.doi.org/10.1007/s00167-016-4303-y.
7. Virolainen P, Lahteenmaki H, Hiltunen A, Sipola E, Meurman O, Nelimarkka O. The reliability of diagnosis of infection during revision arthroplasties. *Scand J Surg* 2002;**91**:178–81.
8. Benito N, Franco M, Ribera A, Soriano A, Rodriguez-Pardo D, Sorli L, et al. Time trends in the aetiology of prosthetic joint infections: a multicentre cohort study. *Clin Microbiol Infect* 2016;**22**:732.e1-8.
9. Minassian AM, Osmon DR, Berendt AR. Clinical guidelines in the management of prosthetic joint infection. *J Antimicrob Chemother* 2014;**69**(Suppl. 1):i29–35.
10. Osmon DR, Berbari EF, Berendt AR, Lew D, Zimmerli W, Steckelberg JM, et al. Diagnosis and management of prosthetic joint infection: clinical practice guidelines by the Infectious Diseases Society of America. *Clin Infect Dis* 2013;**56**:e1–e25.
11. Hughes JG, Vetter EA, Patel R, Schleck CD, Harmsen S, Turgeant LT, et al. Culture with BACTEC Peds Plus/F bottle compared with conventional methods for detection of bacteria in synovial fluid. *J Clin Microbiol* 2001;**39**:4468–71.
12. Fink B, Makowiak C, Fuerst M, Berger I, Schafer P, Frommelt L. The value of synovial biopsy, joint aspiration and C-reactive protein in the diagnosis of late peri-prosthetic infection of total knee replacements. *J Bone Joint Surg (Br)* 2008;**90**:874–8.
13. Yagupsky P, Dagan R, Howard CW, Einhorn M, Kassis I, Simu A. High prevalence of *Kingella kingae* in joint fluid from children with septic arthritis revealed by the BACTEC blood culture system. *J Clin Microbiol* 1992;**30**:1278–81.
14. Font-Vizcarra L, Garcia S, Martinez-Pastor JC, Sierra JM, Soriano A. Blood culture flasks for culturing synovial fluid in prosthetic joint infections. *Clin Orthop Relat Res* 2010;**468**:2238–43.
15. Larsen LH, Lange J, Xu Y, Schonheyder HC. Optimizing culture methods for diagnosis of prosthetic joint infections: a summary of modifications and improvements reported since 1995. *J Med Microbiol* 2012;**61**:309–16.
16. Cobo J, Del Pozo JL. Prosthetic joint infection: diagnosis and management. *Expert Rev Anti Infect Ther* 2011;**9**:787–802.
17. Costerton JW. Biofilm theory can guide the treatment of device-related orthopaedic infections. *Clin Orthop Relat Res* 2005;**437**:7–11.
18. Zimmerli W, Trampuz A, Ochsner PE. Prosthetic-joint infections. *N Engl J Med* 2004;**351**:1645–54.
19. Cune J, Soriano A, Martinez JC, Garcia S, Mensa JA. Superficial swab culture is useful for microbiologic diagnosis in acute prosthetic joint infections. *Clin Orthop Relat Res* 2009;**467**:531–5.
20. Trampuz A, Zimmerli W. Diagnosis and treatment of implant-associated septic arthritis and osteomyelitis. *Curr Infect Dis Rep* 2008;**10**:394–403.
21. Bauer TW, Parvizi J, Kobayashi N, Krebs V. Diagnosis of periprosthetic infection. *J Bone Joint Surg Am* 2006;**88**:869–82.
22. Neut D, van Horn JR, van Kooten TG, van der Mei HC, Busscher HJ. Detection of biomaterial-associated infections in orthopaedic joint implants. *Clin Orthop Relat Res* 2003;**413**:261–8.
23. Esteban J, Pérez-Jorge C, Pérez-Tanoira R, Gómez-Barrena E. Microbiological diagnosis of prosthetic joint infection. In: Trebse R, editor. *Infected total joint arthroplasty: the algorithmic approach.* London: Springer; 2012. p. 165–80.

24. Minassian AM, Newnham R, Kalimeris E, Bejon P, Atkins BL, Bowler IC. Use of an automated blood culture system (BD BACTEC) for diagnosis of prosthetic joint infections: easy and fast. *BMC Infect Dis* 2014;**14**:233.
25. Peel TN, Dylla BL, Hughes JG, Lynch DT, Greenwood-Quaintance KE, Cheng AC, et al. Improved diagnosis of prosthetic joint infection by culturing periprosthetic tissue specimens in blood culture bottles. *MBio* 2016;**7**:e01776-15.
26. Schafer P, Fink B, Sandow D, Margull A, Berger I, Frommelt L. Prolonged bacterial culture to identify late periprosthetic joint infection: a promising strategy. *Clin Infect Dis* 2008;**47**:1403–9.
27. Schwotzer N, Wahl P, Fracheboud D, Gautier E, Chuard C. Optimal culture incubation time in orthopedic device-associated infections: a retrospective analysis of prolonged 14-day incubation. *J Clin Microbiol* 2014;**52**:61–6.
28. Esteban J, Alvarez-Alvarez B, Blanco A, Fernandez-Roblas R, Gadea I, Garcia-Canete J, et al. Prolonged incubation time does not increase sensitivity for the diagnosis of implant-related infection using samples prepared by sonication of the implants. *Bone Joint J* 2013;**95-B**:1001–6.
29. Achermann Y, Goldstein EJ, Coenye T, Shirtliff ME. *Propionibacterium acnes*: from commensal to opportunistic biofilm-associated implant pathogen. *Clin Microbiol Rev* 2014;**27**:419–40.
30. Roux AL, Sivadon-Tardy V, Bauer T, Lortat-Jacob A, Herrmann JL, Gaillard JL, et al. Diagnosis of prosthetic joint infection by beadmill processing of a periprosthetic specimen. *Clin Microbiol Infect* 2011;**17**:447–50.
31. De Vecchi E, Bortolin M, Signori V, Romano CL, Drago L. Treatment with dithiothreitol improves bacterial recovery from tissue samples in osteoarticular and joint infections. *J Arthroplasty* 2016; http://dx.doi.org/10.1016/j.arth.2016.05.008.
32. Drago L, Signori V, De Vecchi E, Vassena C, Palazzi E, Cappelletti L, et al. Use of dithiothreitol to improve the diagnosis of prosthetic joint infections. *J Orthop Res* 2013;**31**:1694–9.
33. Atkins BL, Athanasou N, Deeks JJ, Crook DW, Simpson H, Peto TE, et al. Prospective evaluation of criteria for microbiological diagnosis of prosthetic-joint infection at revision arthroplasty. The OSIRIS Collaborative Study Group. *J Clin Microbiol* 1998;**36**:2932–9.
34. Marin M, Esteban J, Meseguer MA, Sanchez-Somolinos M. Microbiological diagnosis of bone-joint infections. *Enferm Infecc Microbiol Clin* 2010;**28**:534–40.
35. Marculescu CE, Berbari EF, Hanssen AD, Steckelberg JM, Osmon DR. Prosthetic joint infection diagnosed postoperatively by intraoperative culture. *Clin Orthop Relat Res* 2005;**439**:38–42.
36. Peel TN, Buising KL, Choong PF. Diagnosis and management of prosthetic joint infection. *Curr Opin Infect Dis* 2012;**25**:670–6.
37. Renz N, Muller M, Perka C, Trampuz A. Implant-associated infections—diagnostics. *Chirurg* 2016;**87**:813–21.
38. Gristina AG. Biomaterial-centered infection: microbial adhesion versus tissue integration. *Science* 1987;**237**:1588–95.
39. Trampuz A, Zimmerli W. Prosthetic joint infections: update in diagnosis and treatment. *Swiss Med Wkly* 2005;**135**:243–51.
40. Trampuz A, Piper KE, Jacobson MJ, Hanssen AD, Unni KK, Osmon DR, et al. Sonication of removed hip and knee prostheses for diagnosis of infection. *N Engl J Med* 2007;**357**:654–63.
41. Esteban J, Gomez-Barrena E, Cordero J, Martin-de-Hijas NZ, Kinnari TJ, Fernandez-Roblas R. Evaluation of quantitative analysis of cultures from sonicated retrieved orthopedic implants in diagnosis of orthopedic infection. *J Clin Microbiol* 2008;**46**:488–92.
42. Bjerkan G, Witso E, Bergh K. Sonication is superior to scraping for retrieval of bacteria in biofilm on titanium and steel surfaces in vitro. *Acta Orthop* 2009;**80**:245–50.
43. Monsen T, Lovgren E, Widerstrom M, Wallinder L. In vitro effect of ultrasound on bacteria and suggested protocol for sonication and diagnosis of prosthetic infections. *J Clin Microbiol* 2009;**47**:2496–501.
44. Zhai Z, Li H, Qin A, Liu G, Liu X, Wu C, et al. Meta-analysis of sonication fluid samples from prosthetic components for diagnosis of infection after total joint arthroplasty. *J Clin Microbiol* 2014;**52**:1730–6.
45. Liu H, Zhang Y, Li L, Zou HC. The application of sonication in diagnosis of periprosthetic joint infection. *Eur J Clin Microbiol Infect Dis* 2016; http://dx.doi.org/10.1007/s10096-016-2778-6.

46. Esteban J, Sorli L, Alentorn-Geli E, Puig L, Horcajada JP. Conventional and molecular diagnostic strategies for prosthetic joint infections. *Expert Rev Mol Diagn* 2014;**14**:83–96.
47. Holinka J, Bauer L, Hirschl AM, Graninger W, Windhager R, Presterl E. Sonication cultures of explanted components as an add-on test to routinely conducted microbiological diagnostics improve pathogen detection. *J Orthop Res* 2011;**29**:617–22.
48. Portillo ME, Salvado M, Alier A, Martinez S, Sorli L, Horcajada JP, et al. Advantages of sonication fluid culture for the diagnosis of prosthetic joint infection. *J Infect* 2014;**69**:35–41.
49. Holinka J, Pilz M, Hirschl AM, Graninger W, Windhager R, Presterl E. Differential bacterial load on components of total knee prosthesis in patients with prosthetic joint infection. *Int J Artif Organs* 2012;**35**:735–41.
50. Gomez-Barrena E, Esteban J, Medel F, Molina-Manso D, Ortiz-Perez A, Cordero-Ampuero J, et al. Bacterial adherence to separated modular components in joint prosthesis: a clinical study. *J Orthop Res* 2012;**30**:1634–9.
51. Portillo ME, Salvado M, Trampuz A, Siverio A, Alier A, Sorli L, et al. Improved diagnosis of orthopedic implant-associated infection by inoculation of sonication fluid into blood culture bottles. *J Clin Microbiol* 2015;**53**:1622–7.
52. Esteban J, Sandoval E, Cordero-Ampuero J, Molina-Manso D, Ortiz-Perez A, Fernandez-Roblas R, et al. Sonication of intramedullary nails: clinically-related infection and contamination. *Open Orthop J* 2012;**6**:255–60.
53. Yano MH, Klautau GB, da Silva CB, Nigro S, Avanzi O, Mercadante MT, et al. Improved diagnosis of infection associated with osteosynthesis by use of sonication of fracture fixation implants. *J Clin Microbiol* 2014;**52**:4176–82.
54. Esteban J, Gadea I, Perez-Jorge C, Sandoval E, Garcia-Canete J, Fernandez-Roblas R, et al. Diagnosis of spacer-associated infection using quantitative cultures from sonicated antibiotics-loaded spacers: implications for the clinical outcome. *Eur J Clin Microbiol Infect Dis* 2016;**35**:207–13.
55. Perez-Prieto D, Portillo ME, Puig-Verdie L, Alier A, Gamba C, Guirro P, et al. Preoperative antibiotic prophylaxis in prosthetic joint infections: not a concern for intraoperative cultures. *Diagn Microbiol Infect Dis* 2016; http://dx.doi.org/10.1016/j.diagmicrobio.2016.09.014.
56. Levine MJ, Mariani BA, Tuan RS, Booth Jr. RE. Molecular genetic diagnosis of infected total joint arthroplasty. *J Arthroplasty* 1995;**10**:93–4.
57. Bemer P, Plouzeau C, Tande D, Leger J, Giraudeau B, Valentin AS, et al. Evaluation of 16S rRNA gene PCR sensitivity and specificity for diagnosis of prosthetic joint infection: a prospective multicenter cross-sectional study. *J Clin Microbiol* 2014;**52**:3583–9.
58. Riggio MP, Dempsey KE, Lennon A, Allan D, Ramage G, Bagg J. Molecular detection of transcriptionally active bacteria from failed prosthetic hip joints removed during revision arthroplasty. *Eur J Clin Microbiol Infect Dis* 2010;**29**:823–34.
59. Sauer P, Gallo J, Kesselova M, Kolar M, Koukalova D. Universal primers for detection of common bacterial pathogens causing prosthetic joint infection. *Biomed Pap Med Fac Univ Palacky Olomouc Czech Repub* 2005;**149**:285–8.
60. Stoodley P, Kathju S, Hu FZ, Erdos G, Levenson JE, Mehta N, et al. Molecular and imaging techniques for bacterial biofilms in joint arthroplasty infections. *Clin Orthop Relat Res* 2005;**437**:31–40.
61. Vandercam B, Jeumont S, Cornu O, Yombi JC, Lecouvet F, Lefevre P, et al. Amplification-based DNA analysis in the diagnosis of prosthetic joint infection. *J Mol Diagn* 2008;**10**:537–43.
62. De Man FH, Graber P, Luem M, Zimmerli W, Ochsner PE, Sendi P. Broad-range PCR in selected episodes of prosthetic joint infection. *Infection* 2009;**37**:292–4.
63. Fihman V, Hannouche D, Bousson V, Bardin T, Liote F, Raskine L, et al. Improved diagnosis specificity in bone and joint infections using molecular techniques. *J Infect* 2007;**55**:510–7.
64. Mariani BD, Levine MJ, Booth Jr RE, Tuan RS. Development of a novel, rapid processing protocol for polymerase chain reaction-based detection of bacterial infections in synovial fluids. *Mol Biotechnol* 1995;**4**:227–37.
65. Marin M, Garcia-Lechuz JM, Alonso P, Villanueva M, Alcala L, Gimeno M, et al. Role of universal 16S rRNA gene PCR and sequencing in diagnosis of prosthetic joint infection. *J Clin Microbiol* 2012;**50**:583–9.

66. Fenollar F, Roux V, Stein A, Drancourt M, Raoult D. Analysis of 525 samples to determine the usefulness of PCR amplification and sequencing of the 16S rRNA gene for diagnosis of bone and joint infections. *J Clin Microbiol* 2006;**44**:1018–28.
67. Rak M, KavčIč M, Trebse R, Co RA. Detection of bacteria with molecular methods in prosthetic joint infection: sonication fluid better than periprosthetic tissue. *Acta Orthop* 2016;**87**:339–45.
68. Rak M, Barlic-Maganja D, KavčIč M, Trebse R, Cor A. Comparison of molecular and culture method in diagnosis of prosthetic joint infection. *FEMS Microbiol Lett* 2013;**343**:42–8.
69. Gomez E, Cazanave C, Cunningham SA, Greenwood-Quaintance KE, Steckelberg JM, Uhl JR, et al. Prosthetic joint infection diagnosis using broad-range PCR of biofilms dislodged from knee and hip arthroplasty surfaces using sonication. *J Clin Microbiol* 2012;**50**:3501–8.
70. Bjerkan G, Witso E, Nor A, Viset T, Loseth K, Lydersen S, et al. A comprehensive microbiological evaluation of fifty-four patients undergoing revision surgery due to prosthetic joint loosening. *J Med Microbiol* 2012;**61**:572–81.
71. Piper KE, Jacobson MJ, Cofield RH, Sperling JW, Sanchez Sotelo J, Osmon DR, et al. Microbiologic diagnosis of prosthetic shoulder infection by use of implant sonication. *J Clin Microbiol* 2009;**47**:1878–84.
72. Cazanave C, Greenwood-Quaintance KE, Hanssen AD, Karau MJ, Schmidt SM, Gomez Urena EO, et al. Rapid molecular microbiologic diagnosis of prosthetic joint infection. *J Clin Microbiol* 2013;**51**:2280–7.
73. Achermann Y, Vogt M, Leunig M, Wust J, Trampuz A. Improved diagnosis of periprosthetic joint infection by multiplex PCR of sonication fluid from removed implants. *J Clin Microbiol* 2010;**48**:1208–14.
74. Portillo ME, Salvado M, Sorli L, Alier A, Martinez S, Trampuz A, et al. Multiplex PCR of sonication fluid accurately differentiates between prosthetic joint infection and aseptic failure. *J Infect* 2012;**65**:541–8.
75. Esteban J, Alonso-Rodriguez N, del-Prado G, Ortiz-Perez A, Molina-Manso D, Cordero-Ampuero J, et al. PCR-hybridization after sonication improves diagnosis of implant-related infection. *Acta Orthop* 2012;**83**:299–304.
76. Dubouix-Bourandy A, de Ladoucette A, Pietri V, Mehdi N, Benzaquen D, Guinand R, et al. Direct detection of *Staphylococcus* osteoarticular infections by use of Xpert MRSA/SA SSTI real-time PCR. *J Clin Microbiol* 2011;**49**:4225–30.
77. Vasoo S, Cunningham SA, Greenwood-Quaintance KE, Mandrekar JN, Hanssen AD, Abdel MP, et al. Evaluation of the FilmArray blood culture ID panel on biofilms dislodged from explanted arthroplasties for prosthetic joint infection diagnosis. *J Clin Microbiol* 2015;**53**:2790–2.
78. Metso L, Maki M, Tissari P, Remes V, Piiparinen P, Kirveskari J, et al. Efficacy of a novel PCR- and microarray-based method in diagnosis of a prosthetic joint infection. *Acta Orthop* 2014;**85**:165–70.
79. Borde JP, Hacker GA, Guschl S, Serr A, Danner T, Hubner J, et al. Diagnosis of prosthetic joint infec tions using UMD-Universal Kit and the automated multiplex-PCR Unyvero i60 ITI((R)) cartridge system: a pilot study. *Infection* 2015;**43**:551–60.
80. Prieto-Borja L, Rodriguez-Sevilla G, Aunon A, Perez-Jorge C, Sandoval E, Garcia-Canete J, et al. Evaluation of a commercial multiplex PCR (Unyvero i60(R)) designed for the diagnosis of bone and joint infections using prosthetic-joint sonication. *Enferm Infecc Microbiol Clin* 2016; http://dx.doi.org/10.1016/j.eimc.2016.09.007.

CHAPTER 10

Laboratory Detection of the Emergent Pathogen *Kingella kingae* in Children With Septic Arthritis

P. Yagupsky
Ben-Gurion University of the Negev, Beer-Sheva, Israel

1. INTRODUCTION

Septic, pyogenic, and suppurative arthritis are the names given to the inflammation of the joint space caused by the presence of bacteria or fungi. Septic arthritis is more common in childhood than in any other period of human life and more than half of cases are detected in individuals younger than 20 years of age.[1] Because the pathogenesis of pediatric septic arthritis usually involves a hematogenous bacterial spread which is followed by seeding of circulating organisms to the joint tissues, the age distribution of patients is markedly skewed to the left, reflecting the increased incidence of bacteremia in early childhood. In most cases, the etiological agents of septic arthritis in the child are encapsulated organisms that colonize the mucosal surfaces of the upper respiratory tract.[2] The delayed maturity of the T-cell independent arm of the immune system in humans results in an ineffective antibody response to bacterial capsular polysaccharides[1] rendering children below the age of 4 years highly susceptible to invasive bacterial diseases, including those involving the skeletal system.

The etiology of pediatric septic arthritis has markedly evolved over time as the result of improvements in the hygienic living conditions of the population, the advent of antibiotics and the consequent emergence of antimicrobial drug resistance, implementation of effective vaccination programs against common joint pathogens such as *Haemophilus influenzae* type b and *Streptococcus pneumoniae* and, especially, the development of sensitive diagnostic laboratory methods. In the preantibiotic era, Gram-positive bacteria, and particularly *Staphylococcus aureus*, *Streptococcus pyogenes*, and *Streptococcus pneumoniae*, were the main identified causes of skeletal system infections in children.[1] With the advent of penicillin therapy, the role of streptococci diminished while *S. aureus*, which rapidly acquired β-lactamase-mediated resistance to the antibiotic, remained the most common etiology of pediatric osteomyelitis and septic arthritis.[1] In the 1960s, introduction of chocolate-agar medium

The Microbiology of Skin, Soft Tissue, Bone and Joint Infections
ISSN 2451-9006, http://dx.doi.org/10.1016/B978-0-12-811079-9.00010-0

for seeding joint and bone exudates led to the disclosure of *H. influenzae* type b as the prime agent of suppurative arthritis of childhood,[3] and demonstrated the crucial role played by improved laboratory methods in the diagnosis of pediatric osteoarthritis. It should be pointed out that even when synovial fluid aspirates obtained from children with suspected septic arthritis were routinely seeded onto chocolate-agar medium, in up to two-thirds of cases, cultures remained sterile and the diagnosis unconfirmed, suggesting the possibility that a large fraction of joint infections could be caused by pathogens that were not supported by conventional bacteriological methods.[4]

In the late 1980s, pediatricians working at the Soroka University Medical Center in Southern Israel started to inoculate synovial fluid specimens obtained from patients with joint effusions into aerobic blood culture vials (BCVs) from the BACTEC blood culture system, in an attempt to improve the bacteriological diagnosis. Shortly after adoption of this novel approach, a Gram-negative coccobacillus, later identified as *Kingella kingae*, was repeatedly isolated from exudates aspirated from young children, whereas solid media, including chocolate-agar plates, seeded with the same specimens grew no organisms.[5] This observation heralded the recognition of *K. kingae* as an important pathogen of pediatric bone and joint infections, revealed the hitherto hidden role of the organism in the causation of a large fraction of "culture-negative" septic arthritis in early childhood,[6] and emphasized the importance of using improved laboratory methods for confirming the diagnosis of skeletal system infections.

The aims of the present chapter are to describe this emergent pediatric pathogen, to summarize the current knowledge on the epidemiology of carriage and invasive *K. kingae* infections and particularly those of the skeletal system, give an account of the peculiar clinical features of *K. kingae* arthritis, and review the laboratory tools available for its diagnosis.

2. THE ORGANISM

First characterized in the 1960s by Elizabeth King working at the Center for Disease Control (CDC) in Atlanta, the novel bacterium was first baptized *Moraxella kingii* because of the superficial phenotypic resemblance to members of this genus. In-depth bacteriological research resulted in the later assignment of the organism to a new and separate genus, its renaming to *K. kingae*, and placement of the novel species in the Neisseriaceae family.[7] *K. kingae* is an encapsulated facultative anaerobic bacterium that appears as pairs or short chains of short coccobacilli (Fig. 1), displays β-hemolysis and oxidase activity and a negative catalase reaction, acidifies glucose and maltose, and grows on conventional blood-agar and chocolate-agar but not on MacConkey or Kligler media. A wide array of modern bacteriological instruments and technologies, such as VITEK (bioMérieux, Marcy-l'Etoile, France), matrix-assisted laser desorption ionization-time of flight mass spectrometry (MALDI-TOF) or 16S rRNA gene

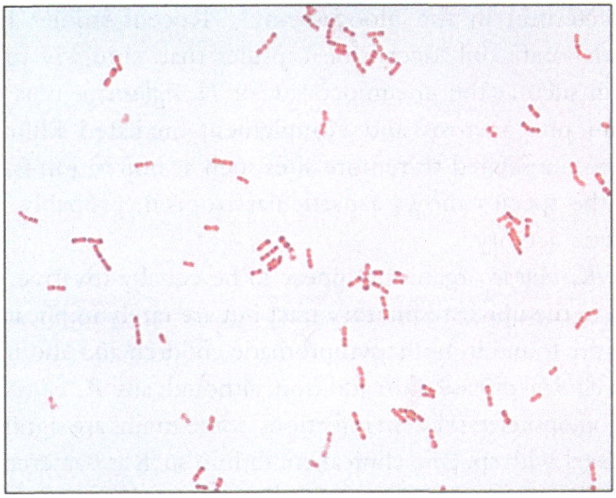

Fig. 1 Typical Gram-stain of *K. kingae* showing plump Gram-negative coccobacilli in pairs and short chains.

sequencing correctly identify the organism. However, clinical microbiology laboratories unfamiliar with the cultural and biochemical features of the species frequently misidentify the bacterium altogether.[8]

3. PATHOGENESIS OF INVASIVE *K. KINGAE* DISEASE

Invasive *K. kingae* infections are thought to occur as a multistep process. Following exposure and acquisition of the organism by intimate child-to-child contact through droplet transmission, the bacterium firmly adheres to the oropharyngeal epithelium by elongated pili to avoid being dislodged by oral secretions, establishing firm mucosal colonization. Carriage of *K. kingae* organisms is a frequent but usually asymptomatic and inconsequential event. The bacterium is first acquired after 6 months of age, its prevalence in the pharynx increases gradually thereafter reaching a prevalence rate of 10%–12% in the second year of life, and declines in older children and adults.[9] This peculiar age-related epidemiological curve shows a remarkable overlap with that of invasive disease.[8] Under favorable conditions, and especially viral infections that damage the upper respiratory and buccal epithelia, such as herpetic gingivostomatitis[8] or hand-foot-and-mouth disease,[10] penetration of carried *K. kingae* organisms into the submucosal layers and invasion of the bloodstream may ensue. Because the *K. kingae* elaborates a potent repeat-in-toxin (RTX), encoded by the *rtx* operon, that is deleterious to epithelial cells, leucocytes, and synoviocytes, it is believed that secretion of this virulence factor facilitates epithelial colonization, tissue invasion, and survival of the bacterium in the bloodstream.[11] Recent studies have shown that *K. kingae* strains elaborate polysaccharide capsules that, similarly to other pathogens of respiratory origin such as the pneumococcus or *H. influenzae* type b, could protect the organism from phagocytosis and complement-mediated killing.[12] Circulating *K. kingae* may, then, be seeded to remote sites such as bones, joints, or the endocardium, for which the species shows a particular tropism, probably mediated by the RTX toxin cytolytic activity.[8]

Not all carried *K. kingae* organisms appear to be equally invasive. Some strains are commonly carried in the upper respiratory tract but are rarely implicated in deep infections, while others are found in both asymptomatic children and also frequently isolated from patients with clinical disease.[13] In addition, although any *K. kingae* strain may cause bloodstream invasion or skeletal system infections, some strains are significantly associated at the population level with specific clinical conditions such as bacteremia with no focus ("occult bacteremia"), osteomyelitis, septic arthritis, or endocarditis.[13]

4. CLINICAL PRESENTATION OF *K. KINGAE* SEPTIC ARTHRITIS

Invasive *K. kingae* infections are almost restricted to otherwise healthy children and >95% of cases are diagnosed between the ages of 6 months to 4 years.[8] Occurrence of the disease in the first semester of life is exceptional, and *K. kingae* infections affecting children aged >4 years and adults are also rare and usually associated with predisposing factors such as malignancies, immunosuppressive conditions, chronic liver or kidney diseases, or antecedent cardiac valve pathology.[8,14]

Involvement of the skeletal system occurs in over one-half of pediatric *K. kingae* infections, followed by bacteremia with no focus in 40%, and endocarditis in <2%.[8,14] Suppurative arthritis constitutes the most common *K. kingae* infection in children and was diagnosed in 63 of the 78 (79.5%) pediatric patients with culture-confirmed skeletal system involvement detected in southern Israel since 1987. Osteomyelitis was observed in five (6.4%), two children (2.6%) had both joint and bone involvement, three (3.8%) children had tenosynovitis, and five (6.4%) children presented with *K. kingae* bacteremia and transient skeletal system symptoms, suggesting an abortive infection.[15] More rare skeletal system manifestations include spondylodiscitis, tenosynovitis, and muscle abscesses.[8]

K. kingae arthritis usually affects the weight-bearing knee, hip or ankle joints and with decreasing frequency those of the upper limbs, wrist, shoulder, or elbow.[8,14] Osteomyelitis generally involves the long tubular bones of the extremities and particularly the femur and tibia.[8] Of notice, the small sternoclavicular and metacarpophalangeal joints, as well as the sternum, clavicle, and tarsal bones, which are rarely infected by traditional pyogenic bacteria, are overrepresented in *K. kingae* septic arthritis and osteomyelitis.[8]

With the notorious exception of cases of endocardial invasion, the clinical presentation of invasive *K. kingae* disease is usually mild and characterized by a limited local and systemic inflammatory response: most patients present in good general condition, have moderate fever, or are even afebrile.[8,14] The blood leukocyte count and acute phase reactants such as C-reactive protein level and erythrocyte sedimentation rate are generally only mildly to moderately elevated and may be entirely normal.[8] However, with the exception of the young age of children with *K. kingae* disease,[6] no clinical or laboratory features appear to be reliable enough to differentiate between *K. kingae* infections and those caused by classic skeletal system pathogens such as *S. aureus*.[16–18]

The joint fluid aspirate shows $<50.0 \times 10^9$ WBC/L in over 20% of children,[8] meeting the laboratory criterion generally employed to rule out suppurative arthritis. Microscopic examination of the exudate rarely reveals the presence of *K. kingae*[8] probably because of the low bacterial load,[19] and the difficulty in visualizing the Gram-negative organism against the pink-stained fibrin background. The seemingly benign clinical and laboratory features of invasive *K. kingae* infections,[8,14] coupled with the suboptimal recovery of the organism in cultures, explain why many cases of joint and bone infections and spondylodiscitis are diagnosed with considerable delay or overlooked altogether.[8] For instance, Kocher's predictive scorings that combine clinical and laboratory data to distinguish between a septic hip from benign transient synovitis, gave to 20 of 28 (71.4%) children with a culture-proven *K. kingae* infection less than a 40% probability of having a bacterial joint infection.[20] Potential reasons for missing the diagnosis of *K. kingae* infections are summarized in Table 1.

Table 1 Potential contributors to the failure to recognize septic arthritis due to *Kingella kingae* in children

Potential contributor	Reason
Failure to suspect a skeletal system infection	Mild clinical presentation; normal WBC count and acute-phase reactants
Failure to suspect bacteremic disease and obtain blood cultures	Lack of high fever and/or leucocytosis
Lack of use of BCVs for seeding synovial fluid aspirates	Unawareness
Gram-stain of synovial fluid aspirate	Usually no organisms seen
Lack of use of NAAAs	Unavailability
Misidentification of the isolate	Unfamiliarity with the cultural and biochemical features of *K. kingae*
Failure to detect *K. kingae* in pharyngeal cultures	Lack of use selective media

5. CULTURE DETECTION OF *K. KINGAE*

5.1 Blood Cultures

Because pediatric septic arthritis generally results from the hematogenic seeding of the organism, routine drawing of blood cultures has been recommended for determining the etiology of the disease. Blood samples are easy to obtain and repeat compared to synovial fluid aspirates or skeletal system tissues and, in many cases, the causative bacterium can be isolated from the blood even when synovial fluid cultures are negative.[21] The diagnostic yield of the blood culture tool varies widely in different series, with an average of 50% in septic arthritis of childhood.[21,22] The sensitivity of blood cultures for diagnosing focal *K. kingae* infections is much lower, and among 81 children with culture-proven skeletal system infections detected in Southern Israel, the organism was recovered from joint and bone exudates in 74 (91.4%), from the blood in 6 (7.4%), and from both, synovial fluid and blood, in a single patient (1.2%), suggesting that, in addition to the fastidious nature of the bacterium, the preceding bacteremic phase of the disease is rather short.[8]

5.2 Synovial Fluid Cultures

For most of the two decades that followed the first description of *K. kingae*, the organism was considered a very rare human pathogen. The unexpected discovery that seeding synovial fluid aspirates from young children with suspected joint infection BCVs substantially improved detection of the bacterium and resulted in the recognition of *K. kingae* as the most common pathogen of septic arthritis in children aged 6–36 months.[8] In a pioneer study, synovial fluid exudates seeded onto routine blood-agar and chocolate-agar plates recovered *K. kingae* in only 2 of 25 children with arthritis, whereas aerobic BACTEC (Becton Dickinson, Cockeysville, Maryland) BCVs inoculated with the same specimens revealed the presence of the organism in all cases.[5] When the positive BCVs were subcultured onto blood-agar and chocolate-agar plates, *K. kingae* grew readily, indicating that routine bacteriological solid media are able to support the nutritional requirements of the organism. In two separate studies conducted in Israel and France in which synovial fluid aspirates from children younger than 3 years with arthritis were routinely inoculated into BACTEC and BacT/Alert BCVs, *K. kingae* was isolated in almost half of the patients with culture-proven disease.[23,24] More recent studies have demonstrated that inoculation of skeletal system exudates in vials of a variety of other commercial blood culture systems such as BACTEC (Becton Dickinson, Cockeysville, Maryland), BacT/Alert (OrganonTeknika Corporation, Durham, North Carolina), Hémoline DUO (bioMérieux, Paris, France), and Isolator 1.5 Microbial Tube (Wampole Laboratories, Cranbury, New Jersey) vials, and an in-house made manual culture method[8] increases the recovery of *K. kingae* compared to conventional bacteriological cultures. These observations confirm that synovial fluid possesses potent antibacterial properties, inhibiting growth of fastidious *K. kingae* organisms.[25] Dilution of the small purulent specimens aspirated from a child's joint in a large volume of nutrient broth decreases the concentration of detrimental factors, improving the recovery of the bacterium.[5]

On average, the automated blood culture systems detect the presence of *K. kingae* after 2–4 days, whereas more prolonged incubation or performance of blind subcultures of negative BCVs does not increase recovery of the organism.[26] Vials detected as positive by the automated instruments are processed following the customary protocol employed for blood cultures. A small volume of the BCV nutrient broth is aspirated, Gram-stained, and subcultured on routine bacteriological media such as blood-agar and chocolate-agar plates, and incubated aerobically at 35°C in a 5% CO_2-enriched atmosphere. Identification of the isolate by appropriate biochemical tests and/or MALDI-TOF technology, and determination of antibiotic susceptibility to relevant antimicrobial drugs and exclusion of β-lactamase production by the nitrocephine assay follow.[27]

5.3 Oropharyngeal Cultures

Confirmation of *K. kingae* as the etiology of childhood septic arthritis is hampered by the poor recovery of the bacterium from blood and synovial fluid specimens, and by the fact that small pediatric joints, which are commonly affected by the organism, are rarely taped.[28,29] Because the colonized oropharynx is the source of bloodstream infection and skeletal system invasion by *K. kingae*,[30] detection of the organism in the oropharynx has been proposed as a means to establish the bacteriological diagnosis of joint infections when cultures are not obtained or yield negative results.[28,29] This approach has the obvious advantages of being noninvasive and easy to perform, which are important considerations in young children. However, it should be pointed out that *K. kingae* is carried in the respiratory tract by 10%–12% of children in the relevant 6 months to 4-year age group[9] and by an even higher percentage of those attending daycare facilities[31,32] and, therefore, the predictive value of a positive pharyngeal culture is limited.

6. DETECTION OF *K. KINGAE* BY NAAAS

Because of the risk for severe morbidity and long-term orthopedic sequelae of septic arthritis, prompt bacteriological confirmation of the diagnosis and early initiation of appropriate antibiotic therapy are critical to prevent acute septic complications and long-term disability. However, even when the BCV method is used, a substantial fraction of presumptive joint infections remains bacteriologically unconfirmed and, in any case, complete identification of the causative agent and antibiotic susceptibility testing require a minimum of an additional 2–3 days.

In recent years, novel culture-independent diagnostic methods based on the enzymatic amplification of nucleic acid sequences have been developed and are gaining increasing acceptance for the diagnosis of infectious conditions in general, and those affecting the skeletal system in particular. These nucleic acid amplification assays (NAAAs) make it possible to detect difficult-to-culture and uncultivable bacteria, exhibit improved sensitivity compared to cultures, shorten the time-to-detection and identification of the etiology of the infection from days to a few hours, and can establish the diagnosis in patients being treated with antibiotics.[33,34] Because of the fastidious nature of *K. kingae*, NAAAs appear to be a natural alternative to culture detection.

The technology of NAAAs has rapidly evolved from the initial conventional PCR to real-time PCR that is less prone to contamination, and from the use of universal primers to primers that amplify species-specific targets and show improved performance. It should be pointed out, however, that although less sensitive than the molecular methods, culture detection of *K. kingae* has the obvious advantages of enabling determination of the antibiotic susceptibility of the isolate as well as allowing typing and the study of its biological features (see Table 2).

6.1 Use of NAAAS in Synovial Fluid Aspirates

The use of universal primers involves extraction of DNA from clinical samples, incubation of the DNA with broad range oligonucleotide primers that anneal to constant regions of the 16S rRNA gene, followed by enzymatic amplification of the intervening variable and species-specific regions.[35] The resulting amplicons are then sequenced and compared with sequences deposited in the Genbank database for species assignment, establishing an accurate identification of the causative organism. Alternatively, the amplicons may be hybridized with species-specific probes

Table 2 Comparative performance of synovial fluid cultures and NAAAs for the detection of *Kingella kingae*

	Detection method			
	Culture		NAAAs	
	Conventional	BCV	16S rRNA	*K. kingae*-specific
Sensitivity[a]	<20%	100%	250%	>400%
Time-to-detection	2–4 days	2–4 days	Same day	Same day
Antibiotic susceptibility	Yes	Yes	No	No
Enables genotyping	Yes	Yes	No	No
Diagnosis of other etiology	Yes	Yes	Yes	No

[a]Relative to BCV cultures.

belonging to the most plausible etiologies of the infection chosen on the basis of clinical and/or epidemiological considerations.

The combined data of the four studies in which broad spectrum NAAAs were compared with the culture results—either conventional, performed by the BCV method, or both—show that the molecular assay improves detection of the organism by >200% [63 of 64 (98.4%) vs 28 of 64 (43.8%), ($P<.001$)].[36–38] These data also imply that the culture recovery of *K. kingae* from skeletal system exudates remains unsatisfactory, even when specimens are inoculated into BCVs. Use of the NAAAs reduces the fraction of culture-negative infection and demonstrates that *K. kingae* is the leading etiology of suppurative arthritis in children aged 6–48 months.

More recently, PCR assays that amplify *K. kingae*–specific targets have been developed and introduced into routine clinical practice demonstrating higher sensitivity and reliability.[6,39–50] The RTX toxin, a virulence factor involved in the colonization of the mucosal surfaces and the survival of the organism in the bloodstream and deep tissues, is produced by all *K. kingae* strains examined so far and, therefore, the RTX locus genes appear to be pertinent targets for molecular diagnosis of the bacterium.[41] The novel NAAAs that amplify the encoding genes allow detection of *K. kingae* strains exhibiting RTX locus polymorphisms including nonsynonymous ones.[6,39–43] Alternatively, the *cpn*60 gene that encodes *K. kingae*'s chaperonin 60 protein, responsible for assisting misfolded or unfolded proteins to reach their native state, has been also successfully employed for this purpose.[44–49]

The performance of universal and species-specific primers for the detection of *K. kingae* was compared in an in vitro study that found that the PCR assay targeting the *rtx*A gene exhibited a sensitivity of 30 colony-forming units (CFU), one order of magnitude higher than the seminested broad range 16S rRNA gene NAAA.[39] These detection thresholds have practical diagnostic implications because the concentration of *K. kingae* organisms in the synovial fluid of children with arthritis is low, ranging between 11 and 300 CFU/mL.[19] The superior sensitivity of the species-specific NAAAs was confirmed in a clinical study that found that amplification with a test that targeted the *cpn*60 gene successfully detected *K. kingae* sequences in 9 of 44 (20.5%) children with culture-negative arthritis, while the universal 16S rRNA primers failed in all cases.[46] In a more recent study *K. kingae* was detected in 56 pediatric patients with arthritis employing both a PCR assay that amplified the 16S rRNA gene and a real-time assay that targeted the *K. kingae*–specific *rtx*A and *rtx*B genes.[50] The broad-spectrum PCR assay was the only positive test in two patients, the species-specific assay was the only positive one in 42 patients, and the organism was detected by both methods in the remaining 12 children. Overall, the 16S rRNA assay detected the presence of the organism in 14 of 56 (25.0%) patients and the *K. kingae*–specific assay in 54 of 56 (96.4%) ($P<.001$), showing the superior performance of the species-specific approach.[50]

The published reports on the comparative performance of *K. kingae*–specific DNA targets and cultures show that the former detected 238 of 239 (99.6%) positive specimens versus 42 of 255 (18.7%) with positive culture isolation.[39–49] It should be pointed out, however, that the broad spectrum NAAAs have the obvious advantage of being capable of detecting any organism, whereas those that target *K. kingae*–specific DNA sequences are limited to diagnosing a single species, and therefore should be used selectively in the well-defined population of children aged 6 months to 4 years. Because *K. kingae* is by far the most common etiology of septic arthritis and osteomyelitis below the age of 4 years, the researchers advised the use of NAAAs that target *K. kingae*–specific sequences as the initial diagnostic test for joint and bone specimens obtained from children in this age interval, reserving the broad-spectrum test that amplifies the 16S rRNA gene for skeletal system exudates in which *K. kingae*–specific assays yield negative results.[46]

Because *K. kingae* is exquisitely susceptible to antimicrobial drugs that are empirically administered to children with suspected invasive infections, previous use of antibiotics has a negative effect upon the culture recovery of the organism from the blood and skeletal system exudates.[4] Recent reports have shown that NAAAs that target the universal rRNA or the species-specific *rtx*A and *rtx*B genes are not negatively affected by previous administration of antibiotics,[50,51] indicating that the molecular diagnosis of *K. kingae* infections may still be possible in patients with partially treated disease.

6.2 Use of NAAAs in Oropharyngeal Specimens

Aspiration of joints in young children frequently poses serious technical difficulties, and therefore these specimens are not always available for culture or NAAA testing. Because colonization of the pharyngeal mucosa by *K. kingae* organisms precedes invasion of the bloodstream and seeding of the skeletal system,[31] search for specific DNA sequences in oropharyngeal specimens has been proposed as a surrogate detection strategy.[28,29,52]

The sensitivity of cultures and NAAAs for detecting *K. kingae* colonization has been compared during the investigation of a large cluster of invasive infections among young attendees to a French daycare facility.[28] Pharyngeal specimens, obtained from close contacts of the index cases, were cultured on a modified BAV medium and analyzed by an NAAA that targeted the *cpn*60 genes.[28] Twelve of 18 (66.7%) individuals were positive using the PCR-based test compared to 6 of 18 (33.3%) detected by culture ($P < .01$), indicating that species-specific NAAAs are more sensitive than cultivation to determine the true carriage state.[28] A more recent study investigated the carriage rate of the organism in New Zealand children aged 6 months to 4 years.[53] Whereas cultures on BAV medium were positive in only 4 out of 176 children (2.3%), a *K. kingae*–specific NAAA detected the organism in 48 children (22.9%), and was positive in all 4 culture-positive specimens, indicating an excellent combination of exquisite sensitivity and specificity. In a study comprising 123 patients aged 6–48 months with skeletal system complaints, the NAAA performed on the oropharyngeal specimen was positive in all 30 children in whom the joint exudate was positive for *K. kingae* by culture and/or by the molecular test.[29] However, 8 pharyngeal samples obtained from 84 children (9.5%) with culture-negative arthritis or osteomyelitis or in whom other organisms were detected, were also positive for *K. kingae*–specific DNA sequences, reflecting the high level of background colonization rate in the young pediatric population.[29] These observations imply that:

- The sensitive species-specific NAAAs have a high negative predictive value, and failure to detect *K. kingae* genomic sequences in an oropharyngeal specimen practically rules out the bacterium as the etiology of a skeletal system infection.
- Failure to isolate *K. kingae* from the oropharynx does not necessarily exclude the organism as the etiology of the disease.
- Because of the high respiratory carriage of *K. kingae* in the relevant children's population aged 6–48 months, and especially among attendees to out-of-home care facilities, the predictive value of a positive pharyngeal NAAA is suboptimal.

7. CONCLUSIONS

In summary, research conducted over the last two decades has firmly established the role of *K. kingae* as an important invasive pathogen of early childhood. Although the use of the BAV method has significantly improved the detection of the organism in synovial fluid exudates, recent data clearly demonstrate that the culture recovery of *K. kingae* remains unsatisfactory. Nucleic acid amplification assays that amplify the 16S rRNA gene sequence double the detection rate of *K. kingae*, and those that target *K. kingae*-specific DNA sequences result in better than fourfold improvement, enabling the bacteriological diagnosis even in antibiotic treated patients. Use of molecular tests reduces the fraction of culture-negative infections, shortens the time to detection, and confirms that *K. kingae* is the leading bacterial etiology of suppurative arthritis in young children.

Although the sensitivity of *K. kingae* cultures is suboptimal compared to that of NAAAs, isolation of the organism has the obvious advantages of enabling antibiotic susceptibility testing and genotyping of the isolate, two important considerations for the patient's clinical management and the study of the epidemiology of the person-to-person transmission of the organism and virulence factors of the causative strains. Therefore, cultures and NAAAs complement each other, and a comprehensive laboratory approach combining both methods appears to offer an optimal answer to both clinical and epidemiological needs (Table 2). The current approach to the microbiological investigation of pediatric septic arthritis, integrating culture and NAAA methods, with special emphasis on *K. kingae* infections, is summarized in Fig. 2.

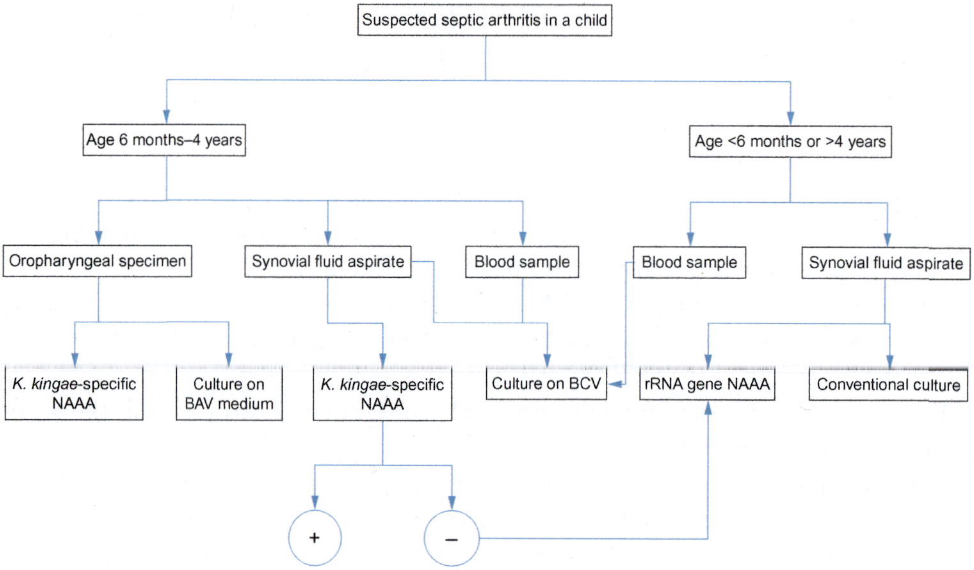

Fig. 2 Integrative approach for the bacteriological diagnosis of pediatric septic arthritis.

REFERENCES

1. Welkon CJ, Long SS, Fisher MC, Alburger PD. Pyogenic arthritis in infants and children: a review of 95 cases. *Pediatr Infect Dis* 1986;**5**(6):669–76.
2. Weintraub A. Immunology of bacterial polysaccharide antigens. *Carbohydr Res* 2003;**338**(23):2539–47.
3. Rotbart HA, Glode MP. *Haemophilus influenzae* septic arthritis in children: report of 23 cases. *Pediatrics* 1985;**75**(2):254–9.
4. Lyon R, Evanich JD. Culture-negative septic arthritis in children. *J Pediatr Orthop* 1999;**19**(5):655–9.
5. Yagupsky P, Dagan R, Howard CW, Einhorn M, Kassis I, Simu A. High prevalence of *Kingella kingae* in joint fluid from children with septic arthritis revealed by the BACTEC blood culture system. *J Clin Microbiol* 1992;**30**(5):1278–81.
6. Slinger R, Moldovan I, Bowes J, Chan F. Polymerase chain reaction detection of *Kingella kingae* in children with culture-negative septic arthritis in eastern Ontario. *Pediatr Child Health* 2016;**21**(2):79–82.
7. Henriksen SD, Bøvre K. Transfer of *Moraxella kingii* Henriksen and Bøvre to the genus *Kingella* gen. nov.in the family Neisseriaceae. *J Syst Bacteriol* 1976;**6**(4):447–50.
8. Yagupsky P. *Kingella kingae*: carriage, transmission, and disease. *Clin Microbiol Rev* 2015;**28**(1):54–79.
9. Amit U, Flaishmakher S, Dagan R, Porat N, Yagupsky P. Age-dependent carriage of *Kingella kingae* in young children and turn-over of colonizing strains. *J Pediatr Infect Dis Soc* 2014;**3**(2):160–2.
10. El Houmami N, Mirand A, Dubourg G, Hung D, Minodier P, Jouve JL, et al. Hand, foot and mouth disease and *Kingella kingae* infections. *Pediatr Infect Dis J* 2015;**34**(5):547–8.
11. Kehl-Fie TE, St Geme 3rd JW. Identification and characterization of an RTX toxin in the emerging pathogen *Kingella kingae*. *J Bacteriol* 2007;**189**(2):430–6.
12. Starr KF, Porsch EA, Heiss C, Black I, Azadi P, St Geme 3rd JW. Characterization of the *Kingella kingae* polysaccharide capsule and exopolysaccharide. *PLoS One* 2013;**8**(9):e75409.
13. Amit U, Porat N, Basmaci R, Bidet P, Bonacorsi S, Dagan R, et al. Genotyping of invasive *Kingella kingae* isolates reveals predominant clones and association with specific clinical syndromes. *Clin Infect Dis* 2012;**55**(8):1074–9.

14. Dubnov-Raz G, Ephros M, Garty BZ, Schlesinger Y, Maayan-Metzger A, Hasson J, et al. Invasive pediatric *Kingellakingae* infections: a nationwide collaborative study. *Pediatr Infect Dis J* 2010;**29**(7):639–43.
15. Yagupsky P, Press J. Unsuspected *Kingellakingae* infections in afebrile children with mild skeletal symptoms: the importance of blood cultures. *Eur J Pediatr* 2004;**163**(9):563–4.
16. Ceroni D, Cherkaoui A, Combescure C, François P, Kaelin A, Schrenzel J. Differentiating osteoarticular infections caused by *Kingella kingae* from those due to typical pathogens in young children. *Pediatr Infect Dis J* 2011;**30**(10):906–9.
17. Basmaci R, Lorrot M, Bidet P, Doit C, Vitoux C, Penneçot G, et al. Comparison of clinical and biologic features of *Kingella kingae* and *Staphylococcus aureus* arthritis at initial evaluation. *Pediatr Infect Dis J* 2011;**30** (10):902–4.
18. Basmaci R, Ilharreborde B, Lorrot M, Bidet P, Bingen E, Bonacorsi S. Predictive score to discriminate *Kingella kingae* from *Staphylococcus aureus* arthritis in France. *Pediatr Infect Dis J* 2011;**30**(12):1121–2.
19. Yagupsky P, Press J. Use of the isolator 1.5 microbial tube for culture of synovial fluid from patients with septic arthritis. *J Clin Microbiol* 1997;**35**(9):2410–2.
20. Yagupsky P, Dubnov-Raz G, Gené A, Ephros M. Israeli-Spanish *Kingella kingae* Research Group. Differentiating *Kingella kingae* septic arthritis of the hip from transient synovitis in young children. *J Pediatr* 2014;**165**(5):985–9.
21. Trujillo M, Nelson JD. Suppurative and reactive arthritis in children. *Semin Pediatr Infect Dis* 1997;**8** (4):242–9.
22. Goldenberg DL, Reed JI. Bacterial arthritis. *N Eng J Med* 1985;**312**(12):764–71.
23. Yagupsky P, Bar-Ziv Y, Howard CB, Dagan R. Epidemiology, etiology, and clinical features of septic arthritis in children younger than 24 months. *Arch Pediatr Adolesc Med* 1995;**149**(5):537–40.
24. Moumile K, Merckx J, Glorion C, Pouliquen JC, Berche P, Ferroni A. Bacterial aetiology of acute osteoarticular infections in children. *Acta Paediatr* 2005;**94**(4):419–22.
25. Gruber BF, Miller BS, Onnen J, Welling R, Wojtys EM. Antibacterial properties of synovial fluid in the knee. *J Knee Surg* 2008;**21**(3):180–5.
26. Baron EJ, Scott JD, Thompkins LS. Prolonged incubation and extensive subculturing do not increase recovery of clinically significant microorganisms from standard automated blood cultures. *Clin Infect Dis* 2005;**41**(11):1677–80.
27. Basmaci R, Bidet P, Jost C, Yagupsky P, Bonacorsi S. Penicillinase-encoding gene bla_{TEM-1} may be plasmid borne or chromosomally located in *Kingella kingae* species. *Antimicrob Agents Chemother* 2015;**59**(2):1377–8.
28. Basmaci R, Ilharreborde B, Bidet P, Doit C, Lorrot M, Mazda K, et al. Isolation of *Kingella kingae* in the oropharynx during *K. kingae* arthritis on children. *Clin Microbiol Infect* 2012;**18**(5):e134–6.
29. Ceroni D, Dubois-Ferrière V, Cherkaoui A, Gesuele R, Combescure C, Lamah L, et al. Detection of *Kingella kingae* osteoarticular infections in children by oropharyngeal swab PCR. *Pediatrics* 2013;**131**(1): e230–5.
30. Yagupsky P, Porat N, Pinco E. Pharyngeal colonization by *Kingella kingae* in children with invasive disease. *Pediatr Infect Dis J* 2009;**28**(2):155–7.
31. Yagupsky P, Dagan R, Prajgrod F, Merires M. Respiratory carriage of *Kingella kingae* among healthy children. *Pediatr Infect Dis J* 1995;**14**(8):673–8.
32. Amit U, Dagan R, Yagupsky P. Prevalence of pharyngeal carriage of *Kingella kingae* in young children and risk factors for colonization. *Pediatr Infect Dis J* 2013;**32**(2):191–3.
33. Fenollar F, Lévy PY, Raoult D. Usefulness of broad-range PCR for the diagnosis of osteoarticular infections. *Curr Opin Rheumatol* 2008;**20**(4):463–70.
34. Ferroni A. Epidemiology and bacteriological diagnosis of paediatric acute osteoarticular infections. *Arch Paediatr* 2007;**14**(Suppl. 2):S91–6.
35. Rosey AL, Albachin E, Quesnes G, Cadilhac C, Pejin Z, Glorion C, et al. Development of a broad-range 16S rDNA real-time PCR for the diagnosis of septic arthritis in children. *J Microbiol Methods* 2007;**68**(1):88–93.
36. Moumile K, Merckx J, Glorion C, Berche P, Ferroni A. Osteoarticular infections caused by *Kingella kingae* in children; contribution of polymerase chain reaction to the microbiologic diagnosis. *Pediatr Infect Dis J* 2003;**22**(9):837–9.

37. Stähelin J, Goledenberger D, Gnehm HE, Altwegg M. Polymerase chain reaction diagnosis of *Kingella kingae* arthritis in a young child. *Clin Infect Dis* 1998;**27**(5):1328–9.
38. Verdier I, Gayet-Ageron A, Ploton C, Taylor P, Benito Y, Freydiere AM, et al. Contribution of a broad range polymerase chain reaction to the diagnosis of osteoarticular infections caused by *Kingella kingae*: description of twenty-four recent pediatric diagnoses. *Pediatr Infect Dis J* 2005;**24**(8):692–6.
39. Cherkaoui A, Ceroni D, Emonet S, Lefevre Y, Schrenzel J. Molecular diagnosis of *Kingella kingae* osteoarticular infections by specific real-time PCR assay. *J Med Microbiol* 2009;**58**(Pt1):65–8.
40. Ceroni D, Cherkaoui A, Ferey S, Kaelin A, Schrenzel J. *Kingella kingae* osteoarticular infections in young children: clinical features and contribution of a new specific real-time PCR assay to the diagnosis. *J Pediatr Orthop* 2010;**30**(3):301–4.
41. Lehours P, Freydière AM, Richer O, Burucoa C, Boisset S, Lanotte F, et al. The rtxA toxin gene of *Kingella kingae*: a pertinent target for molecular diagnosis of osteoarticular infections. *J Clin Microbiol* 2011;**49**(4):1245–50.
42. Juretschko S, Beavers-May T, Hemphill C, Romero J, Stovall S. Superior detection of *Kingella kingae* and *Staphylococcus aureus* in paediatric osteoarticular infections using molecular analysis. In: Abstr 19th Eur Congr Clin Microbiol Infect Dis; 2009. abstr O-469.
43. Williams N, Cooper C, Cundy P. *Kingella kingae* septic arthritis in children: recognizing an elusive pathogen. *J Child Orthop* 2014;**8**(1):91–5.
44. Ilhaerreborde B, Bidet P, Lorrot M, Even J, Mariani-Kurkdjian P, Ligouri S, et al. New real-time PCR-based method for *Kingella kingae* DNA detection: application to samples collected from 89 children with acute arthritis. *J Clin Microbiol* 2009;**47**(6):1837–41.
45. Chometon S, Benito Y, Chaker M, Boisset S, Ploton C, Bérard J, et al. Specific real-time polymerase chain reaction places *Kingella kingae* as the most common cause of osteoarticular infections in young children. *Pediatr Infect Dis J* 2007;**26**(5):377–81.
46. Ferroni A, Al Khoury H, Dana C, Quesne G, Berche P, Glorion C, et al. Prospective survey of acute osteoarticular infections in a French paediatric orthopedic surgery unit. *Clin Microbiol Infect* 2013;**19** (9):822–8.
47. Levy PY, Fournier PE, Fenollar F, Raoult D. Systematic PCR detection in culture-negative osteoarticular infections. *Am J Med* 2013;**126**(12):1143.e25–33.
48. Luegmair M, Chaker M, Ploton C, Berard J. *Kingella kingae* osteoarticular infections of the sternum in children: a report of six cases. *J Child Orthop* 2008;**2**(6):443–7.
49. Haldar M, Butler M, Quinn CD, Stratton CW, Tang YW, Burnham CA. Evaluation of a real-time PCR assay for simultaneous detection of *Kingella kingae* and *Staphylococcus aureus* from synovial fluid in suspected septic arthritis. *Ann Lab Med* 2014;**34**(4):313–6.
50. Filleron A, Michon AL, Jumas-Bilak E, Jeziorski E, Zorngiotti I, Tran TA, et al. What to expect from molecular tools for non-documented pediatric infectious diseases. *Expert Rev Mol Diagn* 2015;**15** (12):1645–56.
51. Grivea IN, Michoula AN, Basmaci R, Dailiana ZH, Tsimitselis G, Bonacorsi S, et al. *Kingella kingae* sequence type-complex 14 arthritis in a 16-month child in Greece. *Pediatr Infect Dis J* 2014;**34**(1):107–8.
52. Ceroni D, Belaieff W, Kanavaki A, Anderson Della Llana R, Lascombes P, Dubois-Ferrière V, et al. Possible association of *Kingella kingae* with infantile spondylodiscitis. *Pediatr Infect Dis J* 2013;**32**:1296–8.
53. Olijve L, Podmore R, Anderson T, Walls T. High rate of oropharyngeal Kingella kingae carriage in New Zealand children. *J Paediatr Child Health* 2016; http://dx.doi.org/10.1111/jpc.13287 [Epub ahead of print].

SECTION III

Alternative and Novel Approaches to the Diagnosis and Treatment of Skin, Soft Tissue, Bone, and Joint Infections

CHAPTER 11

Nanotechnology for the Treatment of Fungal Infections on Human Skin

M. Rai*, A.P. Ingle*, R. Pandit*, P. Paralikar*, I. Gupta*, N. Anasane*, M. Dolenc-Voljč†
*SGB Amravati University, Amravati, India
†University of Ljubljana, Ljubljana, Slovenia

1. INTRODUCTION

Currently, skin diseases are one of the most common among all illnesses in humans and represent a huge public health problem on a global scale.[1,2] They affect 20%–25% of the world's population and are causative agents of infections that vary from country to country.[3] Superficial mycoses involve infections of skin, nails, and hairs which are caused by different groups of fungi.[4] In tropical countries, fungal infections are more common and often reoccur due to favorable climatic conditions. There is a wide range of fungal infections, from superficial, affecting skin, skin appendages and mucous membranes, to systemic infections with invasion of internal organs. Superficial fungal infections are not life threatening, but they spread to other skin regions and may become widespread, be transmitted to other people, and can cause secondary bacterial skin infections and permanent hair loss, negatively affecting the quality of a person's life. They are also disturbing from an aesthetical point of view.[5] The pathogenesis of infection depends on the causative agent, environmental conditions, and immunological status of the host.[6] Mostly, these diseases are spread from person to person. Infective spores can be present on footwear, combs, brushes, towels, and the floor in swimming pools and other sports facilities. Outbreaks of disease can also occur via direct contact with infected animals. Cutaneous mycoses are classified on the basis of the causative fungal agents into dermatophytosis, yeast infections (including *Pityriasis versicolor*, candidiasis and some other yeast infections) and mold infections.[7]

The most common places where skin infection can occur are the skin of the scalp, feet, fingers, mouth, and vagina. Warm and wet places are most favorable for fungal growth. These organisms thrive on skin, especially between toe, armpit, and groin. Tinea and candida are most common in the skin of the toes. Different names are given to tinea infections, depending on the part of the skin where it is affected. *Candida* also causes skin infection on different parts of the body. Fungal skin infection can pass from person to person through

The Microbiology of Skin, Soft Tissue, Bone and Joint Infections
ISSN 2451-9006, http://dx.doi.org/10.1016/B978-0-12-811079-9.00019-7

bathroom floor, towel, clothing, etc. For example, athlete's foot is a fungal infection between toes; if the skin infection is untreated it can spread to the foot. White patches, cracked skin, and red colored marks on the foot are generally observed in the patients. Most common symptoms observed are burning, itching, and unpleasant odor from toes.

"Jock itch" is another skin infection, which occurs mostly in males and affecting the groin and inner thigh region. Red colored patches and swelling are the most common symptoms observed in infected patients.

Ringworm is a term used for infections caused by dermatophytes. Patches of ringworm are red, circular, and itchy. *Candida* causes thrush in mouth and vagina. The common symptoms of vaginal thrush include itching, irritation, and burning sensation. Oral thrush mainly creates white patches on the tongue and is most commonly reported in children.

The present chapter is focused on different nanoparticles and their formulations used for the treatment of various skin infections caused by fungi in humans.

2. DIFFERENT TYPES OF SKIN INFECTIONS

2.1 Dermatophytosis

Dermatophytes are the causative agents of dermatophytosis (tinea). They are the group of parasitizing filamentous fungi that are able to infect the keratinized tissues such as the stratum corneum of the epidermis, nails and hairs.[8–10] Dermatophytes induce a dermal inflammatory response which leads to erythema and scaling, which cause intense itching.[7] The pathogenesis of tinea include many steps, beginning with the contact of infective spores with the skin, adherence to the superficial cells, invasion of keratin layers by the secretion of the keratinases and induction of inflammation.[6] Keratinases degrade the hard keratin into components of low molecular weight which can be utilized by dermatophytes.[11] Dermatophytes are subcategorized into different groups. On the basis of their microscopic characteristics they are classified into *Trichophyton* species, *Epidermophyton* species, and *Microsporum* species. According to their normal habitats, they are classified as anthropophilic species, causing infections only in humans, zoophilic species, primarily pathogenic for different animals, and geophilic species, dwelling in the soil.

They are differentiated by the area of infection. "Tinea" is derived from the Latin, which stands for worm, because formerly it was considered to be an infection caused by worm-like parasites. Different types of tinea infections are presented in Table 1.[12,13]

Dermatophytes are the most common fungi causing superficial skin infections; a frequently isolated pathogen worldwide is *Trichophyton (T.) rubrum*.[14–16]

2.2 Pityriasis Versicolor

Nondermatophytic fungi are currently emerging as important pathogens of superficial mycoses[5] and pityriasis versicolor (also called tinea versicolor) is the next most common fungal disease after dermatophytosis. This disease is caused by the yeast *Malassezia* species,

Table 1 Types of tinea infection and body parts infected[12,13]

Tinea infection	Body parts infected
Tinea faciei	Face
Tinea manuum	Hands (palms and flexoral parts of fingers)
Tinea corporis	Trunk and extremities
Tinea inguinalis	Inguinal fold or groin (genitals)
Tinea pedis	Soles
Tinea capitis	Scalp
Tinea colli	Neck
Tinea unguium	Nails

most commonly by *Malassezia globosa. Malassezia* yeasts are skin commensals that can also be involved in the pathogenesis of *Malassezia* folliculitis, dandruff and seborrhoeic dermatitis. They may cause aggravation of atopic dermatitis and psoriasis on the head and neck. They are lipophilic yeasts that have their natural habitat on the seborrheic parts of the skin. The addition of oil facilitates the growth of these organisms in vitro.[17]

2.3 Candidiasis

Candida species is a group of yeasts that are non-inherently pathogenic and opportunistic. When the host's cellular immune defense is suppressed, they may become pathogenic. Predisposing factors for candidiasis are diabetes, immunosuppression, damage of the stratum corneum, wet conditions, occlusion, obesity, Cushing's syndrome, and some other hormonal diseases. Infants, pregnant women, and elderly are more prone to this disease. Candidiasis usually involves skin folds in axillae, inguinal folds, genital and perigenital areas, skin beneath the breasts in women, hands, fingernails, and mucous membranes. *Candida (C.) albicans* is the most common etiological agent. Some other non-*Candida albicans* yeasts, such as *C. tropicalis*, *C. glabrata*, *C. parapsilosis*, *C. krusei*, and *C. Guilliermondii*, can also cause candidiasis.[7]

2.4 Mold Infections

Superficial mycoses due to molds are less common. Molds are also known as non-dermatophytic filamentous opportunistic fungi that are ubiquitous in the environment. These fungi are not commonly pathogenic in healthy hosts, but are sometimes isolated from skin and nails specimens, most commonly from toenails and feet. They can cause similar clinical characteristics to tinea pedis, tinea manuum, and tinea unguium. Mold infections are more common in tropical climes. *Scopulariopsis brevicaulis*, *Aspergillus* sp., *Acremonium* sp., *Fusarium* sp., *Hendersonula toluroidea*, and *Scytalidium hyalinum* can cause infections in humans. *Aspergillus niger* has been reported as the most frequently occurring species in this group in some geographic regions.[7]

3. DIAGNOSIS OF FUNGAL SKIN INFECTIONS

In clinical practices, diagnosis is often made on the basis of clinical picture. Direct microscopy with KOH examination and cultivation on Sabouraud's agar are used for confirmation of the diagnosis.[18] The molecular markers enable exact identification of the causal dermatophytes. For the identification of the strain from the isolates, different types of PCR techniques can be used which are rapid, simple, and reliable, such as multiplex PCR,[19] ITS based PCR-DGGE[20] and broad-range fungal PCR assays (Taxon-directed PCR assays).[21]

4. TREATMENT

Fungal skin infections are treated either topical or systemical. The way of treatment depends on the severity of the infection, its localization, depth of the skin involved, stage of infection, the causative pathogen, contagiousness, age of the patient, and associated comorbidities. In the great majority of infections, treatment is topical.

Pityriasis versicolor is usually treated with topical azole antifungal drugs in creams, solutions, sprays and shampoos.[22]

Tinea and cutaneous candidiasis are treated with topical antifungal drugs in creams, gels, and solutions. Regular application of topical antifungal drugs for at least a few weeks is needed for cure of the infection.[23] In the case of widespread lesions, deep infections of the scalp and in immunosuppressed patients, systemic antifungal drugs in combination with topical antifungal drugs are needed. In extensive plantar and palmar hyperkeratosis, systemic treatment is preferred as well. Recurrences or reinfections after successful treatment are common. Lack of clinical response may be due to many host and drug-related factors.

A new problem arising with superficial mycoses in general and in yeast infections in particular is the growing resistance to antifungal drugs. Development of resistance is an evolutionary process which is a function of rate of mutation to resistance and the size of the surviving population. It is initiated due to irregular use of the treatment taken by the patient and reinfection with different infectious strains. This resistance is observed due to point mutations occurring in transcriptional regulators of these effector genes or directly in the gene coding for antifungal targets.[24] This may cause failure of conventional antifungal drugs against yeasts and dermatophytes and therefore difficulties in treatment.[25,26]

5. THE NEED OF NANOTECHNOLOGY IN SKIN INFECTIONS

A number of antibiotic drugs have been successfully used in the treatment of a wide range of dermatological diseases, such as acne, rosacea, folliculitis, inflammatory skin conditions and deeper skin infections, such as cellulitis, carbuncles, and furuncles.[27] Resistance to antifungal agents was very rare until the late 1990s. However, the occurrence of fungal yeast infections, including resistant infections, has increased during the last few years. The most common reasons include inadequate, irregular, uncontrolled, and overuse of drugs or increased incidence of immunodeficiency conditions.[27,28]

Recently, multiple drug resistance (MDR) in pathogens to antifungal agents has become a most important worldwide problem, leading to significant negative social, psychological and occupational health effects and worsened quality of life.[28,29]

It is well known that *T. rubrum*, *T. mentagrophytes*, *Microsporum (M) canis*, and *T. tonsurans* are the most common dermatophytes that cause various skin infections.[30] A number of topical and systemic antifungal drugs have been discovered for the control of dermatophytic infections depending upon the specific site of action and target molecules. Antifungal agents are classified into different groups, such as allylamines, azoles (imidazoles and triazoles), morpholines, polyenes, thiocarbamate, and griseofulvin. The details about the site of action and target molecules for different antifungal drugs in dermatophytes are mentioned in Table 2.[28]

The *Candida* species responsible for candidiasis was reported to show resistance to various azole antifungal agents. However, the rate of azole resistance varies from species to species. *C. albicans* showed 1.0%–2.1%, *C. parapsilosis* 0.4%–4.2%, and *C. tropicalis* showed 1.4%–6.6% of resistance.[31,32] Resistance against antifungal drugs used nowadays has been found to be rare among dermatophytes.[33] However, this may change in the future. Terbinafine-resistant *T. rubrum* strains have been reported.[34,35] Recently, Dogra and Uprety[36] reported the threat of chronic and recurrent dermatophytosis in India

Table 2 Antifungal drugs, their site of action and specific target molecules in dermatophytes[28]

Chemical class and antifungal drug	Site of action	Target molecules
Allylamines Terbinafine, naftifine	Ergosterol biosynthesis	Squalene epoxidase
Azoles *Imidazoles* Miconazole, econazole, bifonazole, clotrimazole, ketoconazole *Triazoles* Fluconazole, itraconazole, terconazole, posaconazole	Ergosterol biosynthesis	Cytochrome P450 14α-Lanosterol demethylase
Morpholines Amorolfine	Ergosterol biosynthesis	Sterol reductase and isomerase
Polyenes Amphotericin B, nystatin	Ergosterol biosynthesis	Membrane barrier function
Thiocarbamate Tolnaftate	Ergosterol biosynthesis	Squalene epoxidase
Griseofulvin	Fungal mitotic apparatus	Sliding of microtubules

with special reference to emergence of resistance in *T. rubrum*. Drug resistance, developed to azoles by dermatophytes, is becoming a more prevalent clinical challenge around the globe. Apart from this, resistance to some other antifungal drugs, such as griseofulvin and terbinafine, was also reported.[37]

Moreover, the measures suggested by Ghannoum et al.[38] to avoid and suppress the resistance to antifungal agents include: (i) careful use of antifungals, (ii) appropriate dosing with special emphasis on avoiding treatment with low antifungal dosage, (iii) use of a combination of existing antifungal agents, (iv) treatment with the appropriate antifungals (in cases where the etiological agent is known), and (v) use of surveillance studies to determine the true frequency of antifungal resistance. Some of these strategies are difficult to implement in clinical practice.

Another most important problem associated with treatment of dermatophyte and yeast skin infections are the adverse effects of antifungal agents. Topical antifungal drugs can cause some local side effects in a minority of patients. Azole antifungal drugs like miconazole showed increased burning, itching or irritation of the skin or vagina, and foul-smelling vaginal discharge. Similarly, econazole can cause burning, itching, stinging, and redness, while ketoconazole exhibited adverse effects such as itching, stinging, dry skin, and dry or oily scalp.[39] Nystatin is a member of the polyene group, which exhibits common side effects such as burning, itching, and a rash when applied to the skin. It was also reported to cause eye irritation, redness, and generalized pustular eruption.[40] However, tolnaftate from the thiocarbamate group showed only skin irritation.[41] Terconazole in cream form can sometimes lead to burning or irritation of skin.[42]

Systemic antifungal drugs, used in dermatology, can cause some severe systemic side effects. Antifungal agents were reported to have various adverse effects. Among triazoles, fluconazole, and itraconazole can cause headache, dizziness, diarrhea, stomach pain, heartburn, and changes in the ability to taste food. More severe side effects include excessive tiredness, loss of appetite, upset stomach, vomiting, tingling or numbness in the extremities, fever, chills, medicamentous exanthema, urticaria and angioedema and difficulty in swallowing. Terbinafine, a member of the allylamines, was reported to have very similar adverse effects to those of triazoles. Griseofulvin can cause headache and aplastic anemia.[43] Systemic antifungal drugs are contraindicated in pregnancy and lactation and have to be used cautiously in severe hepatic and renal insufficiency patients and in children.

Overall, it has been observed that patients treated with conventional antifungal drugs, mentioned above, may fail to respond due to drug resistance, leading to spread of the infection, negative social, psychological and occupational health effects and impairment of patient's quality of life. In addition, in some cases antifungal drugs may cause severe adverse effects. Hence, to avoid these unfavorable circumstances, early recognition of infection and effective treatment is needed to reduce morbidity and spreading of infection. Therefore, a new generation of potent broad spectrum antifungals with specific target action against causative agents, with negligible or no side effects in the host would be needed to treat dermatophyte infections in the future.[35] In this context, nanotechnology can help to develop beneficial solution for various skin infections and can provide a new approach to fight drug-resistant fungal infections. Use of nanomaterials with unique biological properties can lead to the formulations of effective antifungal agents for skin infections. The role of nanotechnology and nanomaterials in the management and treatment of skin infections will be discussed here.

5.1 Application of Nanoparticles in Skin Infection

Nanoparticles are nanomaterials measuring <100 nm in one dimension. They have small size and exhibit high surface-to-volume ratio. Due to this special attribute, nanomaterials such as biopolymeric nanoparticles, metal nanoparticles, and inorganic nanoparticles are known for promising antimicrobial activity against fungi causing skin infections.[44] Metal nanoparticles and metal oxide nanoparticles such as silver, gold, and oxide of nanoparticles like zinc, titanium, copper, and iron show a wide range of antifungal activity against dermatophyte infections. Silver compounds have been known for their antimicrobial activity since ancient times. The smaller the size of the nanoparticle, the stronger is the antimicrobial activity. Silver nanoparticles have been emerging as promising antimicrobial agents, which can help in the treatment of fungal infections.[44] Researchers studied the antifungal activity of silver nanoparticles and stated the mechanism for silver nanoparticles (AgNPs) to penetrate the cell membrane of fungi. AgNPs attached to the sulfur-containing proteins of the cell membrane by disrupting the cell membrane.[45] Researchers found that pit formation occurs in the cell membrane of *C. albicans* and formation of pores leads to cell death. AgNPs produced reactive oxygen species (ROS) and hydroxyl radical formation, disrupting mitochondria and forming the apoptotic body, leading to the cell death of fungi.[46–48]

AgNPs are known for their antifungal activity against fungi causing skin infections such as *T. mentagrophytes*, *M. canis*, *T. rubrum*, *C. albicans*, *M. furfur*, *C. tropicalis*, and *C. glabrata*.[45,49–52] AgNPs are used in various cosmetics which helps in prevention of skin infections. There are reports concerning antifungal activity of silver nanoparticles against dermatophytic fungi causing skin infections, such as *E. floccosum*, *M. audouinii*, *M. canis*, *M. gypseum*, *M. nanum*, *T. rubrum*, *T. tonsurans*, *T. violaceum*, *M. furfur*, and *Candida* species.[53,54] Antifungal activity of silver nanoparticles coated with poly vinyl pyrrolidine (PVP) was demonstrated against dermatophytes causing athlete's foot and ringworm infections such as *T. rubrum* and *T. mentagrophytes*.[55] Recently, it was reported that synthesized gold nanoparticles showed activity against dermatophytes, causing athlete's foot. AuNPs have demonstrated antifungal activity against *T. rubrum* and *C. albicans*.[56]

Chitosan is a biopolymer obtained from chitin. Chitosan nanoparticles were synthesized as nanosize chitosan that have smaller size and high surface-to-volume ratio, so their antimicrobial activity is better as compared to chitosan alone.[57] Researchers revealed that chitosan nanoparticles disrupt the cell membrane, may bind with DNA, inhibit mRNA and protein synthesis. Chitosan nanoparticles have superior antimicrobial activity against *Candida* causing skin infections.[57–59] It was reported that a composite of chitosan and silver nanoparticles has excellent antifungal activity against *C. albicans*.[60]

Zinc oxide nanoparticles are known for antifungal activity against dermatophyte infection caused by *M. canis*, *T. verrucosum*, *T. mentagrophyes*.[61,62] Titanium oxide nanoparticles are also known for antifungal activity against skin infections caused by *M. furfur*.[61] The synthesis of sulfur nanoparticles and their antifungal activity against dandruff caused by *Malassezia* yeast was reported.[63]

Paralikar[64] fabricated ketoconazole nanoparticles and demonstrated that the synthesized drug nanoparticles have antifungal activity against *M. furfur*, which causes various skin infections such as pityriasis versicolor, folliculitis, seborrhoeic dermatitis, and dandruff. Ketoconazole nanoparticles can be used in the prevention of fungal skin infections.[64] A ketoconazole-coated silver nanoparticle hybrid was synthesized for the treatment of fungal skin infections. Ketoconazole can be used in the treatment of fungal infections but, due to the drug resistance problem, drug-coated nanoparticles were synthesized as a solution to the problem. Copper oxide is one of the transition elements due to its unique properties. Copper oxide nanoparticles have shown excellent dermatophytic activity against *T. rubrum* and *T. mentagrophytes*.[65] An overview of the use of different nanoparticles against dermatophyte and yeast infections is presented in Table 3.[59–66]

6. ANTIFUNGAL NANOGEL AND FORMULATIONS

Drug-containing formulations are used to treat skin disorders and can be either delivered topically or systematically in skin. In topical drug delivery, drug formulations are applied directly onto the skin, while in systematic delivery, the drug is delivered to the skin via systemic circulation, as in the treatment of dermatological disorders such as acne, psoriasis and eczema.[67,68] Currently, nanoparticles have been used in photoprotector products (titanium oxide and zinc oxide), in cosmetics, antisepsis, phototherapy, treatment of inflammatory diseases and for treatment of sebaceous gland and scalp diseases.[69] Nanoparticulate systems are most widely used for treatment of skin diseases. Nanoparticles have been reported for their antimicrobial properties. Nanoparticle-based formulations such as gels, creams, sunscreens, and ointments are applied on the skin to treat infections. In the field of dermatology, many antibiotics and antifungicidal agents are used for treatment of skin infections but may cause side effects and development of resistance. Nanoparticle formulations have shown higher tolerability, improved safety and excellent effects[69] and can offer an efficient solution to treat difficult disorders.

Table 3 Application of various nanoparticles against fungal infections of skin[45,52–65]

Nanomaterials used	Name of fungi causing skin infections	Skin infections caused by fungus	References
Silver nanoparticles	*Trichophyton rubrum*, *Candida albicans*	Athlete's foot, rashes, swelling, skin candidiasis	45,52,53
Silver nanoparticles	*Malassezia furfur*, *Trichophyton rubrum*, *Candida albicans*, *Candida tropicalis*, *Candida glabrata*	Superficial mycosis, skin candidiasis	54
PVP coated silver nanoparticles	*Trichophyton rubrum* and *Trichophyton mentagrophytes*	Ringworm disease, Athlete's foot	55
Gold nanoparticles	*Trichophyton rubrum*, *Candida albicans*	Athlete's foot, rashes, swelling, skin candidiasis	56
Chitosan nanoparticles	*Candida albicans*	Skin candidiasis	57–59
Chitosan-silver nanoparticles composite	*Candida albicans*, *Candida* spp.	Skin candidiasis	60
Titanium oxide nanoparticles	*Malassezia furfur*	Pityriasis versicolor, folliculitis, seborrhoeic dermatitis, and dandruff	61
Zinc oxide nanoparticles	*Microsporum canis*, *Trichophyton verrucosum*, *Trichophyton mentagrophytes*	Ringworm disease, Athlete's foot	61,62
Sulfur nanoparticles	*Malassezia furfur*	Pityriasis versicolor, folliculitis, seborrhoeic dermatitis, and dandruff	63
Ketoconazole nanoparticles	*Malassezia furfur*	Pityriasis versicolor, folliculitis, seborrhoeic dermatitis, and dandruff	64
Ketoconazole coated silver nanoparticles	*Malassezia furfur*	Pityriasis versicolor, folliculitis, seborrhoeic dermatitis, and dandruff	66
Copper oxide nanoparticles	*Trichophyton rubrum* and *Trichophyton mentagrophytes*	Ringworm, superficial mycosis	65

AgNPs are found to be more effective against various dermatological infections. Silver-loaded silica nanoparticles have a potential to inhibit the growth of microorganisms.[70] This patented study reported that a nanogel formulation of silver metal-based silica nanoparticles showed bactericidal activity against infectious pathogens. AgNPs containing soap also possess microbicidal properties which are found to be useful in treatment of sun-damaged skin and acne.[71] A gel formulation containing AgNPs was found to be effective against multidrug resistant organisms.[72] Toxicity studies on gel containing

AgNPs showed complete safety for topical application in Sprague-Dawley rats. These results suggest that AgNPs could provide a safer alternative to conventional antimicrobial agents in the form of a topical antimicrobial formulation. Nanosilver, used in wound dressing, is already on the market.[73] The antiinflammatory activity of topical nanocrystalline silver cream was evaluated in a guinea pig model for treatment of allergic dermatitis.[74] This study suggests that nanocrystalline silver cream may have therapeutic potential for treatment of inflammatory skin diseases.

Like AgNPs, titanium oxide (TiO_2) was also reported for its wide use in treatment of acne vulgaris, condylomatacuminata, atopic dermatitis, and hyperpigmented skin lesions.[75] A nanoparticle-based nail polish formulation has an advantage in treatment of onychomycosis.[76] Tiwari et al.[62] reported that zinc oxide–based nanoformulation of nail polish showed significant antifungal activity against *T. mentagrophytes* and *M. canis* which are causative agents of onychomycosis. Liposomal nanoparticles in the form of hydrogel were found to be effective against wound infections. They decrease the microbial load and promote wound healing.[77] It was reported that sulfur nanoparticles possess strong antimicrobial activity, and the effect of sulfur nanoparticles on *M. furfur* which is involved in the pathogenesis of seborrheaic dermatitis and dandruff, was also reported.[63]

Mg-Al-citrate-LDH nanohybrid introduced into a body cream formulation containing cocoa-butter showed improved and prolonged activity against *C. albicans* and *C. glabrata.* The resulting body cream formulation showed improved activity and slow release characteristics against *C. albicans* and *C. glabrata.* These results indicate that it could be an alternative to topical antifungal drugs.[78]

Nanoformulations were found to be more permeable through skin and were retained in the skin more effectively than topical antifungal formulations due to less decreased interfacial tension. So, from all the preceding studies, it can be concluded that nanoformulations will be a good alternative to topical formulations for treatment of fungal infection on human skin.

7. NANOTOXICITY TO SKIN

As described in the earlier section, due to their smaller size, nanoparticles have many applications. Particularly in the case of cosmetics and sunscreens, ZnONPs and TiO_2NPs are mainly used.[69] These nanoparticles are intended to protect the skin from damage induced by UV radiation. However, these nano-based products are required to be applied on normal, healthy skin. Moreover, their effectiveness in preventing skin damage is tested under standard laboratory conditions, which mainly utilizes testing on normal, undamaged, healthy skin. But through direct and indirect exposure to the nanoparticles, there is a rise in incidences showing harmful effects to humans and the environment.[79] Much evidence suggests that the nanoparticles used for such purpose are harmful to the user's skin. In reality, it is likely that many users apply these products on damaged skin, as in the case of sunburned skin and dermatitis. Injured skin permits higher penetration and a greater chance of deep penetration of nanoparticles through various layers of skin, thereby leading to harmful effects to the consumers. Drug penetration can increase also in hair follicle openings.[69] Phototoxicity during exposure to UV radiation also has to be considered.[69] The nanoparticles that penetrate inside the skin may cause toxic effects to cells, organs or the entire body of the consumer. Therefore, it is of utmost importance to conduct in vitro and in vivo studies to check the probable harmful effects of nanoparticles.

ZnONPs have been most abundantly used in skin care products and their use for this purpose is rising. But such widespread use of ZnONPs has invited concerns for safety issues. Many studies have reported ZnONPs induced cytotoxic and genotoxic effects on human keratinocytes and fibroblasts.[80–84] The increased oxidative stress and decreased antioxidants play a major role in damaging DNA[80] and inducing other mechanisms resulting in cellular death. It has also been indicated that ZnONPs used in sunscreens induce ROS production.[69,85] The accumulation of ROS inside the cell has a damaging effect, with many targeting the enzymatic system of the cell.

In sunscreen applications, ZnONPs of bigger size do not cross even the stratum corneum of human skin.[86–89] Through a 5 day exposure study performed on human vol unteers with ZnONPs (9 nm) and their bulk counterpart, it has been found that ZnONP exposed persons showed a higher amount of Zn ion in blood and urine as compared to volunteers exposed to the bulk ZnO.[90] This study indicates that small size ZnONPs are being dissolved in the blood, thereby releasing its ion. These ions probably play a part in the harmful effects.

In two different studies involving ZnONPs exposed to sunburned pig skin and mouse skin, it has been found that ZnONPs were able to cross one or two layers of skin whereas, in the case of hairless mice, it penetrated deep inside the stratum corneum.[90] Such a difference in penetration might be due to the thickness of skin. A decrease in collagen content has also been suggested to be an adverse effect of ZnONPs entry inside the skin.[91] Mouse skin is thinner than the pig skin. In humans, in some regions of the body the skin is thin. Therefore, such a body part is at higher risk to the toxic effects of ZnONPs.

TiO_2NPs are also an important component of sunscreen for protecting skin against UV radiations. Similar to ZnONPs, TiO_2NPs also showed cytotoxic effects. By referring to various studies, recently TiO_2NPs have been categorized as a carcinogen by International Agency for Research on Cancer (IARC). The data presented in various studies have shown that TiO_2 dust is the root cause of respiratory tract cancer in rats (Agents Classified by the IARC Monographs). The major reason for this might be the genotoxic activity of TiO_2NPs.[92] Photo-induced chromosomal aberration by TiO_2NPs in the CHO cell line has been reported.[90–92] Under the 10 J cm^{-2} UVA-1 (365 nm) in combination with 4–16 μg mL^{-1} ofanatase particles showed damage to DNA. Likewise,

AgNPs were also reported to show toxic effects on keratinocytes. In this type of nanoparticle surface coating plays an important role in the harmful effect. It was investigated in the exposure of HaCaTcell, a human keratinocytes cell line, to citrate coated 30 nm AgNPs (Cit30 AgNPs), polyethylene glycol coated AgNPs (PEG30 AgNPs). In response they found that both of these types of AgNPs induced apoptotic death, thereby reducing the viable cell count. Particularly Cit30 AgNPs were reported to be more cytotoxic than PEG30 AgNPs.[93]

From this discussion it is evident that there still is a need to know the more accurate mechanism of toxic response of these nanoparticles. Because the aggregation of nanoparticles in skin can also disturb its normal functioning, this can lead to cell damage and diseases. It is also important to focus studies on developing a suitable carrier system, which can check the nanoparticle aggregation in the skin.

8. CONCLUSION

Growth in the field of nanotechnology offers a new approach in the treatment of fungal skin infections. Nanoparticle-based formulations possess a huge potential to combat infections due to their target specific action and increase in bioavailability. Even though the toxic effects of nanoparticles deserve to be carefully highlighted, it is of equal importance to avoid toxicological issues that could spoil the development of potentially beneficial nanotechnologies for human diseases. Together, such studies will help to develop highly biocompatible nano-based products without showing harmful effects or with negligible harmful effects.

ACKNOWLEDGMENTS

The authors are thankful to University Grants Commission, New Delhi for financial assistance under the Special Assistance Programme (SAP) DRS-I. R.P. gratefully acknowledges the Department of Science and Technology, New Delhi, India for providing the INSPIRE fellowship to pursue research.

REFERENCES

1. Dekate S, Padhye S, Gautam A. Identification and characterization of fungi causing superficial mycoses. *Int J Pharm Life Sci* 2011;**2**(5):782–6.
2. Hay RJ, Johns NE, Williams HC, Bolliger IW, Dellavalle RP, Margolis DJ, et al. The global burden of skin disease in 2010: an analysis of the prevalence and impact of skin conditions. *J Investig Dermatol* 2014;**134**(1):1527–34.
3. Gopichand WR, Babulal JU, Madhukar SR. Mycological profile of superficial mycoses in north Maharashtra, India. *Int J Health Sci Res* 2013;**3**(10):90–4.
4. Murtaza M, Rajainthran S, George B. A mycological study of superficial mycoses at the skin clinic in sabah, malaysia. *Int J Pharm Sci Invent* 2013;**2**(3):45–8.
5. Patel P, Mulla S, Patel D, Shrimali G. A study of superficial mycosis in south Gujarat region. *Natl J Community Med* 2010;**1**(2):85–8.

6. Tovar LJ. Pathogenesis of dermatophytosis and tinea versicolor. *Clin Dermatol* 2010;**28**(2):185–9.
7. Grover WCS, Roy CP. Clinico–mycological profile of superficial mycosis in a hospital in North-East India. *Med J Armed Forces India* 2003;**59**(2):114–6.
8. Alshahnl MM, Makimura K, Yamada T, Takatori K, Sawada T. Nourseothricin acetyl transferase: a new dominant selectable marker for the dermatophyte *Trichophyton mentagrophytes*. *Med Mycol* 2010;**48**(4):665–8.
9. Chander J. *Superficial cutaneous mycosis*. 2nd ed. New Delhi: Mehta Publishers; 2009. p. 92–147.
10. Perera W, Rajakulasooriya R, Weerasekera MM, Kottahachchi J, Gunasekera T, Bogahawatta L, et al. Proportion of superficial fungal infections among members of the cleaning staff at the University of Sri Jayewardenepura. *Sri Lankan J Inf Dis* 2014;**4**(1):38–47.
11. Khan AM, Bhadauria S. A review on chemical and molecular characterization of keratinophilic fungi. *Int J Sci Res* 2015;**4**(1):420–3.
12. Das K, Basak S, Ray S. A study on superficial fungal infection from west bengal: a brief report. *J Life Sci* 2009;**1**(1):51–5.
13. Chauhan LK, Varshney M, Pandey V, Sharma P, Verma VK, Kumar P, et al. ROS-dependent genotoxicity, cell cycle perturbations and apoptosis in mouse bone marrow cells exposed to formulated mixture of cypermethrin and chlorpyrifos. *Mutagenesis* 2016;**31**:635–42.
14. Rivera ZS, Losada L, Nierman WC. Back to the future for dermatophyte genomics. *mBio* 2012;**3**(6):1–2.
15. Latka C, Dey SS, Mahajan S, Prabu R, Jangir PK, Gupta C, et al. Genome sequence of a clinical isolate of dermatophyte *Trichophyton rubrum* from India. *FEMS Microbiol Lett* 2015;**362**(8):1–3.
16. Wariso KT, Igunma JA, Oboro IL. Pattern of dermatophytes isolated in the medical microbiology laboratory of the university of port harcourt teaching hospital, rivers state, Nigeria. *Adv Microbiol* 2015;**5**(1):346–50.
17. Oke OO, Onayemi O, Olasode OA, Omisore AG, Oninla OA. The prevalence and pattern of superficial fungal infections among school children in Ile-Ife, South-Western Nigeria. *Dermatol Res Pract* 2014;**2014**:http://dx.doi.org/10.1155/2014/842917.
18. Achkar JM, Fries BC. *Candida* infections of the genitourinary tract. *Clin Microbiol Rev* 2010; **23**(2):253–73.
19. Luo G, Mitchell TG. Rapid identification of pathogenic fungi directly from cultures by using Multiplex PCR. *J Clin Microbiol* 2002;**40**(8):2860–5.
20. Michaelsen A, Pinzari F, Ripka K, Lubitz W, Pinar G. Application of molecular techniques for identification of fungal communities colonising paper material. *Int Biodeterior Biodegrad* 2006;**58**:133–41.
21. Khot P, Fredricks DN. PCR-based diagnosis of human fungal infections. *Expert Rev Anti Infect Ther* 2009;**7**(10):1201–21.
22. Gupta AK, Foley KA. Antifungal treatment for pityriasis versicolor. *J Fungi* 2015;**1**:13–29.
23. Weinstein A, Berman B. Topical treatment of common superficial tinea infections. *Am Fam Phys* 2002;**65**(10):2095–102.
24. Vadeputte P, Ferrari S, Coste AT. Antifungal resistance and new strategies to control fungal infections. *Int J Microbiol* 2011;**85**(1):1–24.
25. Ghelardi E, Celandroni F, Gueye SA, Salvetti S, Senesi S, Bulgheroni A, et al. Potential of ergosterol synthesis inhibitors to cause resistance or cross-resistance in *Trichophyton rubrum*. *Antimicrob Agents Chemother* 2014;**58**(5):2825–9.
26. Mousavi SAA, Salari S, Hadizadeh S. Evaluation of antifungal effect of Silver nanoparticles against *Microsporum canis, Trichophyton mentagrophytes* and *Microsporum gypseum*. *Iran J Biotechnol* 2015; **13**(4):38–42.
27. Chon SY, Doan HQ, Mays RM, Singh SM, Gordon RA, Tyring SK. Antibiotic overuse and resistance in dermatology. *Dermatol Ther* 2012;**25**:55–69.
28. Nigam PK. Antifungal drugs and resistance: current concepts. *Our Dermatol Online* 2015;**6**(2):212–21.
29. Sanglard D. Emerging threats in antifungal-resistant fungal pathogens. *Front Med (Lausanne)* 2016;**3**:11. http://dx.doi.org/10.3389/fmed.2016.00011.
30. Banerjee M, Ghosh AK, Basak S, Das KD, Gangopadhyay DN. Comparative evaluation of effectivity and safety of topical amorolfine and clotrimazole in the treatment of tinea corporis. *Indian J Dermatol* 2011;**56**:657–62.

31. Pfaller MA, Diekema DJ. Twelve years of fluconazole in clinical practice: global trends in species distribution and fluconazole susceptibility of bloodstream isolates of *Candida*. *Clin Microbiol Infect* 2004;**10**(1):11–23.
32. Pfaller MA, Diekema DJ, Rinaldi MG, Barnes R, Hu B, Veselov AV, et al. Results from the ARTEMIS disk global antifungal surveillance study: a 6.5-year analysis of susceptibilities of *Candida* and other yeast species to fluconazole and voriconazole by standardized disk diffusion testing. *J Clin Microbiol* 2005;**43**:5848–59.
33. Silva LB, de Oliveira DBC, da Silva BV, de Souza RA, da Silva PR, Ferreira-Palm K, et al. Identification and antifungal susceptibility of fungi isolated from dermatomycoses. *J Eur Acad Dermatol Venereol* 2014;**28**:633–40.
34. Ghannoum MA, Chaturvedi V, Espinel-Ingroff A, Pfaller MA, Rinaldi MG, et al. Intra- and interlaboratory study of a method for testing the antifungal susceptibilities of dermatophytes. *J Clin Microbiol* 2004;**42**:2977–9.
35. Martinez-Rossi NM, Peres NTA, Rossi A. Antifungal resistance mechanisms in dermatophytes. *Mycopathologia* 2008;**166**:369–83.
36. Dogra S, Uprety S. The menace of chronic and recurrent dermatophytosis in India: is the problem deeperthan we perceive? *Indian Dermatol Online J* 2016;**7**(2):73–6.
37. Peres NTA, Maranhao FCA, Rossi A, Martinez-Rossi NM. Dermatophytes: host-pathogen interaction and antifungal resistance. *An Bras Dermatol* 2010;**85**:657–67.
38. Ghannoum MA, Rice LB. Antifungal agents: mode of action, mechanisms of resistance, and correlation of these mechanisms with bacterial resistance. *Clin Microbiol Rev* 1999;**12**:501–17.
39. Lewis DF, Rodrigues AD, Ioannides C, Parke DV. Adverse reactions of imidazole antifungal agents: computer graphic studies of cytochrome P-450 interactions. *J Biochem Toxicol* 1989;**4**(4):231–4.
40. Rosenberger A, Tebbe B, Treudler R, Orfanos CE. Acute generalized exanthematous pustulosis, induced by nystatin. *Hautarzt* 1998;**49**(6):492–5.
41. http://www.webmd.com/drugs/2/drug-94303/antifungal-tolnaftate-topical/details#side-effects [Accessed 18 October 2016].
42. https://www.drugs.com/sfx/terconazole-topical-side-effects.htm [Accessed 18 October 2016].
43. Myers RS. Antifungal agents. In: Elmer GW, Totah R, Black D, editors. *Immunizing and antimicrobial agents*. Washington, DC: University of Washington; 2006. p. 1–16.
44. Higa LH, Schilrreff P, Perez AP, Morilla MJ, Romero EL. The intervention of nanotechnology against epithelial fungal diseases. *J Biomater Tissue Eng* 2013;**3**(1):1–19.
45. Lara HH, Romero-Urbina DG, Pierce C, Lopez-Ribot JL, Arellano-Jiménez MJ, Jose-Yacaman M. Effect of silver nanoparticles on *Candida albicans* biofilms: an ultrastructural study. *J Nanobiotechnol* 2015;**13**:91. http://dx.doi.org/10.1186/s12951-015-0147-8.
46. Falkiewicz-Dulik M, Macura AB. Nanosilver as substance biostabilising footwear materials in the foot mycosis prophylaxis. *Mikol Lek* 2008;**15**(3):145–50.
47. Kim KJ, Sung WS, Suh BK, Moon SK, Choi JS, Kim JG, et al. Antifungal activity and mode of action of silvernano-particles on *Candida albicans*. *Biometals* 2009;**22**:235–42.
48. Hwang IS, Lee J, Hwang JH, Kim KJ, Lee DG. Silver nanoparticles induce apoptotic cell death in through the increase ofhydroxyl radicals. *FEBS J* 2012;**279**:1327–38.
49. Kim KJ, Sung WS, Moon SK, Choi JS, Kim JG, Lee DG. Antifungal effect of silver nanoparticles on dermatophytes. *J Microbiol Biotechnol* 2008;**18**:1482–4.
50. Gupta A, Bonde SR, Gaikwad S, Ingle A, Gade AK, Rai M. *Lawsonia inermis*-mediated synthesis of silver nanoparticles: activity against human pathogenic fungi and bacteria with special reference to formulation of an antimicrobial nanogel. *IET Nanobiotechnol* 2013;**8**(3):172–8.
51. Mohamed NH, Ismail MA, Abdel-Mageed WM, Mohamed Shoreit AA. Antimicrobial activity of latex silver nanoparticles using *Calotropis procera*. *Asian Pac J Trop Biomed* 2014;**4**(11):876–83.
52. Aghamoosa M, Sabokbar A. Antifungal activity of silver nanoparticle in different sizes against some pathogenic fungi. *J Appl Chem Res* 2014;**8**(4):115–22.
53. Hassanab AA, Mansourb MK, Mahmoud HH. Biosynthesis of silver nanoparticles (Ag-Nps) (a model of metals) by *Candida albicans* and its antifungal activity on some fungal pathogens (*Trichophyton mentagrophytes* and *Candida albicans*). *N Y Sci J* 2013;**6**:27–34.

54. Anasane N, Golinska P, Wypij M, Rathod D, Dahm H, Rai M. Acidophilic actinobacteria synthesized silver nanoparticles showed remarkable activity against fungi-causing superficial mycoses in humans. *Mycoses* 2016;**59**:157–66.
55. Silva E, Saraiva SM, Miguel SP, Correia IJ. PVP coated silver nanoparticles showing antifungal improved activity against dermatophytes. *J Nanopart Res* 2014;**16**:2726. http://dx.doi.org/10.1007/s11051-014-2726-2.
56. Anbarasu R, Selvan G, Baskar S, Raja V. Synthesis of *Evolvulus alsinoides* derived gold nanoparticles for medical applications. *Int J Adv Sci Res* 2016;**2**(01):38–44.
57. Albasarah YY, Somavarapu S, Stapleton P, Taylor KMG. Chitosan-coated antifungal formulations for nebulisation. *J Pharm Pharmacol* 2010;**62**(7):821–8.
58. Kulikov SN, TiurinIu A, Fassakhov RS, Varlamov VP. Antibacterial and antimycotic activity of chitosan: mechanisms of action and role of the structure. *Zh Mikrobiol Epidemiol Immunobiol* 2009;**5**:91–7.
59. Li RC, Guo ZY, Jiang PA. Synthesis, characterization, and antifungal activity of novel quaternary chitosan derivatives. *Carbohydr Res* 2010;**345**(13):1896–900.
60. Kulatunga DCM, Dananjaya SHS, Godahewa GI, Lee J, De Zoysa M. Chitosan silver nanocomposite (CAgNC) as an antifungal agent against *Candida albicans*. *Med Mycol* 2017;**55**:213–22.
61. George SA, Raj MS, Solomon D, Roselin P. Comparative study of the anti-fungal activity of zinc oxide and titanium dioxide nano and bulk particles with anti-fungals against fungi isolated from infected skin and dandruff flakes. *Res Rev J Microbiol Biotechnol* 2014;**3**(3):23–30.
62. Tiwari N, Pandit R, Gaikwad S, Gade A, Rai M. Biosynthesis of zinc oxide nanoparticles by petals extract of *Rosa indica* L., its formulation as nail paint and evaluation of antifungal activity against fungi causing onychomycosis. *IET Nanobiotechnol* 2016;http://dx.doi.org/10.1049/iet-nbt.2016.0003.
63. Baskar S, Pragati P, Chandrababu K. Antimicrobial studies using sulphur nano particles on dandruff causing malassezi yeasts. In: *Proceedings of the World Congress on Engineering, vol. II*, London: WCE; 2015.
64. Paralikar P. Fabrication of ketoconazole nanoparticles and their activity against *Malassezia furfur*. *Nusant Biosci* 2015;**7**(1):43–7.
65. Kumar R, Shukla SK, Pandey A, Srivastava SK, Dikshit A. Copper oxide nanoparticles: an antidermatophytic agent for *Trichophyton* spp. *Nanotechnol Rev* 2015;**4**(5):401–9.
66. Mathew J, Rathod V, Singh D, Ninganagouda S, Singh AK, Kulkarni P. Enhanced efficacy of ketoconazole coated silver nanoparticles against the fungus *Malassezia furfur* a dandruff causing agent. *World J Pharmacy Pharma Sci* 2015;**4**(6):1246–58.
67. Washington N, Washington C, Wilson C. *Transdermal drug delivery. Physiological pharmaceutics: barriers to drug absorption*. New York, NY: CRC Press; 2000. p. 181–195.
68. Baibhav J, Gurpreet S, Rana AC, Saini S, Vikas S. Emulgel: a comprehensive review on the recent advances in topical drug delivery. *Int Res J Pharm* 2011;**2**:66 70.
69. Antonio JR, Antonio CR, Cardeal ILS, Ballavenuto JMA, Oliviera JR. Nanotechnology in dermatology. *An Bras Dermatol* 2014;**89**:126–36.
70. Santra S. Ag loaded silica nanoparticle/nanogel formulation, methods of making, and methods of use. US20130108678; 2013.
71. Lohani A, Verma A, Joshi H, Yadav N, Karki N. Nanotechnology-based cosmeceuticals. *ISRN Dermatol* 2014;http://dx.doi.org/10.1155/2014/843687. Article ID: 843687.
72. Jain J, Arora S, Rajwade JM, Omray P, Khandelwal S, Paknikar KM. Silver nanoparticles in therapeutics: development of an antimicrobial gel formulation for topical use. *Mol Pharm* 2009;**6**(5):1388–401.
73. Papakostas D, Rancan F, Sterry W, Blume-Peytavi U, Vogt A. Nanoparticles in dermatology. *Arch Dermatol Res* 2011;**303**:533–50.
74. Bhol KC, Alroy J, Schechter PJ. Anti-inflammatory effect of topical nanocrystalline silver cream on allergic contact dermatitis in a guinea pig model. *Clin Exp Dermatol* 2004;**29**(3):282–7.
75. Wiesenthal A, Hunter L, Wang S, Wickliffe J, Wilkerson M. Nanoparticles: small and mighty. *Int J Dermatol* 2011;**50**:247–54.
76. Amato SW, Farer A, Hoyte WM, Pavlovsky M, Smith R, Valdiviezo G. Coatings for mammalian nails that include nanosized particles, US Patent 2007/002207; 2007.
77. Sevgi M, Toklu A, Vecchio D, Hamblin MR. Topical antimicrobials for burn infections—an update. *Recent Pat Antiinfect Drug Discov* 2013;**8**(3):161–97.

78. Perera J, Weerasekera M, Kottegoda N. Slow release anti-fungal skin formulations based on citric acid intercalated layered double hydroxides nanohybrids. *Chem Cent J* 2015;**9**:27.
79. Gupta I, Duran N, Rai M. *Nano-silver toxicity emerging concerns and consequences in human health. Nano antimicrobial.* Berlin: Springer; 2012. p. 525–548.
80. Sharma V, Shukla RK, Saxena N, Parmar D, Das M, Dhawan A. DNA damaging potential of zinc oxide nanoparticles in human epidermal cells. *Toxicol Lett* 2009;**185**(3):211–8.
81. Sharma V, Anderson D, Dhawan A. Zinc oxide nanoparticles induce oxidative DNA damage and ROS-triggered mitochondria mediated apoptosis in humanliver cells (HepG2). *Apoptosis* 2010;**17**:852–70.
82. Demir E, Kaya N, Kaya B. Genotoxic effects of zinc oxide and titanium dioxide nanoparticles on root meristem cells of *Allium cepa* by comet assay. *Turk J Biol* 2014;**38**:31–9.
83. Annangi B, Rubio L, Alaraby M, Bach J, Marcos R, Hernandez A. Acute andlong-term in vitro effects of zinc oxide nanoparticles. *Arch Toxicol* 2016;**90**(9):2201–13.
84. Kavithaa K, Paulpandi M, Ponraj T, Murugan K, Sumathi S. Induction of intrinsic apoptotic pathway in human breast cancer (MCF-7) cells through facile biosynthesized zinc oxide nanorods. *Karbala Int J Mod Sci* 2016;**2**(1):46–55.
85. Smijs TG, Bouwstra JA. Focus on skin as a possible port of entry for solid nanoparticles and the toxicological impact. *J Biomed Nanotechnol* 2010;**6**(5):469–84.
86. Martorano LM, Stork CJ, Li YV. UV irradiation-induced zinc dissociation from commercial zinc oxide sunscreen and its action in human epidermal keratinocytes. *J Cosmet Dermatol* 2010;**9**(4):276–86.
87. Cross SE, Innes B, Roberts MS, Tsuzuki T, Robertson TA, McCormick P. Human skin penetration of sunscreen nanoparticles: in vitro assessment of a novel micronized zinc oxide formulation. *Skin Pharmacol Physiol* 2007;**20**(3):148–54.
88. Filipe P, Silva JN, Silva R, de Castro JL Cirne, Marques Gomes M, Alves LC, et al. Stratum corneum is an effective barrier to TiO_2 and ZnO nanoparticles percutaneous absorption. *Skin Pharmacol Physiol* 2009;**22**(5):266–75.
89. Schilling K, Bradford B, Castelli D, Dufour E, Nash JF, Pape W, et al. Human safety review of "nano" titanium dioxide and zinc oxide. *Photochem Photobiol Sci* 2010;**9**(4):495–509.
90. Monteiro-Riviere NA, Wiench K, Landsiedel R, Schulte S, Inman AO, Riviere JE. Safety evaluation of sunscreen formulations containing titanium dioxide and zinc oxide nanoparticles in UVB sunburned skin: an in vitro and in vivo study. *Toxicol Sci* 2011;**123**(1):264–80.
91. Surekha P, Kishore AS, Srinivas A, Selvam G, Goparaju A, Reddy PN, et al. Repeated dose dermal toxicity study of nano zinc oxide with Sprague-Dawley rats. *Cutan Ocul Toxicol* 2010;**31**(1):26–32.
92. Dufour EK, Kumaravel T, Nohynek GJ, Kirkland D, Toutain H. Clastogenicity, photo-clastogenicity or pseudo-photo-clastogenicity: genotoxic effects of zinc oxide in the dark, in pre-irradiated or simultaneously irradiated Chinese hamster ovary cells. *Mutat Res* 2006;**607**:215–24.
93. Bastos V, Ferreira de Oliveira JMP, BrownD JH, Malheiro E, Daniel-da-Silva AL, Duarte IF, et al. The influence of citrate or PEG coating on silver nanoparticle toxicity to a human keratinocyte cell line. *Toxicol Lett* 2016;**249**:29–41.

CHAPTER 12

Alternative Antimicrobials from Natural Products Against Dermatological Infections

S. Suroowan, M.F. Mahomoodally
University of Mauritius, Réduit, Mauritius

1. INTRODUCTION

Dermatological infections (DIs) have afflicted humanity since time immemorial. Even today, DIs continue to prevail as a profound public health concern with national and global dimensions.[1] Globally, it is estimated that around 30%–70% of the earth's population suffer from at least one kind of skin disease, with higher proportions occurring in at-risk subpopulations.[1,2] Indeed, most skin disorders induce a massive economic, physical, psychological, and societal burden amongst infected individuals and their families. The economic and global burden associated with dermatological disorders is enormous, such that in 2010, the number of disability-adjusted life years (DALYs) lost due to skin conditions stood at 36,921,995, and ranked 18th on the DALY's list.[1]

Interestingly, over 2000 dermatological disorders are known to occur worldwide and, in an attempt to relieve sufferers of DIs, a plethora of natural and synthetic conventional antimicrobials have been discovered and developed after the introduction of penicillin in the early 1940s.[3] From the 1940s and up to now a wider range of other antimicrobials belonging to different classes have been developed. Synthetic antimicrobials remain one of the most prescribed drugs in modern medicine and have stood the test of time as significantly relevant in the prophylaxis, management and treatment of DIs.[4]

Nonetheless, the major breakthrough in the discovery of conventional antimicrobials has also accentuated the phenomena of antimicrobial resistance (AMR).[5] Up to 50% of all medicines are prescribed, dispensed or sold inappropriately (medicine use in primary care). In addition, the widespread use of broad-spectrum antimicrobials causes an inclusive harm to infective microbes and skin commensals which naturally maintain the health of the skin, resulting in a weaker defense system of the skin barrier.[6] Besides, nonadherence to treatment regimens has significantly contributed to the occurrence of AMR worldwide, which remains one of the greatest challenges of this century.[7] Every year up to 50,000 deaths result from the devastating effects of AMR across Europe and the United States solely, while hundreds of thousands of death are recorded in other nations,

The Microbiology of Skin, Soft Tissue, Bone and Joint Infections
ISSN 2451-9006, http://dx.doi.org/10.1016/B978-0-12-811079-9.00012-4

Table 1 Deaths resulting from AMR by 2050

Continent	Number of deaths every year[Ref]
Asia	4,730,000[3]
Africa	4,150,000[3]
Latin America	392,000[3]
Europe	390,000[3]
North America	317,000[3]
Oceania	22,000[3]

with estimates of real figures rarely reported and hence known. The number of deaths attributed to AMR by 2050 is also forecasted to escalate up to 10 million deaths per year (Table 1). In addition, fewer antimicrobials are being registered each year, thereby resulting in significant morbidity and mortality numbers from resistant DIs.[8]

Consequently, to overcome the staggering challenge surrounding dermatological infection therapy, the focus has now shifted towards the development of alternative antimicrobials.[9] A plethora of plant species recorded through the conductance of ethnobotanical studies are being investigated *in vitro* and *in vivo* as antimicrobials.[10] Numerous commensal skin microbes have shown to benefit the skin as they synthesize antimicrobial metabolites that can prevent infection from other pathogens.[6] Interestingly, a panoply of biotic resources such as bacteria and fungi have demonstrated notable activity against a number of pathogenic microbes inducing DIs.[4] The quest for the development of novel antimicrobials is geared to increase the existing arsenal of antimicrobials in an attempt to defy existing and new infections as well as improve the quality of life of DIs sufferers.

2. COMMON DERMATOLOGICAL INFECTIONS FROM MICROORGANISMS

Bacteria are microscopic organisms residing primarily on internal and external body surfaces. The colon houses the majority of commensals, estimated at about a hundred trillion bacteria. The skin is the second body part onto which most commensals reside, harboring an estimated trillion bacteria.[11] The skin forms a major protective barrier which protects the body from any bacterial invasion. Nonetheless, upon the occurrence of any kind of trauma, such as lacerations, abrasions, excoriations, inflammatory skin disorders and ulceration, the skin is left vulnerable to bacterial infections. Common primary infections to which the skin is particularly prone include impetigo, folliculitis, furuncles and carbuncles, erysipelas as well as cellulitis. Bacterial pathogens resulting in diverse DIs and which have attracted considerable scientific interest since most of them have developed resistance against existing antimicrobials include *Staphylococcus aureus*, *Staphylococcus epidermidis*, methicillin resistant *Staphylococcus aureus* (MRSA), and *Streptococcus pyogenes*.[12]

Fungi include a diverse range of nonphotosynthetic saprophytes relying on the soil or dead organic matter as habitat. These microorganisms manifest mostly as parasites residing within and on the surface of plants and on animal and human skin. Besides their parasitic nature, they are also members of the normal flora within living organisms, especially in humans. Numerous fungal species induce a wide array of DIs.[13] Common fungal species causing DIs in humans include *Epidermophyton*, *Microsporum* and *Trichophyton* species belonging to the dermatophytes family, and *Malassezia furfur* and *Candida* species. Other fungal species responsible for a lower rate of DIs include *Aspergillus*, *Trichothecium roseum*, *Cladosporium fusarium*, *Curvularia*, *Penicillium*, *Epidermatophyton*, *Drechslera* and *Alternaria* species, most of which cause skin, hair and nail infections.[9,14] Viruses are infectious organisms much smaller than bacteria and fungi. The basic requirement for every virus is the availability of a living cell in order to reproduce. A plethora of virus infections are known to cause DIs. The most common viruses resulting in DIs include the herpes simplex virus (HSV) types 1 and 2.[15]

3. THE SHORTCOMINGS OF CONVENTIONAL ANTIMICROBIALS

The development of a plethora of antibiotic generations over the last half century has enabled control over a vast array of microbial infections. Even with such remarkable development in the field, the curative potential of available antimicrobials remains limited. In addition, the release of novel antimicrobial generations from pharmaceutical industries is stalling,[16] and each year existing and new microbial infections are becoming more difficult to manage and treat. Consequently, people are left infected or die with the infection. An astonishing number of existent microbial infections have developed resistance to available antibiotics, especially in developing and low-income countries. Epitomes of these instances include drug resistance in malaria, *Mycobacterium tuberculosis*, African trypanosomiasis, as well as in a number of fungal species.[17]

All over the world, antimicrobials are either misused or overused. A report from the WHO states that only 50% of patients fully adhere to their treatment regimens. Hence, only 5 patients out of 10 complete their antimicrobial drug course. In developed countries, antimicrobials are even prescribed when unnecessary. In addition, many prescribers fail to adhere to standard prescribing policies, resulting in the irrational use of medicines among patients. Concurrently, the production of counterfeit antimicrobials and their subsequent availability to sufferers leads to treatment failure. All these factors combined enhance the risk of development of resistant microbial strains, which consequently accentuates the AMR phenomena.[18]

4. NATURAL PRODUCTS AS ALTERNATIVES TO CONVENTIONAL ANTIMICROBIALS

It is believed that for every disease affecting mankind, nature has a cure in its storehouse. Indeed, natural products have attracted massive interest during recent years, given the limitations of conventional medicine in managing and/or treating successfully a plethora of ailments. Currently, around 80% of the world's population relies on natural products either fully or to some extent as a form of primary health care.[19,20] Numerous populations worldwide rely to a great extent on natural products as these are perceived to be safer than conventional drugs, besides being economical, effective and easily available.[21]

Since the inception of the antibiotics era, countless antibiotics were derived from natural sources: for example, the isolation of penicillin from *Penicillium* spp., gramicidin and tryocidine from a soil bacterium, *Bacillus brevis* and macrolides originally extracted from various *Streptomyces* species. The golden era of antibiotics resulted due to the isolation of most antibiotics from synthetic methods. Nonetheless, the release and extensive use of newer generations of synthetic antibiotics has contributed to the emergence of newer resistant bacterial species.[22]

With the unprecedented pace of the AMR phenomena coupled with the lateral transfer of resistant genes among bacteria, it is often heard that the antibiotic era's end is not very far from us.[23] The phenomena of antibiotic resistance has led to the development of a diverse range of alternative topical antimicrobials which can serve as an additional line of treatment, as well as potentiate the action of existent antimicrobials when administered concurrently. A plethora of natural products have been studied and are still being investigated extensively for their antimicrobial potential. Panoply of conventional drugs developed during the years 1980–2010 are reported to originate from natural products.[24] Hence, the rejuvenated interest in natural products guarantees that useful resources can still be found in our immediate environment and can be developed into effective antimicrobials.

Over the last few years, focus has shifted back to the study of natural products to attempt to discover antimicrobial extracts as well as to isolate metabolites from natural sources. The biochemical potential offered by nature to researchers in the quest of developing novel antimicrobials is almost infinite. In addition, the use of natural products since time immemorial in the prophylaxis, management and treatment of DIs, highlighted in ethnobotanical studies, also serves as baseline information for scientists to investigate their diverse antimicrobial activities *in vitro, in vivo*, and *in silico*.[25]

5. ETHNOBOTANICAL STUDIES

Ethnobotanical studies aim at investigating the traditional relationship between humans and plants. Many plants and spices have long-standing uses as foodstuff. In addition, plants have been used since time immemorial in the prevention and treatment of minor and chronic ailments by indigenous people of all cultures worldwide.[26] Plants were identified through the doctrine of signatures whereby any herb bearing similar morphology to a body part was assigned to treat any ailment related to the given body part.[27] Another common method was to experiment on oneself by trial and error of the utility of the plants against diseases. The use of plant species as medicines has been successful throughout time and, as stated, they are still currently being used to a great extent by various cultures worldwide. Several hundreds of plant species are still being recorded in ethnobotanical studies in the management and treatment of DIs, most of which have not been investigated *in vitro* and *in vivo* against DIs. Traditionally, common plants species used in the management of dermatological infections include *Aristea ecklonii* Baker., *Cissampelos capensis* Thunb., *Chenopodium ambrosioides* Bert. Ex Steud., *Argyrolobium argenteum* (Jacq.) Eckl. & amp; Zeyh., the tuber is used in combination with *Elephantorrhiza elephantina* to treat acne. *Ekebergia capensis* Sparrm., *Eucalyptus camaldulensis* Dehnh., Harpephyllum caffrum Bernh. ex Krauss, *Diospyros mespiliformis* Hochst. ex A.DC., and *Solanum incanum* L. These plant species are notably used in the management of DIs such as acne and ringworm.[26]

6. PLANT-DERIVED ANTIMICROBIALS

Plant-derived antimicrobials are simply the secondary metabolites that the plant harnesses as part of its metabolism. A multiplicity of plant-based metabolites bear notable effects against a wide variety of pathogens. Indeed, diverse classes of plant secondary metabolites such as phenolics, phenolic acids, quinones, saponins, flavonoids, tannins, coumarins, terpenoids, and alkaloids are known to possess a certain level of antimicrobial activity. Given the immense chemical and hence structural diversity of these secondary metabolites, each one possesses a unique antimicrobial activity.[28]

Many plants are known to harness cinnamic and caffeic acids, which are common derivatives of phenylpropane, as part of their metabolism.[29] Indeed, common herbs such as tarragon and thyme which biosynthesize cinnamic and caffeic acids have been shown to possess noteworthy antibacterial, antifungal and antiviral agents.[30] An assorted number of other compounds derived from this class are also known to be bactericidal: for example, catechol and pyrogallol which are hydroxylated phenols containing two and three hydroxyl groups respectively.[31] The number of hydroxyl groups they possess is thought to determine their relative toxicity against microorganisms. Among the phenolic compounds, eugenol is widely known to possess antimicrobial activity.[30] Quinones remain an interesting secondary metabolite class to explore in the hope of elucidating novel antimicrobials. This secondary metabolite class is believed to target surface exposed adhesins, cell wall polypeptides, and membrane bound enzymes, as well as render substrates unavailable to the microorganisms.[32] Upon microbial infection, most plants synthesize flavones, flavonoids, and flavonols. Catechins are secondary metabolites that fall within this class and have demonstrated remarkable *in vitro* activity against *Vibrio cholerae*, *Streptococcus mutans*, and *Shigella* as well as against other microorganisms.[33–35] Hence, exploring this metabolite class remains an interesting avenue. Tannins have been postulated to have the ability of inactivating microbial adhesins, enzymes, and cell envelope proteins, among many other antimicrobial activities. Several coumarin derivatives have been found to be effective against different microorganisms. Interestingly, coumarin has been shown to possess *in vitro* inhibitory activity against *Candida albicans*. Terpenes 3 and terpenoids, in addition, have been shown to be active against bacteria, viruses, protozoa and fungi. Solamargine, a glycoalkaloid forming part of the alkaloids, is considered to be useful against HIV and other AIDS-related infections. The antimicrobial activity of peptides was first reported in 1942 and a plethora of plants are known to synthesize these secondary metabolites.[34,35]

7. *IN VITRO* STUDIES OF PLANT SPECIES AND THEIR METABOLITES AGAINST DIs

In vitro assays allow for validation of plant species activity against given bacterial pathogens. A common plant that has attracted common interest as a source of antimicrobial agents against DIs is *Aegle marmelos*. It grows widely on the Asian continent with most distribution in Northern India, Bangladesh, Thailand and Indochina.[36] This plant is well known in traditional Indian medicine, given its many medicinal properties. Throughout time, this plant has been utilized to a great extent for its antidiarrheal, antimicrobial, antipyretic, ulcer healing, antifertility and antiinflammatory properties. Its essential oil is believed to inhibit the Ca^{2+} dipicolinic acid metabolism in dermatophytic fungi and hence delays or prevents spore germination.[37,38] The aqueous and ethanolic extracts of the plant are also known to be active against a wide range of bacteria causing DIs, such as *S. aureus* and *Pseudomonas aeruginosa*.[39]

Laurus nobilis is commonly grown in various warm parts of the world such as southern Europe, around the shores of the Mediterranean Sea, which includes Lebanon. Its essential oil has been shown to possess both antibacterial and antifungal properties. A study investigated the antibacterial extracts of *L. nobilis* against *S. aureus*, *Enterococcus faecalis*, *S. epidermidis*, *E. coli* and *P. aeruginosa*. The extracted constituents included alginate, fucoidan, essential and fatty oil. All extracts demonstrated good antibacterial activity against the different bacterial strains, with MIC values ranging from 60.97 to 78.84 mg/mL.[40]

A study investigated the antimicrobial activities of skin care preparations from plant extracts of *Tithonia diversifolia* Helms. (A Gray), *Aloe secundiflora* (Miller) and *Azadirachta indica* (A. Juss) against *E. coli* and *C. albicans*. Herbal soaps and lotions were prepared as a means of local administration of the plant extracts onto the skin surface. Results of this study showed that *T. diversifolia* herbal soap inhibited *E. coli* to a greater extent than other plants, but at the same time was poorly active against *C. albicans*. Soap made from *Azadirachta indica* extract was most effective against *C. albicans* while *Thevetia peruviana* oil extract was more active against *S. aureus* in comparison to *E. coli*.[41] Enthrallingly, the results of this study confirmed the ethnomedicinal application of these plants in skin disorder management and treatment.

Olive leaf water extract (15% w/v) was found to display fungicidal activity against *C. albicans* within 3 h. The antimicrobial property of olive leaf water extract is due to the presence of alkaloids, tannins, saponins and flavonoids. Leaf extract from *Cassia alata* has been shown to inhibit *S. aureus*. The antimicrobial activity of crude water-soluble arrowroot, *Puerariae radix* tea, against *E. coli* and *S. aureus* has also been reported.[42]

In a study performed in South Africa, 132 organic and aqueous extracts were prepared from 47 plant species. This study endeavored to screen the antimicrobial properties of these plants widely claimed to be of utility against DIs. The following plant species – *Aristea ecklonii*, *Chenopodium ambrosioides*, *Diospyros mespiliformis*, *Elephantorrhiza elephantina*, *Eucalyptus camaldulensis*, *Gunnera perpensa*, *Harpephyllum caffrum*, *Hypericum perforatum*, *Melianthus comosus*, *Terminalia sericea* and *Warburgia salutaris* – demonstrated broad-spectrum activity with minimum inhibitory concentration (MIC) $\leq$1.00 mg/mL when assayed against the disease-causing pathogens. The results retrieved for the organic extract of *Elephantorrhiza elephantine*, a plant known traditionally for treating acne vulgaris, were intriguing, with an MIC value of 0.05 mg/mL against *Propionibacterium acnes*, hence supporting the traditional claim behind the use of this plant. Another interesting result was recorded for *Diospyros mespiliformis*, traditionally known to treat ringworm infection, which demonstrated an MIC value of 0.10 mg/mL against *Trichophyton mentagro-phytes* and *Microsporum canis* (MIC 0.50 mg/mL). Other results that attracted particular attention in this study were from a combination of the aqueous root extracts of *Pentanisia prunelloides* and *Elephantorrhiza elephantina* in a 1:1 ratio: a synergistic antimicrobial effect was demonstrated and the sum of the fractional inhibitory concentration ranged between 0.31–0.38 against *S. aureus*, gentamycin–methicillin resistant *S. aureus*, *S. epidermidis* and *C. albicans*.[43]

Tannins are polyphenolic compounds that can be found in common foods such as tea, red fruits, vegetables and cocoa.[44,45] The antimicrobial properties of tannins were appraised long ago by the scientific community. Current research depicts that tannins can exert both a bacteriostatic and a bactericidal effect. In addition, a study suggested that tannins can inhibit the coagulase activity of *S. aureus*; this is due to its ability to bind iron, form a complex and hence inactivate the protein. Also, a study has witnessed the synergistic activity of tannins in combination with conventional antibiotics assayed against *S. aureus*. Interestingly, the MIC of several conventional antibiotics was drastically reduced. Among this class of polyphenolic compounds, tannic acid has been found to display the most potent antimicrobial activity. Besides its antimicrobial properties, tannins possess antiinflammatory as well as healing properties, which could be of great benefit to patients suffering from burn infections.[46]

8. FRUITS AND VEGETABLES AS A SOURCE OF ANTIMICROBIALS

While herbals and spices have been in the scientific limelight and investigated for their antimicrobial properties for centuries, recently much attention has shifted to fruits and vegetables.[47] Several hundreds of fruits and vegetables are known to possess antimicrobial properties. A study investigating the antibacterial properties of common fruits and vegetables included aronian berry, asparagus, bell pepper, beet, blackberry, blueberry, broccoli, carrot, cucumber, cherry, cranberry, garlic, ginger, grape, red onion, red cabbage, rhubarb, rutabaga, raspberry, pomegranate, spinach, strawberry, and green tea.[47] Interestingly, the authors highlighted that no green vegetable used in the study possessed bactericidal or bacteriostatic activity against *S. epidermidis*, an important pathogen involved in DIs. On the other hand, the methanolic extract of pomegranate peels displayed a broad-spectrum activity against three bacterial species involved in DIs: *E. coli*, *Proteus vulgaris* and *S. aureus*. *Pomegranate juice*, direct extract from guava and xoconostle pears inhibited the growth of *E. coli* while *C. annuum* extract was active against *S. aureus*.[48–50] Garlic, one of the most widely studied plants today, demonstrated a bacteriostatic effect on *E. coli*, *Streptococcus* and *Staphylococcus species*.[51] Red cabbage has also been shown to have significant antibacterial and antifungal activities against MRSA *S. aureus*, *E. coli* and *S. aureus*.[42] Other fruits and vegetables metabolizing secondary metabolites that have been found to inhibit dermatological pathogens are listed in Table 2.

Table 2 Plant metabolites as targets for dermatological infection-causing microorganisms

Plant product	Secondary metabolite	Target microorganism[Ref]
Pomegranate fruit peels	Phenolics and flavonoids	*L. monocytogenes*, *S. aureus*, *E. coli*[67]
Apple peels	Polyphenolic compounds	*S. aureus*[67]
Almond skin extracts	Polyphenols	*S. aureus*, *L. monocytogenes*[68]
Coconut husk	Phytochemical including phenolics and tannins	*L. monocytogenes*, *S. aureus* and *V. cholera*[69]
Green tea waste	Tannins	*S. aureus*, *E. coli*[70]
Acorn, chestnut, and persimmon hull	Tannins	*S. aureus*, *E. coli*, *L. monocytogenes*[70]
Tomato seeds	Metabolites such as fatty acids, carotenoids, saponins, phenolic compounds	*S. aureus*, *S. epidermidis*, *Micrococcus luteus*, *E. faecalis*, *B. cereus*, *C. albicans*[71]
Quince fruit peel	Polyphenolic compounds such as chlorogenic acid, catechin, quercetin and kaempferol	*S. aureus*, *P. aeruginosa*, *E. coli*, *C. albicans*[72]
Walnut green husk	Antioxidants such as phenolic compounds	*S. aureus*, *B. cereus*, *B. subtilis*[73]
Grape pomace	Phenolic acids, flavonoids, stilbenes	*S. aureus*[74]
Olive pomace	Phenolic compounds including oleocanthal, deoxyloganic acid lauryl ester	*E. coli* O157:H7, *S. enterica*, *L. monocytogenes*, and *S. aureus*[75]
Beet root pomace extract	Phenolics, flavonoids betacyanins, betaxanthins	*S. aureus*, *B. cereus*, *E. coli*, *P. aeruginosa*[76]
Legume hulls (*Vigna radiate*, *Cicer arietinum* and *Cajanus cajan*)	Polyphenolic compounds, flavonoids	*B. cereus*, *S. aureus*[77]

9. ESSENTIAL OILS

Essential oils are volatile substances extracted from plants mostly by distillation methods. A number of essential oils are known to be natural antimicrobials and could be of great interest for helping to manage and treat DIs.[52] Numerous assays performed using tea tree oil (TTO) extracted from *Melaleuca alternifolia* have resulted in positive antimicrobial activities against *S. aureus* as well as against MRSA. Another essential oil that has shown significant activity *in vitro* against *S. aureus* is Valencia orange oil. Interestingly, when tested on keratinocyte cell lines, it displayed no toxicity.[53]

Ocimum basilicum is a herb renowned in traditional medicine notably for its exceptional medicinal virtues. It has a long-standing use as a tonic, vermifuge, diuretic, antispasmodic and for lower and upper respiratory tract infections.[54,55] Since it possesses a strong flavor, it is widely used in the food industry as an aromatic and a flavoring agent. Given its traditional use to treat or manage acne, it has attracted much interest from researchers in the field of cosmetics. In addition, its essential oil has been under rigorous investigation in the hope of deciphering novel antimicrobial agents. A total of 36 constituents were identified from its essential oil, out of which the major ones were linalool, methyl chavicol, methyl cinnamate, methyleugenol, eugenol and geraniol.[56] In a study, its essential oil was found to inhibit both *B. cereus* and *S. aureus* with *B. cereus* being the most susceptible bacteria.[57]

A study investigated the effects of three different concentrations (25%, 50% and 100%) cinnamon oil against *S. aureus* ATCC 25923, *S. pyogenes* ATCC 19615, *S. epidermidis* ATCC 12228 and *P. aeruginosa* ATCC 10145 using the disk diffusion method. Most bacterial species were found sensitive to cinnamon oil, including *S. aureus, S. pyogenes* and *S. epidermidis* while *P. aeruginosa* demonstrated a high resistance. Even at low concentration the essential oil was found to demonstrate significant antimicrobial activity against *S. pyogenes* (MIC = 1.6%), *S. aureus* (MIC = 6.3%) and *S. epidermidis* (MIC = 12.5%).[58]

10. ANTIMICROBIALS OF ANIMAL ORIGIN

The medicinal application of animal species, their parts and products has been appraised in early texts such as those found in archives, including papyrus, alongside other ancient medical texts. The utilization of animal species to treat minor to chronic maladies is considered an important cultural asset and has emerged as a field known as zootherapy.[59] The worldwide use of animals as a source of medicine is as significant as the use of herbals. A number of animal-derived products are of relevance in DIs. Indeed, a wide array of peptides possessing noteworthy antimicrobial activity from animal sources, including pleurocidin, lactoferrin, defensins and protamine, have been documented over recent years to be active against *E. coli*. Chitosan, derived from the exoskeletons of crustaceans and arthropods, has been recorded to possess antibacterial activity namely against *S. aureus* and *E. coli*.[42] Other animal products effective against *S. aureus* are recorded in Table 3.

Table 3 Animal products as targets of dermatological infection organisms

Animal product	Constituent	Found in	Target microorganism[Ref]
Chitosan	Polycationic biopolymer Compound	Exoskeletons of crustaceans and arthropods	*S.aureus*[78]
Lactoperoxidase	Glycoprotein	Raw milk, colostrum, saliva and other biological secretions	*S. aureus*[79]
Ovotransferrin	Glycoprotein (peptide OTAP-92)	Egg albumin	*S. aureus*[79]
Casein and whey	Bioactive peptides (Isracidin as1 casein f1-23; Casocidin-I f150-188; Kappacin)	Milk protein	*S. aureus*[80]

11. ANTIMICROBIALS OF MARINE ORIGIN

Marine organisms also provide a potential source of novel antimicrobials. Indeed, marine products such as alginates, carrageenan and agar as phycocolloids have a long-standing use in both traditional and modern medicine and hence are included in a variety of pharmaceutical products. During the last few decades, primary and secondary metabolites derived from marine organisms have attracted major interest among pharmaceutical companies, leading to the isolation and investigation of numerous compounds. Several marine products (Table 4) and isolated metabolites (Table 5) from marine organisms assayed *in vitro* have proven to be antibacterial, antifungal and antiviral.[60,61]

12. ANTIBIOTICS OF BACTERIAL ORIGIN

The antibiotics produced by bacteria have garnered an increased interest among researchers during recent years. A number of antibiotics produced by bacteria are being employed as a means of fungal control. Alongside, a plethora of bacterial microorganisms reside in the mouth and in the soil and are known to have antimicrobial properties. A study by Sheikh 2010 investigated the antimicrobial potential of soil and saliva microorganisms. A bacterial and a fungal filtrate were prepared from a soil and saliva mixed sample. The bacterial and fungal filtrates were tested against the following pathogenic species: *S. aureus*, methicillin resistant *S. aureus* (MRSA) and *A. niger*. The bacterial filtrate was ineffective against *A. niger* but exhibited a considerable antagonistic effect on *S. aureus* and MRSA. Similarly, the fungal filtrate inhibited the growth of *A. niger* and was ineffective against *S. aureus* and MRSA bacterial strains. Interestingly, it was found that

Table 4 Marine products as targets against dermatological pathogens

Algae	Species	Metabolites	Target[Ref]
Brown algae	*Colpomenia sinuosa, Dictyota dichotoma, Dictyota dichotoma var. implexa, Petalonia fascia and Scytosiphon lomentaria*	Hydrocarbons, terpenes, acids, phenols, sulfur-containing compound, aldehydes	*S. aureus*[81]
Green algae	*Chlorococcum humicola*	Phytochemicals such as saponins, tannins, carotenoids, flavonoids, alkaloids, glycosides	*S. aureus*[82]
Brown algae	*Himanthalia elongate*	Phytol, fucosterol, neophytadiene or palmitic, palmitoleic and oleic acids	*S. aureus*[83]
Marine brown algae	*Ecklonia cava, E. kurome, E. stolonifera, Eisenia aborea, Eisenia bicyclis, Ishige okamurae, Pelvetia siliquosa*	Phlorotannins	*S. aureus, MRSA*[84]
Macrofungi	*Fomitopsis lilacinogilva, Psathyrella* spp., *Ramaria* spp., *Hohenbuehelia* spp., *Agaricus* spp., *Lentinus* spp., *Strobilomyces* spp.	Grifolin, pleuromutilin, striatins A, B & C, oudemansin, scorodonin, ganomycins, retapamulin	*S. aureus, B. cereus*[85]
Nongilled fungal group	*Lycoperdon perlatum, Cantharellus cibarius, Clavaria vermiculris, Ramaria Formosa Quel, Marasmius oreades, Pleurotus pulmonarius*	Total phenols, flavonoid and ascorbic acid	*S. aureus (ATCC 25923)*[86]
Edible mushroom fungi	*C. sinensis, P. australis*	Polysaccharides, triterpenes	*S. epidermidis S. aureus, B. cereus*[87]

B. subtilis demonstrated a potent inhibitory effect on all three pathogens, followed by *E. coli*. In contrast, *Bacillus megaterium* strongly inhibited the growth of the bacterial species solely. The fungal filtrate of *Fusarium oxysporum* and *Emericella nidulans* successfully inhibited the growth of *S. aureus*, while the growth of MRSA was inhibited only by the filtrate of *E. nidulans*.[4]

A wealth of literature is available regarding the antimicrobial properties of diverse bacteria species. Strains of *Pediococcus acidilactici* and *P. pentosaceus* synthesize pediocin, which can suppress *S. aureus*. *Lactobacillus curvatus* CRL705 harnesses lactocin as a target for *E. coli* strains. Bacterial species such as *Lactobacillus bulgaricus*, *Lactobacillus plantarum* and *Lactobacillus reuteri* biosynthesize bulgaricin, plantaricin and reuterin, respectively, for which the common target is *S. aureus*.[28]

Table 5 Marine metabolites as inhibitors of dermatological pathogens

Metabolites	Effect[Ref]
Seven sesterterpenes sulfates from the tropical sponge *Dysidea* sp.	Inhibitory effect against isocitrate lysate from *C. albicans*. Antibacterial effects against *Bacillus subtilis* and *Proteus vulgaris*[47]
Puupehanol/marine sponge *Hyrtios* sp.	Inhibitory activity against *Cryptococcus neoformans* and *Candida krusei*[88]
Fascioquinol A and B/*Fasciospongia* sp.	*S. aureus* (inhibitory concentration 50 IC50 0.9–2.5 μM) and *B. subtilis* (IC50 0.3–7 μM)[89]
Sargafuran/*Sargassum macrocarpum*	Bactericidal and completely killed *Propionibacterium acnes* by lysing bacterial cells[90]
Dehydroxychlorofusarielin B/marine-derived fungus *Aspergillus* sp.	Antibacterial activity against *S. aureus* and methicillin- and multidrug-resistant *S. aureus*[91]
Eurysterols A and B/*Euryspongia*	Antifungal activity against amphotericin B-resistant and wild-type strains of *C. albicans*, with MIC values in turn of 15.6 and 62.5 μg/mL[92]
2-(2′,4′-dibromophenoxy)-4,6-dibromophenol/*Dysidea granulosa*	*In vitro* antibacterial activity, especially against methicillin-resistant and sensitive *S. aureus*, vancomycin-resistant and -sensitive *Enterococci* and *Bacillus* sp[93]
4,4′,6-tribromo-2,2′-biphenol/ *Pseudoalteromonas* sp. CMMED 290	Significant antibacterial activity against methicillin-resistant *S. aureus*[94]
2,2′,3-tribromo-biphenyl-4,4′-dicarboxylic acid/*the marine bacterium Pseudoalteromonas phenolica* O-BC30T	Exhibited anti-methicillin-resistant *S. aureus* activity against all 10 clinical isolates of these microorganisms, with MIC values between 1 and 4 μg/mL. Also highly active against *B. subtilis* and *Enterococcus serolicida*[95]
5-hydroxyindole-type Alkaloids/*Hyrtios* sp.	*C. albicans* isocitrate lyase inhibitory activity[96]
Nostoflan/*Nostoc flagelliforme*	Anti-HSV-1[97]

13. CONTROLLED CLINICAL TRIALS OF HERBAL MEDICINES AGAINST DERMATOLOGICAL INFECTIONS

Many plant extracts have demonstrated significant antimicrobial activity *in vitro* but only a few are subjected to rigorous controlled clinical trials (CCTs). While the perception that antimicrobials derived from herbal medicines are safer than conventional drugs is not necessarily true, CCTs allow for validation of these claims. There is no escaping the fact that CCTs are the sole way through which evidence regarding the therapeutic effects of herbal medicines can be generated. It allows for validation of therapeutic doses, knowledge of therapeutic, side and adverse effects, as well as clinical inefficacy.[62]

In a CCT regrouping 104 patients with impetigo contagiosa, two different formulations of crude extracts of black tea were administered to patients. An aqueous lotion (1%) and a Vaseline base (5%) prepared from *Thea assamica* were prepared and administered as the patients were separated into two distinct groups. The healing rate of patients under herbal treatment was compared with patients under framycetin and gramicidin ointment treatment. It was recorded that the number of patients cured in the group receiving the Vaseline base (5%) prepared from *Thea assamica* healed at the same rate as patients under antibiotic therapy.[63]

In another CCT involving 30 inpatients either infected or colonized with MRSA, the patients were divided into two groups. The first group received tea tree oil (TTO) 4% nasal ointment and 5% body wash for 1–34 days; this constituted the experimental group (EG). The control group (CG) was treated with 2% mupirocin nasal ointment and triclosan body wash for 2–14 days. All the recruited patients also received intravenous vancomycin and were routinely screened for MRSA 48 and 96 hours after cessation of topical therapy. Only 18 patients, 8 from the EG and 10 from the CG, completed the trial. More patients from the EG cleared the infection: 5 versus 2 from the CG.[64]

A single-blind randomized clinical trial compared the effects of using 5% TTO versus 5% benzoyl peroxide among 124 patients suffering from mild acne. The number of inflamed and noninflamed lesions were counted and recorded at baseline and then monthly. After 3 months, both treatments were found to be effective, despite the fact that patients under benzyl peroxide treatment improved at a more significant rate. However, lower cases of skin discomfort were recorded in patients receiving TTO (44%) as compared to benzyl peroxide (79%). Recently, a study extending over 4 weeks compared the effects of *Ocimum gratissimum* oil against benzoyl peroxide and a placebo. Students suffering from acne lesions were recruited and four different concentrations of *O. gratissimum* oil were prepared (0.5%, 1%, 2% and 5% v/v) and incorporated into distinct bases: polysorbate 80, cetomacrogol, petrolatum and alcohol. The time taken for half the number of lesions to disappear was taken as a reference to evaluate the efficacy of the natural oil extracts. Indeed, it was found that preparations containing 2% and 5% *Ocimum* oil in alcohol and 5% in cetomacrogol were significantly more active than benzoyl peroxide ($P < 0.05$), while 2% oil in cetomacrogol had similar activity to the reference product.[65]

14. DISCUSSION

The idea of man looking around him and exploiting the available natural resources to identify cures to ailments is an ancient practice. It is believed that nature is a storehouse in which a remedy exists for every ailment.[19,20] The production of synthetic drugs derived from chemical processes on the one hand slowed the tendency for man to look into his immediate natural environment to identify new and different natural cures, and on the other hand has led to a greater dependence on conventional drugs. The discovery of penicillin in the early 1940s paved the way to the development of more and more new classes and generations of antimicrobials during the past half century. While this was considered the golden era of antibiotics, resistance to novel antimicrobials developed soon after among microorganisms, mainly through the lateral transfer of resistant genes.

In addition to the antibiotic resistance phenomena, the use of antimicrobials is accompanied by a plethora of adverse or side effects, resulting in a number of cases of noncompliance among patients with their medication regimen. Concurrently, the overuse of antimicrobials by prescribing when not necessary has accelerated the AMR phenomena.[66] The realization by scientists that the curative potential of antibiotics is obviously limited has again fueled research into the area of natural products. With mounting mortality numbers resulting from DIs every year, probing natural products as novel antimicrobials has become more important than ever.

Indeed, natural products provide an extraordinarily vast array of plant, animal, marine and microbial resources that can be exploited as novel antimicrobials. Plants have always been the first alternative for both primitive and modern man to exploit as medicine. A multiplicity of plants have been identified as antimicrobials by primitive people and recorded in ancient literature and citations. Indeed over 100 South African plants have been mentioned in ancient texts as being antimicrobials. Ethnobotanical studies have revealed myriads of plants, most of which have not been assayed *in vitro* for their claimed effects.[26] A multitude of herbals, fruits and legumes have been assayed against various types of microorganisms, mostly *S. aureus, S. epidermidis, E. coli, Klebsiella pneumoniae, Pseudomonas aeruginosa and C. albicans*. A wide array of plant extracts have produced interesting results against the most resistant dermatological pathogens, such as *S. epidermidis, S. aureus*, and MRSA, among several other bacterial species. Interestingly, some studies have highlighted the synergistic antimicrobial activity resulting from the combination of two different plant species extracts. Other studies depict that the activity of available antibiotics can be made more potent by combining with herbal extracts.

The potential that products of animal origin possess also cannot be neglected. Similar to plants, animals have been at the immediate reach of humans since time immemorial and many animal species are still employed as medicine by various cultures of the world. A number of *in vitro* investigations performed on some animal products, such as pleurocidin, lactoferrin, defensins, protamine, casein and whey depict their efficacy against *E. coli* and *S. aureus*.

During recent years, marine organisms have attracted significant interest from pharmaceutical companies in the hope of discovering novel therapeutic agents. There is no denying that marine organisms present an enormous potential as a drug source against DIs of bacterial, fungal and bacterial origin. The multiplicity of marine sponges investigated against DIs have produced interesting results against *B. subtilis, P. acnes*, MRSA and *S. aureus*. In addition, a plethora of compounds have been isolated from these sponges for the first time. Some marine fungus species, notably *Aspergillus* spp., were found to be effective against a multidrug-resistant strain of MRSA and hence could provide better alternatives for the management and treatment of DIs resulting from MRSA infection.

Highlighting useful plants through ethnobotanical studies against DIs and assaying them *in vitro* are only the baseline steps in the process of unveiling to the world novel antimicrobial extracts, compounds and products. Not every plant from ethnobotanical studies is subjected to *in vitro* assays and few plants producing significant antimicrobial activity against the pathogens involved in DIs are tested clinically. Conducting clinical trials allows for dosage validation, testing for side or adverse effects, comparison of the effects of the natural product against conventional antimicrobials and most importantly allows the investigation of the efficacy of the product in humans. Despite the importance of clinical trials, only a few are performed, particularly with respect to natural products against DIs. The efficacy of vaseline base (5%) prepared from *Thea assamica* was equal to that of framycetin and gramicidin ointment in patients suffering from impetigo contagiosa. A higher proportion of patients colonized with MRSA receiving tea tree oil 4% nasal ointment and 5% body wash for 1–34 days cleared the infection compared to those receiving 2% mupirocin nasal ointment and triclosan body wash for 2–14 days. In patients with acne, the use of TTO cleared fewer lesions than benzoyl peroxide, but skin discomfort from using this natural product was much lower.

15. CONCLUSION

Natural products provide a diversified source of antimicrobial compounds in the quest for developing novel antimicrobials for DIs. Most plant species recorded in ethnobotanical studies should be studied further for their purported and time-tested effects in managing different DIs. Other natural products derived from marine, animal and microbial sources provide interesting incentives for the development of newer classes and generations of dermatological antimicrobials. Plant, marine, animal and microbial source products should be evaluated clinically for distinct dermatological infections whereby their safety and efficacy can be determined. Extracts derived from natural products following controlled clinical trials can be developed for use, either singly or in combination with other conventional antimicrobials to relieve sufferers and improve their quality of life, as well as to defy resistant microorganisms.

REFERENCES

1. Ndihokubwayo JB, Yahaya AA, Desta AT, Ki-Zerbo G, Odei EA, Keita B, et al. Antimicrobial resistance in the African Region: Issues, challenges and actions proposed. *African Health Monitor* 2013;(16):27–30.
2. Hay RJ, Johns NE, Williams HC, Bolliger IW, Dellavalle RP, Margolis DJ, Marks R, et al. The global burden of skin disease in 2010: an analysis of the prevalence and impact of skin conditions. *J Invest Dermatol* 2014;**134**:1527–34. http://dx.doi.org/10.1038/jid.2013.446.
3. CDC. Antibiotic resistance threats in the United States. <http://www.cdc.gov/drugresistance/threat-report-2013/pdf/ar-threats-2013-508.pdf>; 2013 [Accessed 29 september 2016].

4. Sheikh HMA. Antimicrobial activity of certain bacteria and fungi isolated from soil mixed with human saliva against pathogenic microbes causing dermatological diseases. *Saudi J Biol Sci* 2010;**17**:331–9. http://dx.doi.org/10.1016/j.sjbs.2010.06.003.
5. Motta AS, Olivera FC, Brandelli A. Screening antimicrobial activity among bacteria isolated from the Amazon Basin. *Braz J Microbiol* 2004;**35**:307–10. http://dx.doi.org/10.1590/s1517-83822004000300007.
6. Nakatsuji T, Gallo RL. Dermatological therapy by topical application of non-pathogenic bacteria. *J Invest Dermatol* 2014;**134**(1):11–4. http://dx.doi.org/10.1038/jid.2013.379.
7. Cooke J, Stephens P, Ashiru-Oredope D, Johnson AP, Livermore DM, Sharland M. Antibacterial usage in English NHS hospitals as part of a national Antimicrobial Stewardship Programme. *Public Health* 2014;**128**:693–7. http://dx.doi.org/10.1016/j.puhe.2014.06.023.
8. Review on Antimicrobial Resistance. Antimicrobial Resistance: *Tackling a Crisis for the Health and Wealth of Nations* 2014.
9. Gupta C, Garg AP, Uniyal RC, Kumari A. Comparative analysis of the antimicrobial activity of cinnamon oil and cinnamon extract on some food-borne microbes. *Afr J Microbiol Res* 2008;**2**(9):247–51.
10. Bradacs G, Heilmann J, Weckerle CS. Medicinal plant use in Vanuatu: a comparative ethnobotanical study of three Islands. *J Ethnopharmacol* 2011;**137**:434–48. http://dx.doi.org/10.1016/j.jep.2011.05.050.
11. Sender R, Fuchs S, Milo R. Revised estimates for the number of human and bacteria cells in the body. *PLoS Biol* 2016;**14**(8)http://dx.doi.org/10.1371/journal.pbio.1002533. e1002533.
12. Riain UN. Recommended management of common bacterial skin infections. *Prescriber* 2013;**22**(15–16):14–24. http://dx.doi.org/10.1002/psb.1143.
13. Oninla OA. Superficial fungi skin infections: the bane of dermatoses in Nigeria. *J Med Microb Diagn* 2014;**3**:152. http://dx.doi.org/10.4172/2161-0703.1000152.
14. Roberts DT, Taylor WD, Boyle J. Guidelines for treatment of onychomycosis. *Br J Dermatol* 2003;**148**:402–10. http://dx.doi.org/10.1046/j.1365-2133.2003.05242.x.
15. Aly R. Microbial infections of skin and nails. In: Baron S, editor. *Medical microbiology*. Galveston, TX: University of Texas Medical Branch at Galveston; 1996 [chapter 98].
16. Kaplan W, Laing R. *Priority medicines for Europe and the world.* Geneva: World Health Organization; 2004.
17. Sosa A de J, Byarugaba DK, Ama'bile-Cuevas CF, Hsueh P-R, Kariuki S, Okeke IN, editors. *Antimicrobial resistance in developing countries.* New York, Dordrecht, Heidelberg, London: Springer; 2010.
18. Medicines use in primary care in developing and transitional countries: Fact book summarizing results from studies reported between 1990 and 2006. Geneva: World Health Organization. <http://www.who.int/medicines/publications/primary_care_8April09.pdf>; 2009 [Accessed 29 September 2016].
19. Gangadhar M, Shraddha K, Ganesh M. Antimicrobial screening of garlic (*Allium sativum*) extracts and their effect on glucoamylase activity in-vitro. *J Appl Pharm Sci* 2012;**2**(01):106–8.
20. Rahman MS, Salehin MF, Jamal M, Parvin A, Alam MK. Antibacterial activity of *Argemone mexicana* L. against water borne microbes. *Res J Med Plant* 2011;**5**(5):621–6. http://dx.doi.org/10.3923/rjmp.2011.621.626.
21. Beidokhti MN, Prakash D. Antioxidant and anti-inflammatory potential of selected medicinal plants of Lamiaceae family. *Int J Pharm Pharm Sci* 2013;**5**(Suppl. 1):100–4.
22. Van Boeckel TP, Gandra S, Ashok A, Caudron Q, Grenfell BT, Levin SA, et al. Global antibiotic consumption 2000 to 2010: an analysis of national pharmaceutical sales data. *Lancet Infect Dis* 2014;**14**(8):742–50. http://dx.doi.org/10.1016/s1473-3099(14)70780-7.
23. Ashkenazi S. Beginning and possibly the end of the antibiotic era. *J Paediatr Child Health* 2013;**49**(3):E179–82. http://dx.doi.org/10.1111/jpc.12032.
24. Newman DJ, Cragg GM. Natural products as sources of new drugs over the 30 years from 1981 to 2010. *J Nat Prod* 2012;**75**(3):311–35. http://dx.doi.org/10.1021/np200906s.
25. Suroowan S, Mahomoodally F, Ragoo L. Management and treatment of dengue and chikungunya—natural products to the rescue. *Comb Chem High Throughput Screen* 2016;**19**(7):554–64. http://dx.doi.org/10.2174/1386207319666160506123401.
26. Asowata-Ayodele AM, Afolayan AJ, Otunola GA. Ethnobotanical survey of culinary herbs and spices used in the traditional medicinal system of Nkonkobe Municipality, Eastern Cape, South Africa. *S Afr J Bot* 2016;**104**:69–75. http://dx.doi.org/10.1016/j.sajb.2016.01.001.

27. Kaefer CM, Milner JA. The role of herbs and spices in cancer prevention. *J Nutr Biochem* 2008;**19**:347–61. http://dx.doi.org/10.1016/j.jnutbio.2007.11.003.
28. Gyawali R, Ibrahim SA. Natural products as antimicrobial agents. *Food Control* 2014;**46**:412–29. http://dx.doi.org/10.1016/j.foodcont.2014.05.047.
29. Cowan MM. Plant products as antimicrobial agents. *Clin Microbiol Rev* 1999;**12**:564–82.
30. Thomson WAR. *Medicines from the Earth*. Maidenhead: Mc Graw-Hill Book Co.; 1978.
31. Geissman TA. Flavonoid compounds, tannins, lignins and related compounds. In: Florkin M, Stotz EH, editors. *Pyrrole pigments, isoprenoid compounds and phenolic plant constituents*. New York, NY: Elsevier; 1963. p. 285.
32. Stern JL, Hagerman AE, Steinberg PD, Mason PK. Phlorotannin-protein interactions. *J Chem Ecol* 1996;**22**:1887–99. http://dx.doi.org/10.1007/bf02028510.
33. Batista O, Duarte A, Nascimento J, Simones MF. Structure and antimicrobial activity of diterpenes from the roots of *Plectranthus hereroensis*. *J Nat Prod* 1994;**57**:858–61. http://dx.doi.org/10.1021/np50108a031.
34. Borris RP. Natural products research: perspectives from a major pharmaceutical company. *J Ethnopharmacol* 1996;**51**:29–38. http://dx.doi.org/10.1016/0378-8741(95)01347-4.
35. Vijaya K, Ananthan S, Nalini R. Antibacterial effect of theaflavin, polyphenon 60 (*Camellia sinensis*) and *Euphorbia hirta* on *Shigella* spp.—a cell culture study. *J Ethnopharmacol* 1995;**49**:115–8. http://dx.doi.org/10.1016/0378-8741(95)90039-x.
36. Brijesh S, Daswani P, Tetali P, Antia N, Birdi T. Studies on the antidiarrhoeal activity of *Aegle marmelos* unripe fruit: validating its traditional usage. *BMC Complement Altern Med* 2009;**9**(1):47. http://dx.doi.org/10.1186/1472-6882-9-47.
37. Dhankhar S, Ruhil S, Balhara M, Dhankhar S, Chhillar A. *Aegle marmelos* (Linn.) Correa: a potential source of phytomedicine. *J Med Plant Res* 2011;**5**(9):1497–507.
38. Rana B, Singh U, Taneja V. Antifungal activity and kinetics of inhibition by essential oil isolated from leaves of *Aegle marmelos*. *J Ethnopharmacol* 1997;**57**(1):29–34.
39. Venkatesan D, Karunakaran M, Kumar SS, Palaniswamy P, Ramesh G. Antimicrobial activity of *Aegle marmelos* against pathogenic organism compared with control drug. *Ethnobotanical Leaflets* 2009;**13**: 968–74.
40. Chmit M, Kanaan H, Habib J, Abbass M, Mcheik A, Chokr A. Antibacterial and antibiofilm activities of polysaccharides, essential oil, and fatty oil extracted from *Laurus nobilis* growing in Lebanon. *Asian Pac J Trop Med* 2014;**7**(Suppl. 1):S546–52. http://dx.doi.org/10.1016/s1995-7645(14)60288-1.
41. Kareru PG, Keriko JM, Kenji GM, Thiong'o GT, Gachanja AN, Mukiira HN. Antimicrobial activities of skincare preparations from plant extracts. *Afr J Tradit Complement Altern Med* 2010;**7**:214–8. http://dx.doi.org/10.4314/ajtcam.v7i3.54777.
42. Hayek SA, Gyawali R, Ibrahim SA. Antimicrobial natural products. In: Méndez-Vilas A, editor. *Microbial pathogens and strategies for combating them: science, technology and education*. Spain: Formatex Research Centre; 2013. p. 910–21.
43. Mabona U, Viljoen A, Shikanga E, Marston A, Vuuren SV. Antimicrobial activity of southern African medicinal plants with dermatological relevance: from an ethnopharmacological screening approach, to combination studies and the isolation of a bioactive compound. *J Ethnopharmacol* 2013;**148**:45–55. http://dx.doi.org/10.1016/j.jep.2013.03.056.
44. Ikeda I, Imasato Y, Sasaki E, Nakayama M, Nagao H, Takeo T, et al. Tea catechins decrease micellar solubility and intestinal absorption of cholesterol in rats. *Biochim Biophys Acta* 1992;**1127**(2):141–6. http://dx.doi.org/10.1016/0005-2760(92)90269-2.
45. Barrett A, Ndou T, Hughey CA, Straut C, Howell A, Dai Z, et al. Inhibition of alpha-amylase and glucoamylase by tannins extracted from cocoa, pomegranates, cranberries, and grapes. *J Agric Food Chem* 2013;**61**(7):1477–86. http://dx.doi.org/10.1021/jf304876g.
46. Hupkens P, Boxma H, Dokter J. Tannic acid as a topical agent in burns: historical considerations and implications for new developments. *Burns* 1995;**21**(1):57–61. http://dx.doi.org/10.1016/0305-4179(95)90784-w.
47. Lee D, Shin J, Yoon KM, Kim TI, Lee SH, Lee HS, et al. Inhibition of *Candida albicans* isocitrate lyase activity by sesterterpene sulfates from the tropical sponge Dysidea. *Bioorg Med Chem Lett* 2008;**18**:5377–80. http://dx.doi.org/10.1016/j.bmcl.2008.09.059.

48. Ibrahim SA, Bor T, Song D, Tajkarimi M. Survival and growth characteristics of *Escherichia coli* O157: H7 in pomegranate-carrot and pomegranate-apple blend juices. *Food Nutr Sci* 2011;**2**:844. http://dx.doi.org/10.4236/fns.2011.28116.
49. Al-Zoreky NS. Antimicrobial activity of pomegranate (*Punica granatum* L.) fruit peels. *Int J Food Microbiol* 2009;**134**:244–8. http://dx.doi.org/10.1016/j.ijfoodmicro.2009.07.002.
50. Özkan G, Sagdiç O, Göktürk Baydar N, Kurumahmutoglu Z. Antibacterial activities and total phenolic contents of grape pomace extracts. *J Sci Food Agric* 2004;**84**:1807–11. http://dx.doi.org/10.1002/jsfa.1901.
51. Lv F, Liang H, Yuan Q, Li C. In vitro antimicrobial effects and mechanism of action of selected plant essential oil combinations against four food-related microorganisms. *Food Res Int* 2011;**44**:3057–64. http://dx.doi.org/10.1016/j.foodres.2011.07.030.
52. Sevgi M, Toklu A, Vecchio D, Hamblin MR. Topical antimicrobials for burn infections—an update. *Recent Pat Antiinfect Drug Discov* 2013;**8**(3):161–97. http://dx.doi.org/10.2174/1574891x08666131112143447.
53. Muthaiyan A, Martin EM, Natesan S, Crandall PG, Wilkinson BJ, Ricke SC. Antimicrobial effect and mode of action of terpeneless cold-pressed Valencia orange essential oil on methicillin resistant *Staphylococcus aureus*. *J Appl Microbiol* 2012;**112**(5):1020–33. http://dx.doi.org/10.1111/j.1365-2672.2012.05270.x.
54. Srivastava S, Cahill DM, Conlan XA, Adholeya A. A novel in vitro whole plant system for analysis of polyphenolics and their antioxidant potential in cultivars of *Ocimum basilicum*. *J Agric Food Chem* 2014;**62**(41):10064–75. http://dx.doi.org/10.1021/jf502709e.
55. Ramesh B, Satakopan VN. In vitro antioxidant activities of *Ocimum* species: *Ocimum basilicum* and *Ocimum sanctum*. *J Cell Tissue Res* 2010;**10**(1):2145–50.
56. Padalia RC, Verma RS, Chauhan A, Goswami P, Chanotiya CS, Saroj A, et al. Compositional variability and antifungal potentials of *Ocimum basilicum*, *O. tenuiflorum*, *O. gratissimum* and *O. kilimandscharicum* essential oils against *Rhizoctonia solani* and *Choanephora cucurbitarum*. *Nat Prod Commun* 2014;**9**(10):1507–10.
57. Al Abbasy DW, Pathare N, Al-Sabahi JN, Kha SA. Chemical composition and antibacterial activity of essential oil isolated from Omani basil (*Ocimum basilicum* Linn.). *Asian Pac J Trop Dis* 2015;**5**(8):645–9. http://dx.doi.org/10.1016/s2222-1808(15)60905-7.
58. Shabani NRM, Ismail Z, Ismail WI, Zainuddin N, Rosdan NH, Roslan MNF, et al. Antimicrobial activity of cinnamon oil against bacteria that cause skin infections. *JSRAD* 2016;**3**(2):1–6.
59. Lev E. Traditional healing with animals (zootherapy): Medieval to present-day Levantine practice. *J Ethnopharmacol* 2003;**86**:107–18. http://dx.doi.org/10.1016/s0378-8741(02)00377-x.
60. Ely R, Supriya T, Naik CG. Antimicrobial activity of marine organisms collected off the coast of South East India. *J Exp Mar Biol Ecol* 2004;**309**:121–7. http://dx.doi.org/10.1016/j.jembe.2004.03.010.
61. Abad MJ, Bedoya LM, Bermejo P. Marine compounds and their antimicrobial activities. In: Méndez-Vilas A, editor. *Science against microbial pathogens: communicating current research and technological advances*. Spain: Formatex Research Centre; 2011. p. 1293–306.
62. Martin KW, Ernst E. Herbal medicines for treatment of bacterial infections: a review of controlled clinical trials. *J Antimicrob Chemother* 2003;**51**:241–6. http://dx.doi.org/10.1093/jac/dkg087.
63. Sharquie KE, Al-Turfi IA, Al-Salloum SM. The antibacterial activity of tea in vitro and in vivo (in patients with impetigo contagiosa). *J Dermatol* 2000;**27**:706–10. http://dx.doi.org/10.1111/j.1346-8138.2000.tb02263.x.
64. Caelli M, Porteous J, Carson CF, Heller R, Riley TV. Tea tree oil as an alternative topical decolonization agent for methicillin-resistant *Staphylococcus aureus*. *J Hosp Infect* 2000;**46**:236–7. http://dx.doi.org/10.1016/s0962-4562(01)80024-2.
65. Orafidiya LO, AgbaniE O, Oyedele AO, Babalola OO, Onayemi O. Preliminary clinical tests on topical preparations of *Ocimum gratissimum* Linn leaf essential oil for the treatment of acne vulgaris. *Clin Drug Investig* 2002;**22**:313–9. http://dx.doi.org/10.2165/00044011-200222050-00005.
66. Alumran A, Hou XY, Hurst C. Validity and reliability of instruments designed to measure factors influencing the overuse of antibiotics. *J Infect Public Health* 2012;**5**(3):221–32. http://dx.doi.org/10.1016/j.jiph.2012.03.003.

67. Agourram A, Ghirardello D, Rantsiou K, Zeppa G, Belviso S, Romane A, et al. Phenolic content, antioxidant potential and antimicrobial activities of fruit and vegetable by-product extracts. *Int J Food Prop* 2013;**16**(5):1092–104. http://dx.doi.org/10.1080/10942912.2011.576446.
68. Mandalari G, Bisignano C, D'Arrigo M, Ginestra G, Arena A, Tomaino A, et al. Antimicrobial potential of polyphenols extracted from almond skins. *Lett Appl Microbiol* 2010;**51**(1):83–9. http://dx.doi.org/10.1111/j.1472-765x.2010.02862.x.
69. Wonghirundecha S, Sumpavapol P. Antibacterial activity of selected plant by-products against food-borne pathogenic bacteria. In: International conference on nutrition and food sciences. IPCBEE 2012, vol.39: p. 116-120; 2012.
70. Sung SH, Kim KH, Jeon BT, Cheong SH, Park JH, Kim DH, et al. Antibacterial and antioxidant activities of tannins extracted from agricultural by-products. *J Med Plants Res* 2012;**6**(15):3072–9. http://dx.doi.org/10.5897/jmpr11.1575.
71. Taveira M, Silva LR, Vale-Silva LA, Pinto E, Valentaeo P, Ferreres F, et al. Lycopersicon esculentum seeds: an industrial byproduct as an antimicrobial agent. *J Agric Food Chem* 2010;**58**(17):9529–36. http://dx.doi.org/10.1021/jf102215g.
72. Fattouch S, Caboni P, Coroneo V, Tuberoso CI, Angioni A, Dessi S, et al. Antimicrobial activity of Tunisian quince (*Cydonia oblonga* Miller) pulp and peel polyphenolic extracts. *J Agric Food Chem* 2007;**55**(3):963–9. http://dx.doi.org/10.1021/jf062614e.
73. Oliveira I, Sousa A, Ferreira IC, Bento A, Estevinho L, Pereira JA. Total phenols, antioxidant potential and antimicrobial activity of walnut (*Juglans regia* L.) green husks. *Food Chem Toxicol* 2008;**46**(7):2326–31. http://dx.doi.org/10.1016/j.fct.2008.03.017.
74. Sagdic O, Ozturk I, Yilmaz MT, Yetim H. Effect of grape pomace extracts obtained from different grape varieties on microbial quality of beef patty. *J Food Sci* 2011;**76**(7):M515–21. http://dx.doi.org/10.1111/j.1750-3841.2011.02323.x.
75. Friedman M, Henika PR, Levin CE. Bactericidal activities of health promoting, food-derived powders against the foodborne pathogens *Escherichia coli*, Listeria monocytogenes, *Salmonella enterica*, and *Staphylococcus aureus*. *J Food Sci* 2013;**78**(2):M270–5. http://dx.doi.org/10.1111/1750-3841.12021.
76. Djilas S, Markov S. Antioxidant and antimicrobial activities of beet root pomace extracts. *Czech J Food Sci* 2011;**29**(6):575–85. http://dx.doi.org/10.1016/j.jff.2012.04.008.
77. Kanatt SR, Arjun K, Sharma A. Antioxidant and antimicrobial activity of legume hulls. *Food Res Int* 2011;**44**(10):3182–7. http://dx.doi.org/10.1016/j.foodres.2011.08.022.
78. Leleu S, Herman L, Heyndrickx M, De Reu K, Michiels C, De Baerdemaeker J, et al. Effects on Salmonella shell contamination and trans-shell penetration of coating hens' eggs with chitosan. *Int J Food Microbiol* 2011;**145**(1):43–8. http://dx.doi.org/10.1016/j.ijfoodmicro.2010.11.023.
79. Perez KL, Taylor TM. Competitive research and development on antimicrobials and food preservatives. In: Taormina PJ, editor. *Microbiological research and development for the food industry*. New York, NY: CRC Press Taylor and Francis Group; 2013. p. 109.
80. Korhonen HJ, Rokka S. Properties and applications of antimicrobial proteins and peptides from milk and eggs. In: Hettiarachchy NS, Sato K, Marshall MR, Kannan A, editors. *Bioactive food proteins and peptides: applications in human health*. New York, NY: CRC Press Taylor and Francis Group; 2012. p. 49.
81. Demirel Z, Yilmaz-Koz FF, Karabay-Yavasoglu UN, Ozdemir G, Sukatar A. Antimicrobial and antioxidant activity of brown algae from the Aegean Sea. *J Serb Chem Soc* 2009;**74**(6):619–28. http://dx.doi.org/10.2298/jsc0906619d.
82. Bhagavathy S, Sumathi P. Jancy Sherene Bell I. (2011). Green algae *Chlorococcum humicola* a new source of bioactive compounds with antimicrobial activity. *Asian Pac J Trop Biomed* 2011;**1**(1):S1–7. http://dx.doi.org/10.1016/s2221-1691(11)60111-1.
83. Plaza M, Santoyo S, Jaime L, García-Blairsy Reina G, Herrero M, Senorans F, et al. Screening for bioactive compounds from algae. *J Pharm Biomed Anal* 2010;**51**(2):450–5. http://dx.doi.org/10.1016/j.jpba.2009.03.016.
84. Eom SH, Kim YM, Kim SK. Antimicrobial effect of phlorotannins from marine brown algae. *Food Chem Toxicol* 2012;**50**(9):3251–5. http://dx.doi.org/10.1016/j.fct.2012.06.028.
85. Bala N, Aitken EA, Cusack A, Steadman KJ. Antimicrobial potential of Australian macro fungi extracts against foodborne and other pathogens. *Phytother Res* 2012;**26**(3):465–9. http://dx.doi.org/10.1002/ptr.3563.

86. Ramesh C, Pattar MG. Antimicrobial properties, antioxidant activity and bioactive compounds from six wild edible mushrooms of western ghats of Karnataka, India. *Pharmacognosy Res* 2010;**2**(2):107. http://dx.doi.org/10.4103/0974-8490.62953.
87. Ren L, Hemar Y, Perera CO, Lewis G, Krissansen GW, Buchanan PK. Antibacterial and antioxidant activities of aqueous extracts of eight edible mushrooms. *Bioact. Carbohydr. Diet Fibre* 2014;**3**:41–51. http://dx.doi.org/10.1016/j.bcdf.2014.01.003.
88. Xu WH, Ding Y, Jacob MR, Agarwal AK, Clark AM, Ferreira D, et al. Puupehanol, a sesquiterpene hydroquinone derivative from the marine sponge *Hyrtios* sp. *Bioorg Med Chem Lett* 2009;**19**:6140–3. http://dx.doi.org/10.1016/j.bmcl.2009.09.015.
89. Zhang H, Khalil ZG, Capon RJ. Fascioquinols A-F: bioactive meroterpenes from a deep-water Southern Australian marine sponge, Fasciospongia sp. *Tetrahedron* 2011;**67**:2591–5. http://dx.doi.org/10.1016/j.tet.2011.02.015.
90. Kamei Y, Sueyoshi M, Hayashi K, Terada R, Nozaki H. The novel anti-propionibacterium acnes compound, sargafuran, found in the marine brown algae *Sargassum macrocarpum*. *J Antibiot* 2009;**62**:259–63. http://dx.doi.org/10.1038/ja.2009.25.
91. Nguyen HP, Zhang D, Lee V, Kang JS, Choi HD, Son BW. Dehydrochlorofusarielin B, an antibacterial polyoxigenated decalin derivative from the marine-derived fungus *Aspergillus* sp. *J Nat Prod* 2007;**70**:1188–90. http://dx.doi.org/10.1021/np060552g.
92. Boonlarppradab C, Faulkner DJ. Eurysterols A and B, cytotoxic and antifungal steroidal sulfates from a marine sponge of the genus *Euryspongia*. *J Nat Prod* 2007;**70**:846–8. http://dx.doi.org/10.1021/np060472c.
93. Shridhar DM, Mahajan GB, Kamat VP, Naik CG, Parab RR, Thakur NR, Mishra PD. Antibacterial activity of 2-(2′,4′-dibromophenoxy)-4,6-dibromophenol from *Dysidea granulosa*. *Mar Drugs* 2009;**7**:464–71. http://dx.doi.org/10.3390/md7030464.
94. Feher D, Barlow R, McAtee J, Hemscheidt TK. Highly brominated antimicrobial metabolites from a marine *Pseudoalteromonas* sp. *J Nat Prod* 2010;**73**:1963–6. http://dx.doi.org/10.1021/np100506z.
95. Isnansetyo A, Kamei Y. Anti-methicillin-resistant *Staphylococcus aureus* (MRSA) activity of MC21-B, an antibacterial compound produced by the marine bacterium *Pseudoalteromonas phenolica* O-BC30T. *Int J Antimicrob Agents* 2009;**34**:131–5.
96. Lee HS, Yoon KM, Han YR, Lee RJ, Chung SC, Kim TI, et al. 5-Hydroxyindole-type alkaloids as *Candida albicans* isocitrate lyase inhibitors, from the tropical sponge Hyrtios. *Bioorg Med Chem Lett* 2009;**19**:1051–3.
97. Kanekiyo K, Hayashi K, Takenaka H, Lee JB, Hayashi T. Anti-herpes simplex virus target of an acidic polysaccharide, nostoflan, from the edible blue-green alga *Nostoc flagelliforme*. *Biol Pharm Bull* 2007;**30**:1573–5.

CHAPTER 13

Medicinal Plants Versus Skin Disorders: A Survey From Ancient to Modern Herbalism

A. Parsaeimehr, S.O. Martinez-Chapa, R. Parra-Saldívar
School of Engineering and Science, Tecnologico de Monterrey, Campus Monterrey, Mexico

1. INTRODUCTION

Since ancient times, mankind has searched for cures of diseases in nature. It has been discovered that plants and their extracts are effective for treating skin diseases and disorders. The archeological evidence shows that use of medicinal plants dates back to the Paleolithic era, although the Sumerians created the first list of medicinal plants 5000 years ago.[1,2]

Many cultures have a solid historical background of using plants to treat and cure skin disorders. For example, the Mayans have used peppers (Solanaceae), cacao (Malvaceae), agave (Asparagaceae), and tobacco (Solanaceae) to treat skin disorders, while the Chinese have used *Cassia angustifolia* (Caesalpiniaceae), *Rheum palmatum* (Polygonaceae), *Coptis chinensis* (Ranunculaceae), and *Phellodendron amurense* (Rutaceae).[3,4] The Ethiopians have also applied a range of medicinal plants, such as *Lippia adoensis* (Verbenaceae), *Olinia rochetiana* (Oliniaceae), *Verbascum sinaiticum* (Scrophulariaceae), *Malva parviflora* (Malvaceae), *Acokanthera schimperi* (Apocynaceae), *Euclea schimperi* (Ebenaceae), *Inula confertiflora* (Asteraceae), *Melilotus elegans* (Leguminosae), and *Plumbago zeylanica* (Plumbaginaceae), to treat skin disorders.[5] It is interesting that people from different regions of the world have successfully found efficient endemic medicinal plants for their endemic skin diseases. For example, the Assamese people have consumed milled seeds of *Cajanus cajan* (Fabaceae) and heated oil of *Brassica juncea* (Brassicaceae) mixed with *Cynodon dactylon* (Poaceae), and *Curcuma longa* (Zingiberaceae) juices to cure skin disorders.[6] A survey from the northeastern part of India has shown how medicinal plants such as *Artemisia nilagirica* (Asteraceae), *Calotropis gigantea* (Apocynaceae), *Cannabis sativa* (Cannabaceae), *Cassia alata* (Fabaceae), *Centella asiatica* (Apiaceae), *Datura metal* (Solanaceae), *Drymaria cordata* (Caryophyllaceae), *Jatropha* spp. (Euphorbiaceae), *Litsea cubeba* (Lauraceae), *Mimosa pudica* (Fabaceae), *Plantago major* (Plantaginaceae), and *Plumeria acutifolia* (Apocynaceae) have been widely used against skin infections.[7] In Trás-os-Montes e Alto Douro (northern part of Portugal), *Arctium minus* (Asteraceae) and *Calendula officinalis* (Asteraceae) have been used against skin inflammation,

The Microbiology of Skin, Soft Tissue, Bone and Joint Infections
ISSN 2451-9006, http://dx.doi.org/10.1016/B978-0-12-811079-9.00013-6

callus, and warts. These plants have antiinflammatory, antispasmodic, keratolytic, phlebotonic, and sedative activities. Similarly, *Borago officinalis* (Boraginaceae) and *Erythraea centaurium* (Gentianaceae), having emollient, dermatologically protective, detoxifying, antidysenteric, and antipyretic activities, have been considered antieczema agents.[8] The Dutch have also used medicinal plants, including *Hypericum perforatum* (Hypericaceae), *Matricaria recutita* (Asteraceae), *Thymus vulgaris* (Lamiaceae), *Urtica dioica* (Urticaceae), *Valeriana officinalis* (Caprifoliaceae), *Achillea millefolium* (Asteraceae), *Alchemilla vulgaris* (Rosaceae), *Allium cepa* (Amaryllidaceae), *Melissa officinalis* (Lamiaceae), *Mentha piperita* (Lamiaceae), and *Sambucus nigra* (Adoxaceae) against skin disorders.[9] In South Africa, plants, such as *Croton sylvaticus* (Euphorbiaceae), *Warburgia salutaris* (Canellaceae), *Pentanisia prunelloides* (Rubiaceae), and *Melianthus comosus* (Melianthaceae), which are known for their antioxidant, antiinflammatory, and 5-lipoxygenase inhibitory activities, have been also used to cure skin disorders.[10]

Researchers have applied various engineering tools to discover and isolate bioactive molecules from pharmaceutical plants. From 1950 to 1990 around 100 new plant-based medicines, such as deserpidine, rescinnamine, and vincristine, have been introduced.[11] Contemporary statistics reveal that about 25% of medicines in the world are originally derived from plants and many bioactive compounds, such as eguggulsterone, teniposide, nabilone, plaunotol, z-guggulsterone, lectinan, artemisinin, and ginkgolides, have been widely used for pharmaceutical purposes.[12] Medicinal plants possess a great therapeutic potential against various types of skin disorders, and in comparison to synthetic drugs, they are relatively cheap, have fewer side effects, and are more acceptable to the patients.[13] Nowadays, the bioactive compounds from medicinal plants have attracted much attention as alternatives to synthetic drugs. In this chapter, we describe the use of medicinal plants and their bioactive molecules to fight against skin disorders.

2. MEDICINAL PLANTS VERSUS SKIN DISORDERS

2.1 Acne

Acne typically affects skins with a higher number of oil glands and causes inflammation, papules, skin lesions, and postinflammatory hyperpigmentation. Evidence shows that acne has genetic, hormonal, psychological, and infectious backgrounds.[14] *Staphylococcus epidermidis* and *Propionibacterium* have been identified as bacteria which trigger the inflammation in acne.[15] The Thai medicinal plants, such as *Senna alata* (Fabaceae), *Eupatorium odoratum* (Asteraceae), *Garcinia mangostana* (Clusiaceae), and *Barleria lupulina* (Acanthaceae), have been known to have strong inhibitory effects on these bacteria.[16] Furthermore, it has been known that ethanol extracts of four medicinal plants (*Mollugo pentaphylla* (Molluginaceae), *Matteuccia orientalis* (Onocleaceae), *Angelica anomala* (Apiaceae), and *Orixa japonica* (Rutaceae)) inhibit the growth of these pathogens, and decrease the secretion of interleukin-8 (IL-8) and tumor necrosis factor-alpha

(TNF-α) in THP-1 cells (human monocytic cell line derived from an acute monocytic leukemia patient). Considerably, a low or moderate cytotoxicity in human dermal fibroblasts and HaCaT cells (spontaneously transformed aneuploid immortal keratinocyte cell line from human adult skin) has been reported from the ethanol extracts of *Mollugo pentaphylla*, *Matteuccia orientalis*, and *Angelica anomala*.[17] *Psoralea corylifolia* (Fabaceae) and *Magnoliae officinalis* (Magnoliaceae) have also shown antimicrobial activities 2.8 and 2.5 times higher than salicylic acid, respectively, against *Propionibacterium* acne. The extracts also displayed inhibitory effects on the secretion of IL-8 and TNF-α from THP-1 cells.[18] Seed oil of *Punica granatum* (Lythraceae) has also been used against skin disorders such as acne, since it has stimulatory effects on keratinocyte proliferation, the type I procollagen synthesis, the interstitial collagenase, and the regeneration of epidermis.[19] Another medicinal plant that can be used against acne is *Thymus pulegioides* (Lamiaceae), which is native to Europe and grown in alkaline or neutral soil. The compounds thymol, linalool, fenchone, carvacrol, γ-terpinene, *p*-cymeme, linalool, thymol, β-caryophyllene, hymol, carvacrol, geraniol, linalool, and alpha-terpenyl acetate were identified in this plant.[20,21]

Safaei-Ghomi and colleagues reported that the polar subfraction of methanol extracts of *Thymus pulegioides* has an antioxidant activity higher than butylated hydroxytoluene, and the essential oil of this plant has significant antiviral and antimicrobial activities.[22] Thymus essential oil inhibits the growth of *Propionibacterium* and the activation of nuclear factor κB (NF-κB) involved in transcription of cytokines related to inflammatory responses.[23] *Candida albicans* is a dimorphic fungus responsible for skin infections and the essential oil of this plant has been known to have anti-*Candida* activities. Furthermore, the essential oil of *Thymus pulegioides* has been reported to have a synergistic effect with fluconazol and amphotericin B against multidrug resistant *Candida albicans*.[24] *Ocimum gratissimum* (Lamiaceae) is a medicinal plant that grows in Africa, Southern Asia, and Latin America. It has been reported that *Ocimum gratissimum* oil with *Aloe vera* (Asphodelaceae) gel works against acne. *Ocimum gratissimum* contains bioactive molecules including thymol, eugenol, methyl chavicol, linalool, limonene, β-pinene, and α-terpineol which fight against acne.[25,26] Orafidiya and colleagues reported that the *Ocimum* oil (2%) in cetomacrogol blend base was more efficient against acne compared to benzoyl peroxide 10% lotion (an organic compound used for curing acne). [27] *Vitex agnus-castus* (Verbenaceae) has also been used against acne. This plant contains the compounds artemetin, casticin, 3-hydroxy-5,6,7,4-tetramethoxy flavone, penduletin, *p*-hydroxybenzoic acid, methyl 3,4-dihydroxybenzoate, methyl isovanillate, vanillic acid, 3,4-dihydroxybenzoic acid, 1,8-cineole, caryophyllene, terpinen-4-ol, 1,8-cineole, cedrelanol, and 6a,11a-dihydro-6H-[1]benzofuro[3,2-c][1,3]dioxolo[4,5-g]chromen-9-ol. Vitex extracts have been known to affect the pituitary glands to balance secretion of luteinizing and follicle-stimulating hormones, and to regulate estrogen and progesterone levels in hormonal acne (Fig. 1).[28,29]

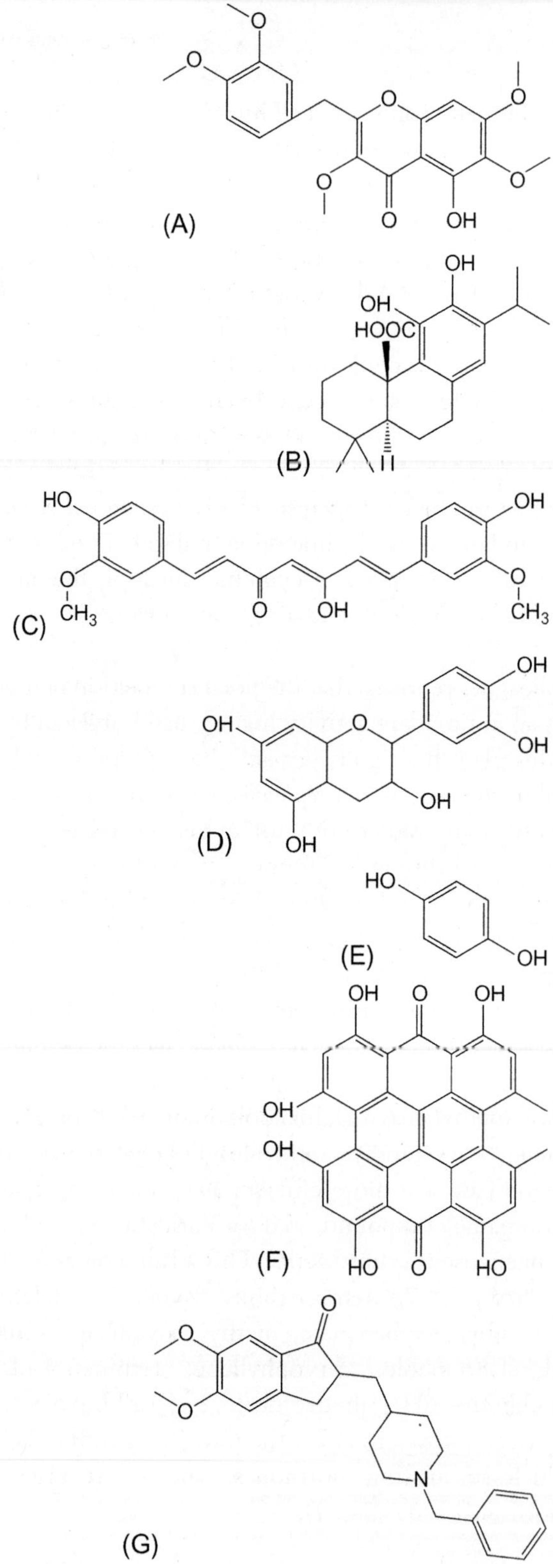

Fig. 1 See figure legend on opposite page.

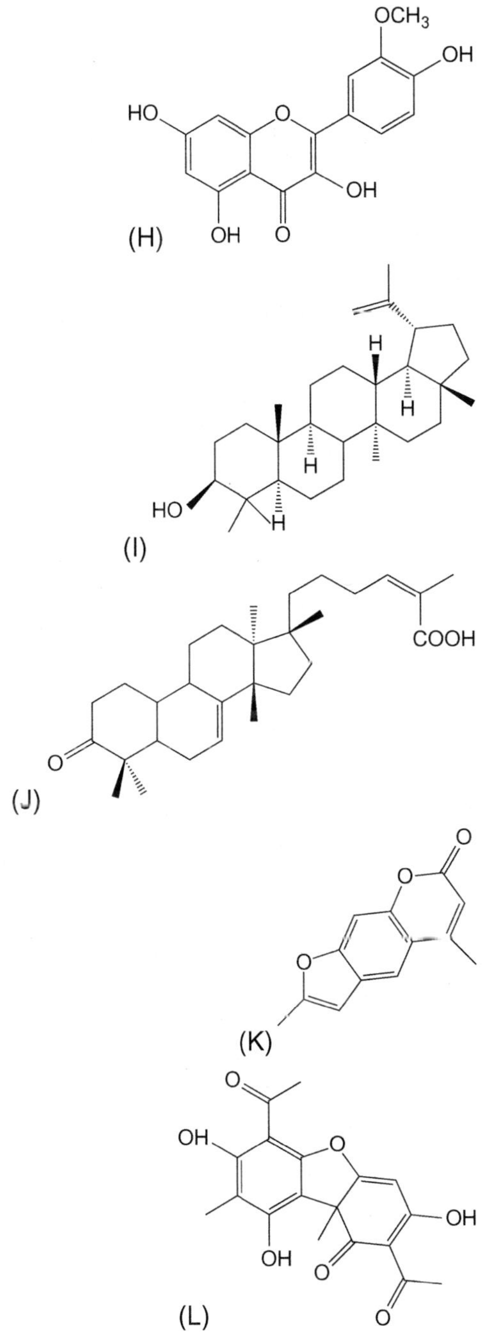

Fig. 1 (A–L) Chemical structures of some bioactive compounds mentioned in the chapter. (A) artemetin, (B) carnosic acid, (C) curcumin, (D) epicatechin, (E) hydroquinone, (F) hypericin, (G) isaindigotone, (H) isorhamnetin, (I) lupeol, (J) masticadienonic acid, (K) trioxsalen, (L) usnic acid.

2.2 Atopic Dermatitis

Atopic dermatitis (AD) is a type of inflammation of skin, and the affected people have an increased risk of infections. Genetic, immune system, and environmental exposures have been believed to play major roles in AD.[30] It has been reported that the ethanol extract of *Impatiens balsamina* (Balsaminaceae) at the dosage of 100 mg/kg significantly inhibits the scratching behavior of NC mice (a mice model for testing AD). Two compounds (kaempferol 3-rutinoside and 2-hydroxy-1, 4-naphthoquinone) isolated from *Impatiens balsamina* have also been identified as inhibiting the scratching behavior of NC mice.[31] *Rheum tanguticum* (Polygonaceae) is another medicinal plant whose rhizomes have inhibitory effects on AD. The oral consumption of ethanol extract of the rhizomes of *Rheum tanguticum* at a dosage of 30–300 mg/kg/day for a duration of 5 weeks inhibited hapten-induced dermatitis in NC/Nga mice, and at the dosage of 100 mg/kg/day decreased the numbers of white blood cells, neutrophils, and eosinophils in the blood. It also significantly reduced the level of immunoglobulin E (IgE) and inhibited 5-lipoxygenase (5-LOX)-catalyzed leukotriene production.[32] The bark of *Betula platyphylla* (Betulaceae) has been applied to cure a range of inflammatory diseases such as AD. Oral consumption of *Betula platyphylla* at a dosage from 25 to 250 mg/kg showed inhibitory influences on lesions, severity of scores, hypertrophy, and inflammation of skin. Further studies showed that the IgE level was significantly decreased by *Betula platyphylla*, even though no change in expression of interferon-γ mRNA was observed. The researchers have concluded that the extracts of *Betula platyphylla* inhibits the development of AD-like skin lesions in NC/Nga mice through the suppression of the T-helper 2 cell response.[33] In another study, *Scutellariae radix* (Lamiaceae) and *Aloe vera* gel were used against AD.

Scutellariae radix decreased the expression level of interleukin 5 (IL-5), but had no influence on IgE. *Aloe vera* upregulated the expression of IgE, and decreased both IL-5 and interleukin 10 (IL-10), and it regulated the immunological responses in AD.[34] In Chinese traditional medicine, the pentaherbs formulation (PHF) has been reported to possess immunomodulatory properties that can be used to treat skin disorders, including AD. The formulation is composed of Herba menthae (1 g), Cortex moutan (2 g), Rhizoma atractylodis (2 g), Flos lonicerae (2 g), and Cortex phellodendri (2 g). It has been shown that consumption of the PHF twice a day at a dosage up to 1 mg/mL is potent enough to reduce the proliferation of peripheral blood mononuclear cells, and supernatant concentrations of brain-derived neurotrophic factor, interferon (IFN)-γ, and TNF-α.[35,36] A study on the use of virgin coconut against AD and *Staphylococcus aureus* infection on 26 patients showed that only 5% of subjects who received twice-daily treatment of virgin coconut for 4 weeks had positive *Staphylococcus aureus* colonies.[37] There are solid pieces of evidence showing Herpes simplex virus (HSV) infects the skin epidermis in AD patients. It has also been demonstrated that HSV infection can be treated with the extracts of medicinal plants, such as *Bidens pilosa* (Asteraceae) and *Houttuynia cordata* (Saururaceae).[38]

2.3 Psoriasis

The inflammatory characteristic of psoriasis alters the epidermis, and it destructively affects the wound-healing process based on the abnormal differentiation of the cells and excessive hyperproliferation of keratinocytes. The extracts of plants such as *Psoralea corylifolia* (Fabaceae) and *Centella asiatica* (Apiaceae) block the keratinocyte replication (IC_{50} values of 18.4 ± 0.6 and 209.9 ± 9.8 mg/mL, respectively), while the antipsoriasis activities of the plant *Centella asiatica* are suggested due to the existence of triterpenoid glycosides madecassoside and asiaticoside compounds (IC_{50} values of 8.6 ± 0.1 and 8.4 ± 0.6 μM respectively) in the extracts.[39,40]

Alpinia galanga (Zingiberaceae), *Curcuma longa*, and *Annona squamosa* have been evaluated for antipsoriasis activity using the HaCaT keratinocyte cell line, and it was found out that each of the selected plants has a specific influence on psoriasis activity. For instance, *Curcuma longa* extract reduced the expression of colony stimulating factor 1 (CSF-1), interleukin 8 (IL-8), nuclear factor NF-kappa-B p100 subunit (NF-kB2), nuclear factor NF-kappa-B p105 subunit (NF-kB1) and nuclear factor NF-kappa-B p65 subunit (RelA). However, *Alpinia galanga* promoted the expression of tumor necrosis factor alpha-induced protein 3 (TNFAIP3), and it inhibited the expression of (CSF-1) and nuclear factor NF-kappa-B p100 subunit (NF-kB2). *Annona squamosa* (Annonaceae) extract decreased the expression of cluster of differentiation 40 (CD40) and NF-kB1.[41]

The isolated flavonoid quercetin from the rhizome of *Smilax china* (Smilacaceae) Linn has been tested against psoriasis using mouse tail (epidermal thickness and percentage orthokeratotic values), and at the dosage of 50 mg/kg significantly inhibited leukocyte migration and showed antiproliferant activity, decreased the percentage of orthokeratotic values, epidermal thickness, and expressed orthokeratosis.[42]

Infection is a part of psoriasis. Previous reports suggested the psoriatic infection can be controlled using 20 mL of the mixed extracts of *Olax scandens* (Olacaceae) and *Holarrhena pubescens* (Apocynaceae) for a duration of 20 days, or using a mix of *Phyllanthus reticulatus* (Euphorbiaceae) and *Ricinus communis* (Euphorbiaceae).[43]

Leukotrienes are known as mediators of inflammatory reactions, and 5-lipoxygenase is known as the key enzyme of leukotriene biosynthesis; thus molecules with anti-5-lipoxygenase activities are valuable for treating psoriasis.[44,45] Reviews by Schneider and colleagues indicated that the plants *Atractylodes lancea* (Asteraceae), *Isatis tinctoria* (Brassicaceae), *Linaria saxatilis* (scrophulariaceae), *Haplophyllum hispanicum* (Rutaceae), *Plantago asiatica* (Plantaginaceae), *Chelidonium majus* (Papaveraceae), *Artocarpus communis* (Moraceae), *Salvia aethiopis* (Lamiaceae), *Belamcanda chinensis* (Iridaceae), *Glycyrrhiza inflate* (Fabaceae), *Dalbergia odorifera* (Fabaceae), and *Asiasarum sieboldii* (Aristolochiaceae) have compounds with lipoxygenase inhibitory activity (IC50 values <1 μm).[46,47] Evidence shows most of the compounds with antilipoxygenase activity belong to the class of phenolic compounds, such as quercetin, magnolol, apigenin, luteolin, siderito-flavone, gnaphalin, silibinin, centaureidin, belamcandol, baicalein.

Boswellia serrata (Burseraceae) and *Pistacia terebinthus* (Anacardiaceae) were also reported as a source of triterpencarboxylic acids, such as acetyl-11-keto-β-boswellic acid, masticadienolic acid, moronic acid, and oleanolic acid, which have inhibitor activities against the 5-lipoxygenase enzyme.[48]

Allium species can play a modulatory role in immune responses. They are potent in inducing lymphocyte proliferation and macrophage phagocytosis, stimulating release of interleukin-2, tumor necrosis factor-alpha and interferon-gamma.[49,50] Extracts of *Allium* inhibit the phase of activation of the macrophage in leishmaniasis disease. The evidence shows a single amount of 20 mg/kg garlic extract or its protein fraction (0.04 mg/kg) inhibits the macrophage in Balb/c mice. More studies explained that *Allium* extracts may have the ability to alter the transferring process of proteins such as f-actin, talin, and paxillin, which are important at the engulfment processes of phagocytosis.[51] Compounds such as diallyl sulfide, S-allylmercaptocysteine, and ajoene are also reported to inhibit the nuclear transcription factor kappaB linked to the activation of psoriasis; these molecules can be isolated from *Allium* species.[52] *Isatis tinctoria* (Brassicaceae) and *Chelidonium majus* are also reported to contain alkaloids including isaindigotone, tryptanthrin and chelerythrine, sanguinarine with anti 5-lipoxygenase activities.[53]

2.4 Skin Cancer

Skin cancer is related to the growth and progress of abnormal cells with the ability to invade and spread; ultraviolet radiation, human papillomavirus, chemicals, and genetic syndromes are some main causes of skin cancer.[54]

Lupeol is a triterpenoid, which has strong antioxidant, antimutagenic, antiinflammatory, and antiarthritic characteristics, and it can be used in pharmaceutical applications. Lupeol can be found in different plants, such as *Senegalia visco* (Fabaceae), *Abronia villosa* (Nyctaginaceae), *Gossampinus malabarica* (Malvaceae), *Strobilanthes callosa* (Acanthaceae), and dandelion coffee.[55,56] Lupeol is reported to have anticarcinogenic activities and can inhibit the expression of nitric oxide synthase, ornithine decarboxylase, and cyclooxygenase-2. It also inhibits the activation of nuclear factor kappa B (NF-κB), and suppresses the activation of the inhibitor of NF-κB kinase subunit alpha (IKK-α) and phosphatidyl inositol 3-kinase (PI3K).[57,58]

Zingiber officinale (Zingiberaceae) is reported to have bioactive compounds, including zerumbone, shogaols, zingerone, vanilloids, [6]-gingerol and [6]-paradol, (3*R*,5*S*)-3,5-diacetoxy-1,7-bis(4-hydroxy-3-methoxyphenyl)heptane, (5*S*)-5-acetoxy-1,7-bis(4-hydroxy-3-methoxyphenyl)heptan-3-one, 3-acetoxy-l,5-epoxy-1-(3,4-dihydroxy-5-methoxyphenyl)-7-(4-hydroxy-3-methoxy-phenyl)heptane, 1,7-bis(4-hydroxy-3-methoxyphenyl)hept-4-en-3-one, which can be used against skin cancer.[59] For example, [6]-gingerol inhibits the angiogenesis of human endothelial cells, synthesis of nitric oxide (NO), epidermal ornithine decarboxylase activity, 12-*O*-tetradecanoylphorbol

13-acetate (TPA)-skin tumor promotion, and NF-κB DNA binding.[60] Zerumbone has also been reported to possess significant anticancer activities, and it is potent in preventing tumor initiation and the promotion phases by induction of antioxidative and reducing inflammatory responses.[61]

Ursolic acid and oleanolic acid are two other bioactive molecules isolated from *Glechoma hederacea* (Lamiaceae) with anticancer activities comparable to retinoic acid, a metabolite of vitamin A and an inhibitor of tumor promotion.[62]

Saraca asoka (Fabaceae) is also reported to have significant amounts of flavonoids, which can be used against skin cancer. Based on the evidence, the flavonoids extracted from *Saraca asoka* have been used to treat 7,12 dimethyl benz(a)anthracene (DMBA) induced skin cancer in mice models. The results showed that the expression of markers including β-D-glucuronidase, myeloperoxidase, sialic acid, hexokinase, and caspase 3 were returned to the normal range by using *Saraca asoka* extracts.[63]

3′-Methoxy-3,4′,5,7-tetrahydroxyflavone (isorhamnetin) is a plant flavonoid with anticancer activities. This molecule can be found in plants such as *Hippophae rhamnoides* (Elaeagnaceae), *Radix Astragali* (Fabaceae), *Rhazya stricta* (Apocynaceae), *Rumex nervosus* (Polygonaceae), *Gomphrena martiana* (Amaranthaceae), *Callianthemum taipaicum* (Ranunculaceae), *Oenanthe javanica* (Apiaceae), and *Arthrocnemum glaucum* (Chenopodiaceae).[64,65] Isorhamnetin targets the epidermal growth factor (EGF) induced neoplastic cell transformation, anchorage dependent and independent growth of A431 human epithelial carcinoma cell (epidermoid carcinoma), kinas activity of phosphoinositide 3-kinase (PI3-K), extracellular signal regulated kinase (MEK1), and mitogen-activated protein (MAP). [66,67]

Epicatechin and its derivatives ((−)epigallocatechin-3-gallate (EGCG), 3,3′,4′,5,7-*O*-penta-acetyl(−)-epicatechin (PEPI), 3′-*O*-ethyl(−)-epicatechin (ethyl EPI)) can be found in a range of medicinal plants; *Acacia catechu* (Fabaceae), *Spatholobus suberectus* (Fabaceae), *Crataegus pinnatifida* (Rosaceae), *Camellia sinensis* (Theaceae), *Vaccinium angustifolium* (Ericaceae), *Scurrula atropurpurea* (Loranthaceae), *Pterocarpus marsupium* (Fabaceae) have all been shown to have anticancer activities and are potent to be used against a range of skin tumor promoters.[68,69]

Curcumin is a phenol compound from the group of curcuminoids and it is isolated from the rhizomes of *Curcuma longa*. The literature shows that curcumin inhibits the lipid peroxidation, glutathione S-transferase activity (GST), thiobarbituric acid reactive substance (TBARS), COX-2 and lipoxygenase activities, and MAPK/extracellular signal-regulated protein kinase 1/2 (MAPK/ERK1/2) levels, which are typically high in skin cancer.[70] Curcumin also induces the cellular tumor antigen p53, blocks the NF-κB cell survival pathway, and increases the level of c-Jun N-terminal kinases (JNK) and P38 mitogen-activated protein kinases.[71]

Silybum marianum (Asteraceae) is a medicinal plant originally from the south of Europe and Asia; this plant contains flavonolignans, including silibinin, isosilibinin, silicristin, and silidianin, which are of importance for treating skin cancers. Evidence shows the flavonolignans of this plant increase the expression of MAPK/ERK1/2, JNK, p38, inhibit SHC-transforming protein (Shc), and the epidermal growth factor receptor (EGF) to reduce cell proliferation and induce apoptosis in cancerogenic cells.[72,73]

2.5 Skin Pigmentation Disorders

The pigment melanin is responsible for giving the skin its color. Different skin disorders can inhibit the synthesis of melanin and damage melanocytes, the melanin-producing cells. Vitiligo is one of the main skin pigmentation disorders (SPDs), and even though the epidermis and dermis are normal in vitiligo patients, the absence of melanin and melanocytes are the main cause for this SPD.[74] The tested extracts of the plants *Paeonia lactiflora* (Paeoniaceae), *Ligusticum chuanxiong* (Apiaceae), *Scutellaria baicalensis* (Lamiaceae), *Ligustrum lucidum* (Oleaceae), *Prunus mume* (Rosaceae), and *Dalbergia odorifera* against vitiligo demonstrated tyrosinase activity (an enzyme that controls the production of melanin). Further experiments showed that melanin was resynthesized from dopachrome by autoxidation and the effect of the medicinal plant extracts was similar to 8-methoxypsoralen (xanthotoxin), a bioactive molecule isolated from *Ammi majus* (Apiaceae) that is used to treat vitiligo.[75]

Psoralens belong to the furocoumarins group of natural products, and they can be isolated from plants such as *Psoralea corylifolia* (Fabaceae), *Ficus carica* (Moraceae), *Apium graveolens* (Apiaceae), *Petroselinum crispum* (Apiaceae), and citrus fruits. Trioxsalen (4,5,8 trimethylpsoralen), a furanocoumarin and a psoralen derivative, has been effectively used against vitiligo. Typically, the patients treated with trioxsalen show the repigmentation of vitiliginous parts and sunlight tolerance.[76]

Brosimum gaudichaudii (Moraceae) is another medicinal plant that has been used for photochemotherapy against psoriasis and vitiligo. *Brosimum gaudichaudii* is a source of compounds such as xanthyletin, and coumarins gaudichaudine, psoralen, luvangetin, and bergapten.[77]

Hypericin is an anthraquinone derivative from *Hypericum perforatum* (Hypericaceae) with antiviral, antibacterial, and nonspecific kinase inhibitor activities that has also been used against vitiligo, and the molecules xanthotoxin (Ammoidin) and imperatorin (Ammidin) from the plant *Ammi majus* (Apiaceae) are other samples which have been applied against vitiligo. The results showed that more than 50% of the subjects were completely treated, while the others had moderate or slight repigmentation.[78,79]

Khellin is a crystalline extract from the plant *Ammi visnaga* (Apiaceae), which has been used by Egyptians to treat ureteral colic. It is suggested that khellin has the potential to be used against vitiligo; this molecule has no considerable photochemical effect on DNA.[80]

Hydroquinone is an aromatic organic compound and a well-known molecule used in the treatment of pigmentation disorders. This molecule can be found in the plants

Vaccinium microcarpum (Ericaceae) and *Arctostaphylos alpine* (Ericaceae). In tests by Clarys and Barel, the efficacy of an ascorbate-phytohydroquinone complex that inhibits the synthesis of melanin and promotes the degradation of the existing melanin was analyzed. After 1 month of treatment, a clear depigmentation of the macules was measured, and none of the volunteers reported any side effects from the prolonged treatment with the hydroquinone-containing product.[81]

Tyrosinase is a copper-containing enzyme and a key enzyme in melanin biosynthesis. Different plants, such as *Heterotheca inuloides* (Asteraceae), *Sophora flavescens* (Fabaceae), *Anacardium occidentale* (Anacardiaceae), *Mentha rotundifolia* (Lamiaceae), *Artocarpus heterophyllus* (Moraceae), *Avocado proanthocyanidins* (Lauraceae), *Broussonetia papyrifera* (Moraceae), *Cornus officinalis* (Cornaceae), *Rhus javanica* (Anacardiaceae), and *Pinus densiflora* (Pinaceae) were reported to have inhibitor activities on tyrosinase.[82]

Hwang and Lee demonstrated that the extracts from the plants *Acorus gramineus* (Acoraceae), *Capsella bursa-pastoris* (Brassicaceae), *Morus bombycis* (Moraceae), *Perilla frutescens* (Lamiaceae), *Quercus dentate* (Fagaceae), *Rhus javanica* (Simaroubaceae), *Schizopepon bryoniaefolius* (Cucurbitaceae), and *Sophora flavescens* (Legumes) have inhibitory effects on melanogenesis (($>$50%) melanin synthesis at 50 μg/mL).[83]

Salidroside is a glucoside of tyrosol that has been isolated from the plant *Rhodiola rosea* (Crassulaceae), and it is reported as a protective agent against damaging UV. Studies are suggesting the potential of this compound for use against depigmentation; a significant inhibition of salidroside (33.0%) and paeonol (22.2%–30.9%), a bioactive molecule isolated from the Paeoniaceae family, was observed on the tyrosinase activity. The potential of salidroside was measured slightly lower than arbutin (18.4%–44.7%), a glycosylated hydroquinone molecule which can be found in the *Arctostaphylos uva-ursi* plant. However, salidroside exhibited a dose-dependent inhibition characteristic in melanin synthesis at a low concentration of 100 μM. The paeonol and arbutin expressed inhibition rates of 27.4%–37.2% and 25.8%–45.6% within 500–1000 μM.

(2*Z*,8*Z*)-Matricaria acid methyl ester is another bioactive molecule isolated from the methanol extract of *Erigeron breviscapus* (Asteraceae). This molecule demonstrated significant depigmentation ($IC_{50}=25.4$ μM), and antityrosinase activities in B16F10 melanoma cells.[84,85]

3. MICROBIAL INFECTIONS OF SKIN

The literature reveals that a diverse range of medicinal plants have effectively been used against microbial infections of skins. For example, *Terminalia avicennioides* (Combretaceae), *Phyllanthus discoideus* (Phyllanthaceae), *Bridelia ferruginea* (Euphorbiaceae), *Ageratum conyzoides* (Asteraceae), *Ocimum gratissimum* (Lamiaceae), and *Acalypha wilkesiana* (Euphorbiaceae) have been used against *Staphylococcus aureus*, a Gram-positive coccal bacterium responsible for skin infections. Additionally, *L. adoensis* and *Olinia rochetiana* have been used for treatment of different infections caused by *Staphylococcus aureus* and the fungi *Trichophyton mentagrophytes* (Arthrodermataceae).[86]

Usnic acid is a dibenzofuran derivative that can be found in the *Usnea barbata* (Parmeliaceae) plant. This bioactive compound has proven to possess a negative influence on Gram-positive bacteria such as *Staphylococcus aureus* (including methicillin-resistant strains—MRSA), the *Propionibacterium* and the *Corynebacterium* species which are responsible for skin infections.[87,88]

It is also suggested that resveratrol, a plant defense molecule, can be used against skins infections. Based on the evidence this molecule has an inhibitory influence on *Staphylococcus aureus*, *Enterococcus faecalis*, and *Pseudomonas aeruginosa* at a dosage from 171 to 342 μg/mL, and at a dosage of 25–50 μg/mL on the dermatophytes fungi such as *Trichophyton mentagrophytes*, *Trichophyton tonsurans*, *Trichophyton rubrum*, *Epidermophyton floccosum*, and *Microsporum gypseum*.[89]

Carnosic acid is known as a phenolic diterpene that can be found in *Rosmarinus officinalis* (Lamiaceae) and *Salvia pachyphylla* (Lamiaceae). This compound has shown anti-inflammatory properties, and can suppress the production of IL-6, IL-8, and MCP-1 from keratinocyte HaCaT cells, which was stimulated with sodium lauryl sulfate and retinoic acid.[90]

Carnosic acid is also reported to be potent in downregulating the expression level of NO, TNF-α, and the prostaglandin E_2 (PGE_2) from RAW264.7 cells activated by the toll-like receptor (TLR)-2 ligands, and blocks the synthesis of lipopolysaccharide, peptidoglycan, pam3CSK, and TLR4 ligand in bacteria.[91] Glycyrrhetinic acid is a triterpenoid component of *Glycyrrhiza glabra* (Fabaceae) that demonstrated its effectiveness against HSV-1 skin infections.[92] The isolated bioactive molecule "prodelphinidin B-2 3,3′-di-*O*-gallate" (PB233′OG), from the bark of *Myrica rubra* (Myricaceae), has also been reported to have antiviral activities with no cellular cytotoxic effects. Studies demonstrated that PB233′OG, on the one hand, inhibits the HSV-2 attachment to Vero cells, interfering with the penetration of HSV-2 into cells, and on the other hand affects the late stage(s) of the HSV-2 infection cycle, reducing the viral infectivity at high viral loading, which can be used against viral infection of skins.[93]

4. CONCLUSION AND PERSPECTIVES

Nature has always generously provided mankind with products that can be used for our health care. Undoubtedly the identification and isolation of bioactive molecules remains a vital key factor in searching for new medicines. Today's bioactive molecules from medicinal plants, such as lupeol, epicatechin, curcumin, and zerumbone, were identified, isolated, and used for pharmaceutical purposes, and some are under biological testing for entering the pharmaceutical markets. Nevertheless, not all the potential of medicinal plants has come to practical usage yet, and substantial numbers of bioactive compounds from plant sources remain undiscovered. Moreover, the mode of action of most of these molecules is not yet well understood; thus usage of new technologies, revolutionary methods for screening secondary metabolites, and more investment in this area of science seem necessary.

ACKNOWLEDGMENT

The constructive comments provided by Dr. Hyundoo Hwang from the School of Engineering and Sciences, Tecnologico de Monterrey, Campus Monterrey, Ave. Eugenio Garza Sada 2501, Monterrey, N.L. 64849, México is thankfully acknowledged.

REFERENCES

1. Srinivasan D, Nathan S, Suresh T, Perumalsamy PL. Antimicrobial activity of certain Indian medicinal plants used in folkloric medicine. *J Ethnopharmacol* 2001;**74**:217–20.
2. Pan SY, Litscher G, Gao SH, Zhou SF, Yu ZL, Chen HQ, et al. Historical perspective of traditional indigenous medical practices: the current renaissance and conservation of herbal resources. *Evid Based Complement Alternat Med* 2014;**27**:1–20.
3. Duke JA. *Duke's handbook of medicinal plants of Latin America*. Taylor & Francis group, Boca Raton London New York: CRC Press; 2008.
4. Cuellar MJ, Giner RM, Recio MC, Manez S, Rios JL. Topical anti-inflammatory activity of some Asian medicinal plants used in dermatological disorders. *Fitoterapia* 2001;**72**:221–9.
5. Solanki R. Treatment of skin diseases through medicinal plants in different regions of the world. *Int J Biomed Res* 2011;**2**:73–88.
6. Saikia AP, Ryakala VK, Sharma P, Goswami P, Bora U. Ethnobotany of medicinal plants used by Assamese people for various skin ailments and cosmetics. *J Ethnopharmacol* 2006;**106**:149–57.
7. Begum D, Nath SC. Ethnobotanical review of medicinal plants used for skin diseases and related problems in Northeastern India. *J Herbs Spices Med Plants* 2000;**7**:55–93.
8. Neves JM, Matos C, Moutinho C, Queiroz G, Gomes LR. Ethnopharmacological notes about ancient uses of medicinal plants in Trás-os-Montes (northern of Portugal). *J Ethnopharmacol* 2009;**124**:270–83.
9. van Asseldonk AG. *Traditional and modern herbalism in The Netherlands*. DG Beek-Ubbergen: Institute for Ethnobotany and Zoopharmacognosy (IEZ); 2001.
10. Frum Y, Viljoen AM. In vitro 5-lipoxygenase and anti-oxidant activities of South African medicinal plants commonly used topically for skin diseases. *Skin Pharmacol Physiol* 2006;**19**:329–35.
11. Verma S, Singh SP. Current and future status of herbal medicines. *Vet World* 2008;**1**:347–50.
12. Rates SM. Plants as source of drugs. *Toxicon* 2001;**39**:603–13.
13. Bandaranayake WM. Quality control, screening, toxicity, and regulation of herbal drugs. In: Ahmad I, Aqil F, Owais M, editors. *Modern phytomedicine: turning medicinal plants into drugs*. KGaA, Weinheim, Germany: Wiley-VCH Verlag GmbH & Co; 2006. p. 25–58.
14. Cordain L, Lindeberg S, Hurtado M, Hill K, Eaton SB, Brand-Miller J. Acne vulgaris: a disease of Western civilization. *Arch Dermatol* 2002;**138**:1584–90.
15. Nishijima S, Kurokawa I, Katoh N, Watanabe K. The bacteriology of acne vulgaris and antimicrobial susceptibility of Propionibacterium acnes and *Staphylococcus* epidermidis isolated from acne lesions. *J Dermatol* 2000;**27**:318–23.
16. Chomnawang MT, Surassmo S, Nukoolkarn VS, Gritsanapan W. Antimicrobial effects of Thai medicinal plants against acne-inducing bacteria. *J Ethnopharmacol* 2005;**101**:330–3.
17. Kim SS, Kim JY, Lee NH, Hyun CG. Antibacterial and anti-inflammatory effects of Jeju medicinal plants against acne-inducing bacteria. *J Gen Appl Microbiol* 2008;**54**:101–6.
18. Lee EJ, Bae SY, NamKung W, Lee YH. Antibacterial and anti-inflammatory effects of medicinal plants against acne-inducing bacteria. *J Soc Cosmet Sci Korea* 2010;**36**:57–63.

19. Kanlayavattanakul M, Lourith N. Therapeutic agents and herbs in topical application for acne treatment. *Int J Cosmet Sci* 2011;**33**:289–97.
20. Oh TH, Kim SS, Yoon WJ, Kim JY, Yang EJ, Lee NH, et al. Chemical composition and biological activities of Jeju Thymus quinquecostatus essential oils against *Propionibacterium* species inducing acne. *J Gen Appl Microbiol* 2009;**55**:63–8.
21. Figueiredo AC, Barroso JG, Pedro LG, Salgueiro L, Miguel MG, Faleiro ML. Portuguese Thymbra and Thymus species volatiles: chemical composition and biological activities. *Curr Pharm Des* 2008;**14**:3120–40.
22. Safaei-Ghomi J, Ebrahimabadi AH, Djafari-Bidgoli Z, Batooli H. GC/MS analysis and in vitro antioxidant activity of essential oil and methanol extracts of *Thymus caramanicus* Jalas and its main constituent carvacrol. *Food Chem* 2009;**115**:1524–8.
23. Zu Y, Yu H, Liang L, Fu Y, Efferth T, Liu X, et al. Activities of ten essential oils towards *Propionibacterium acnes* and PC-3, A-549 and MCF-7 cancer cells. *Molecules* 2010;**15**:3200–10.
24. Khan MS, Malik A, Ahmad I. Anti-candidal activity of essential oils alone and in combination with amphotericin B or fluconazole against multi-drug resistant isolates of *Candida albicans. Med Mycol* 2012;**50**:33–42.
25. Orafidiya LO, Agbani EO, Oyedele AO, Babalola OO, Onayemi O, Aiyedun FF. The effect of aloe vera gel on the anti-acne properties of the essential oil of *Ocimum gratissimum* Linn leaf—a preliminary clinical investigation. *Int J Aromather* 2004;**14**:15–21.
26. Telci I, Bayram E, Yılmaz G, Avcı B. Variability in essential oil composition of Turkish basils (*Ocimum basilicum* L.). *Biochem Syst Ecol* 2006;**34**:489–97.
27. Orafidiya LO, Agbani EO, Oyedele AO, Babalola OO, Onayemi O. Preliminary clinical tests on topical preparations of *Ocimum gratissimum* Linn leaf essential oil for the treatment of acne vulgaris. *Clin Drug Investig* 2002;**22**:313–9.
28. Yarnell E, Abascal K. Herbal medicine for acne vulgaris. *Altern Complement Ther* 2006;**12**:303–9.
29. Vimal A, Vikram L, Singhal S, Anil B. *Vitex negundo*: a Chinese chaste tree. *Int J Pharm Innov* 2011;**5**:9–20.
30. Palmer CN, Irvine AD, Terron-Kwiatkowski A, Zhao Y, Liao H, Lee SP, et al. Common loss-of-function variants of the epidermal barrier protein filaggrin are a major predisposing factor for atopic dermatitis. *Nat Genet* 2006;**38**:441–6.
31. Oku H, Ishiguro K. Antipruritic and antidermatitic effect of extract and compounds of *Impatiens balsamina* L. in atopic dermatitis model NC mice. *Phytother Res* 2001;**15**:506–10.
32. Jin JH, Ngoc TM, Bae K, Kim YS, Kim HP. Inhibition of experimental atopic dermatitis by rhubarb (rhizomes of *Rheum tanguticum*) and 5-lipoxygenase inhibition of its major constituent, emodin. *Phytother Res* 2011;**25**:755–9.
33. Kim EC, Lee HS, Kim SK, Choi MS, Lee S, Han JB, et al. The bark of *Betula platyphylla* var. japonica inhibits the development of atopic dermatitis-like skin lesions in NC/Nga mice. *J Ethnopharmacol* 2008;**116**:270–8.
34. Kim J, Seok Lee I, Park S, Choue R. Effects of *Scutellariae radix* and *Aloe vera* gel extracts on immunoglobulin E and cytokine levels in atopic dermatitis NC/Nga mice. *J Ethnopharmacol* 2010;**132**:529–32.
35. Leung TF, Wong KY, Wong CK, Fung KP, Lam CW, Fok TF, et al. In vitro and clinical immunomodulatory effects of a novel Pentaherbs concoction for atopic dermatitis. *Br J Dermatol* 2008;**158**:1216–23.
36. Chan BC, Hon KL, Leung PC, Sam SW, Fung KP, Lee MY, et al. Traditional Chinese medicine for atopic eczema: PentaHerbs formula suppresses inflammatory mediators release from mast cells. *J Ethnopharmacol* 2008;**120**:85–91.
37. Verallo-Rowell VM, Dillague KM, Syah-Tjundawan BS. Novel antibacterial and emollient effects of coconut and virgin olive oils in adult atopic dermatitis. *Dermatitis* 2008;**19**:308–15.
38. Chiang LC, Chang JS, Chen CC, Ng LT, Lin CC. Anti-Herpes simplex virus activity of *Bidens pilosa* and *Houttuynia cordata. Am J Chin Med* 2003;**31**:355–62.
39. Rahman M, Alam K, Zaki Ahmad M, Gupta G, Afzal M, Akhter S, et al. Classical to current approach for treatment of psoriasis: a review. *Endocr Metab Immune Disord Drug Targets* 2012;**12**: 287–302.

40. Sampson JH, Raman A, Karlsen G, Navsaria H, Leigh IM. In vitro keratinocyte antiproliferant effect of *Centella asiatica* extract and triterpenoid saponins. *Phytomedicine* 2001;**8**:230–5.
41. Saelee C, Thongrakard V, Tencomnao T. Effects of Thai medicinal herb extracts with anti-psoriatic activity on the expression on NF-κB signaling biomarkers in HaCaT keratinocytes. *Molecules* 2011;**16**:3908–32.
42. Vijayalakshmi A, Ravichandiran V, Velraj M, Nirmala S, Jayakumari S. Screening of flavonoid "quercetin" from the rhizome of *Smilax china* Linn. for anti-psoriatic activity. *Asian Pac J Trop Biomed* 2012;**2**:269–75.
43. Subbaiah KV, Savithramma N. Bio-prospecting and documentation of traditional medicinal plants used to treat itching, psoriasis and wounds by ethnic groups of Kurnool District, Andhra Pradesh, India. *Asian J Pharm Clin Res* 2012;**5**:127–31.
44. Rådmark O, Samuelsson B. Regulation of the activity of 5-lipoxygenase, a key enzyme in leukotriene biosynthesis. *Biochem Biophys Res Commun* 2010;**396**:105–10.
45. Dixon RAF, Diehl RE, Opas E, Rands E, Vickers PJ, Evans JF, et al. Requirement of a 5-lipoxygenase-activating protein for leukotriene synthesis. *Nature* 1990;**343**:282–4.
46. Schneider I, Bucar F. Lipoxygenase inhibitors from natural plant sources. Part 2: medicinal plants with inhibitory activity on arachidonate 12-lipoxygenase, 15-lipoxygenase and leukotriene receptor antagonists. *Phytother Res* 2005;**19**:263–72.
47. Schneider I, Bucar F. Lipoxygenase inhibitors from natural plant sources. Part 1: medicinal plants with inhibitory activity on arachidonate 5-lipoxygenase and 5-lipoxygenase [sol] cyclooxygenase. *Phytother Res* 2005;**19**:81–102.
48. Braca A, Dal Piaz F, Marzocco S, Vassallo A, De Tommasi N. Triterpene derivatives as inhibitors of protein involved in the inflammatory process: molecules interfering with phospholipase A2, cycloxygenase, and lipoxygenase. *Curr Drug Targets* 2011;**12**:302–21.
49. Clement F, Pramod SN, Venkatesh YP. Identity of the immunomodulatory proteins from garlic (*Allium sativum*) with the major garlic lectins or agglutinins. *Int Immunopharmacol* 2010;**10**:316–24.
50. Hodge G, Hodge S, Han P. *Allium sativum* (garlic) suppresses leukocyte inflammatory cytokine production in vitro: potential therapeutic use in the treatment of inflammatory bowel disease. *Cytometry* 2002;**48**:209–15.
51. Ghazanfari T, Hassan ZM, Khamesipour A. Enhancement of peritoneal macrophage phagocytic activity against *Leishmania* major by garlic (*Allium sativum*) treatment. *J Ethnopharmacol* 2006;**103**:333–7.
52. Aggarwal BB, Shishodia S. Suppression of the nuclear factor-κB activation pathway by spice-derived phytochemicals: reasoning for seasoning. *Ann N Y Acad Sci* 2004;**1030**:434–41.
53. Werz O. Inhibition of 5-lipoxygenase product synthesis by natural compounds of plant origin. *Planta Med* 2007;**73**:1331–57.
54. Diepgen TL, Mahler V. The epidemiology of skin cancer. *Br J Dermatol* 2002;**146**:1–6.
55. Saleem M. Lupeol, a novel anti-inflammatory and anti-cancer dietary triterpene. *Cancer Lett* 2009;**285**:109–15.
56. Wal A, Srivastava RS, Wal P, Rai A, Sharma S. Lupeol as a magical drug. *Pharm Biol Eval* 2015;**2**:142–51.
57. Saleem M, Afaq F, Adhami VM, Mukhtar H. Lupeol modulates NF-κB and PI3K/Akt pathways and inhibits skin cancer in CD-1 mice. *Oncogene* 2004;**23**:5203–14.
58. Lee TK, Poon RT, Wo JY, Ma S, Guan XY, Myers JN, et al. Lupeol suppresses cisplatin-induced nuclear factor-κB activation in head and neck squamous cell carcinoma and inhibits local invasion and nodal metastasis in an orthotopic nude mouse model. *Cancer Res* 2007;**67**:8800–9.
59. Jiang H, Xie Z, Koo HJ, McLaughlin SP, Timmermann BN, Gang DR. Metabolic profiling and phylogenetic analysis of medicinal *Zingiber* species: tools for authentication of ginger (*Zingiber officinale* Rosc.). *Phytochemistry* 2006;**67**:1673–85.
60. Kim EC, Min JK, Kim TY, Lee SJ, Yang HO, Han S, et al. [6]-Gingerol, a pungent ingredient of ginger, inhibits angiogenesis in vitro and in vivo. *Biochem Biophys Res Commun* 2005;**335**:300–8.
61. Mashhadi NS, Ghiasvand R, Askari G, Hariri M, Darvishi L, Mofid MR. Anti-oxidative and anti-inflammatory effects of ginger in health and physical activity: review of current evidence. *Int J Prev Med* 2013;**4**:1–7.

62. Tokuda H, Ohigashi H, Koshimizu K, Ito Y. Inhibitory effects of ursolic and oleanolic ancid on skin tumor promotion by 12-O-tetradecanoylphorbol-13-acetate. *Cancer Lett* 1986;**33**:279–85.
63. Cibin TR, Devi DG, Abraham A. Chemoprevention of skin cancer by the flavonoid fraction of *Saraca asoka*. *Phytother Res* 2010;**24**:666.
64. Yang RY, Lin S, Kuo G. Content and distribution of flavonoids among 91 edible plant species. *Asia Pac J Clin Nutr* 2008;**17**:275–9.
65. Saud SM, Young MR, Jones-Hall YL, Ileva L, Evbuomwan MO, Wise J, et al. Chemopreventive activity of plant flavonoid isorhamnetin in colorectal cancer is mediated by oncogenic Src and β-catenin. *Cancer Res* 2013;**73**:5473–84.
66. Kim JE, Lee DE, Lee KW, Son JE, Seo SK, Li J, et al. Isorhamnetin suppresses skin cancer through direct inhibition of MEK1 and PI3-K. *Cancer Prev Res* 2011;**4**:582–91.
67. Sun B, Sun GB, Xiao J, Chen RC, Wang X, Wu Y, et al. Isorhamnetin inhibits H_2O_2-induced activation of the intrinsic apoptotic pathway in H9c2 cardiomyocytes through scavenging reactive oxygen species and ERK inactivation. *J Cell Biochem* 2012;**113**:473–85.
68. Agarwal R, Katiyar SK, Zaidi SI, Mukhtar H. Inhibition of skin tumor promoter-caused induction of epidermal ornithine decarboxylase in SENCAR mice by polyphenolic fraction isolated from green tea and its individual epicatechin derivatives. *Cancer Res* 1992;**52**:3582–8.
69. Pourcel L, Routaboul JM, Cheynier V, Lepiniec L, Debeaujon I. Flavonoid oxidation in plants: from biochemical properties to physiological functions. *Trends Plant Sci* 2007;**12**:29–36.
70. F'guyer S, Afaq F, Mukhtar H. Photochemoprevention of skin cancer by botanical agents. *Photodermatol Photoimmunol Photomed* 2003;**19**:56–72.
71. Narayan S. Curcumin, a multi-functional chemopreventive agent, blocks growth of colon cancer cells by targeting β-catenin-mediated transactivation and cell–cell adhesion pathways. *J Mol Histol* 2004;**35**:301–7.
72. Singh RP, Agarwal R. Flavonoid antioxidant silymarin and skin cancer. *Antioxid Redox Signal* 2002;**4**:655–63.
73. Ramasamy K, Agarwal R. Multitargeted therapy of cancer by silymarin. *Cancer Lett* 2008;**269**:352–62.
74. Plensdorf S, Martinez J. Common pigmentation disorders. *Am Fam Physician* 2009;**79**:109–16.
75. Tiechi L, Wenyuan Z, Mingyu X, Meihua Z, Weixin F. Studies on upregulation of 89 Chinese medicinal plant extracts on melanin biosynthesis. *J Clin Dermatol* 1999;**3**.
76. Forrest JB, Forrest HJ. Case report: malignant melanoma arising during drug therapy for vitiligo. *Eur J Surg Oncol* 1980;**13**:337–40.
77. Vilegas W, Pozetti GL, Harumi Yariwake Vilegas J. Coumarins from *Brosimum gaudichaudii*. *J Nat Prod* 1993;**56**:416–7.
78. Ortonne JP. Psoralen therapy in vitiligo. *Clin Dermatol* 1989;**7**:120–35.
79. Fujii HM, Pedroza AR, González W. Vitiligo: its treatment with hypericin. *Prensa Méd Argent* 1966;**53**:1812–7.
80. Abdel-Fattah A, Aboul-Enein MN, Wasset GM, El-Menshawi BS. An approach to the treatment of vitiligo by khellin. *Dermatologica* 1982;**165**:136–40.
81. Clarys P, Barel A. Efficacy of topical treatment of pigmentation skin disorders with plant hydroquinone glucosides as assessed by quantitative color analysis. *J Dermatol* 1998;**25**:412–4.
82. Jung SW, Lee NK, Kim SJ, Han DS. Screening of tyrosinase inhibitor from plants. *Korean J Food Sci Technol* 1995;**27**:891–6.
83. Hwang JH, Lee BM. Inhibitory effects of plant extracts on tyrosinase, L-DOPA oxidation, and melanin synthesis. *J Toxicol Environ Health* 2007;**70**:393–407.
84. Peng LH, Liu S, Xu SY, Chen L, Shan YH, Wei W, et al. Inhibitory effects of salidroside and paeonol on tyrosinase activity and melanin synthesis in mouse B16F10 melanoma cells and ultraviolet B-induced pigmentation in guinea pig skin. *Phytomedicine* 2013;**20**:1082–7.
85. Perfumi M, Mattioli L. Adaptogenic and central nervous system effects of single doses of 3% rosavin and 1% salidroside *Rhodiola rosea* L. extract in mice. *Phytother Res* 2007;**21**:37–43.
86. Akinyemi KO, Oladapo O, Okwara CE, Ibe CC, Fasure KA. Screening of crude extracts of six medicinal plants used in South-West Nigerian unorthodox medicine for anti-methicillin resistant *Staphylococcus aureus* activity. *BMC Complement Altern Med* 2005;**5**:1–7.

87. Segatore B, Bellio P, Setacci D, Brisdelli F, Piovano M, Garbarino JA, et al. *In vitro* interaction of usnic acid in combination with antimicrobial agents against methicillin-resistant *Staphylococcus aureus* clinical isolates determined by FICI and ΔE model methods. *Phytomedicine* 2012;**19**:341–7.
88. Weckesser S, Engel K, Simon-Haarhaus B, Wittmer A, Pelz K, Schempp CM. Screening of plant extracts for antimicrobial activity against bacteria and yeasts with dermatological relevance. *Phytomedicine* 2007;**14**:508–16.
89. Chan MM. Antimicrobial effect of resveratrol on dermatophytes and bacterial pathogens of the skin. *Biochem Pharmacol* 2002;**63**:99–104.
90. Oh J, Yu T, Choi SJ, Yang Y, Baek HS, An S, et al. Syk/Src pathway-targeted inhibition of skin inflammatory responses by carnosic acid. *Mediat Inflamm* 2012;**2012**:1–13.
91. Hortelano S. Molecular basis of the anti-inflammatory effects of terpenoids. *Inflamm Allergy Drug Targets* 2009;**8**:28–39.
92. Khan MT, Ather A, Thompson KD, Gambari R. Extracts and molecules from medicinal plants against herpes simplex viruses. *Antivir Res* 2005;**67**:107–19.
93. Cheng HY, Lin TC, Ishimaru K, Yang CM, Wang KC, Lin CC. In vitro antiviral activity of prodelphinidin B-2 3,3′-di-*O*-gallate from *Myrica rubra*. *Planta Med* 2003;**69**:953–6.

CHAPTER 14

Natural Products as Source of New Antimicrobial Compounds for Skin Infections

A. Pasdaran[*,†], **A. Hamedi**[†,‡]
[*]School of Pharmacy, Research and Development Center of Plants and Medicinal Chemistry, Guilan University of Medical Sciences, Rasht, Iran
[†]Medicinal Plants Processing Research Center, Shiraz University of Medical Sciences, Shiraz, Iran
[‡]School of Pharmacy, Shiraz University of Medical Sciences, Shiraz, Iran

1. INTRODUCTION

Skin, as the outer covering tissue of the human body, creates a protective barrier surface against loss of water and other important metabolites. The skin performs various crucial physiological functions, demonstrating the importance of this organ in human health.[1] Overall, these functions include heat regulation, synthesis and storage of metabolites, absorption, excretion, and more.[2] In addition, the skin as a live and active tissue plays an important role in the prevention of pathogens entering into the body. It also acts as part of the adaptive immune system. Therefore, anatomical or physiological health of skin tissue is very important for human immunity.[3] Based on its anatomical location, the skin is exposed to a large volume of microorganisms. Among the colonizing bacteria, coagulase-negative *staphylococci*, *Staphylococcus epidermidis*, *Corynebacterium*, *Propionibacterium*, *Micrococcus*, and some of the *Acinetobacter* species are the most important microorganisms isolated from the skin.[4,5] Another prevalent microorganism that colonizes skin tissue is the *Malassezia* fungi.[6] A balance of the physical structures and biological parameters of the skin is necessary to maintain human health, and any disruption in this equation can result in skin disorders or infections. In addition to infectious microorganisms, various factors can affect skin susceptibility to infections. The patient's age, immunodeficiencies, chronic underlying disease, wounds, trauma, and location of the affected part are the most important parameters.[7] Diabetic or immunodeficient patients and children are more susceptible to infections from Gram-negative microorganisms.[8,9]

The aim of this chapter is to comprehensively review natural compounds used for the treatment of skin infection, as well as to provide guidelines for selection of suitable sources for screening of new antibacterials based on natural sources.

The Microbiology of Skin, Soft Tissue, Bone and Joint Infections
ISSN 2451-9006, http://dx.doi.org/10.1016/B978-0-12-811079-9.00014-8

2. CLINICAL AND INVESTIGATIONAL APPROACH TO SKIN BACTERIAL INFECTIONS

Although a wide range of microorganisms can be pathogenic to the skin, bacterial and viral infections are the more prevalent. In hospitalized patients, bacterial skin infections are of major concern.[10] Some of the most common clinical forms of these infections are cellulitis, erysipelas, impetigo, folliculitis, furuncles, and carbuncles.[11] Table 1 shows some of the most common clinical forms of skin infections, involved locations and treatment protocols.

With antibiotics losing effectiveness on pathogens over the last decade, drug resistance will be a crucial problem in the near future. On the other hand, several problems associated with the search for novel antimicrobial compounds are impediments to developing new antibiotics.[42]

Some of these determining parameters are as follows:

1. The antibacterial spectrum of a novel compound must be broad enough to cover high diversity of pathogenic microbial species.
2. In view of the molecular aspect, an antimicrobial compound must be highly potent for the microbial targets with little or no harm or side effects on humans. Ideally, these must be different, or absent, in humans. In this situation, novel compounds having specific locking with bacterial receptors present future options on structure relationship activity for optimization of the potency, safety, and efficacy of these compounds for antibiotic development.[43,44]
3. The physicochemical properties of newly found active molecules or "lead" compounds are crucial and have to be optimized for better pharmacokinetic and pharmacodynamics properties.[45] This optimization process for drug candidates may involve molecular weight, lipophilicity, polarity, changing number of hydrogen bond donors and acceptors.[46–48] Any changes in these physicochemical properties may affect the potency or safety of the molecules.
4. Despite the large biodiversity of microbes, one of the main factors for the properties of antibiotic candidates is the cell wall penetration ability of these compounds. The outer membrane, in addition to the peptidoglycan layer and inner membrane in Gram-negative bacteria, serves as an impervious barrier for many chemical compounds.[49,50] Although porins mainly serve as entrance channels for many of the antibacterial compounds, different families of efflux pumps such as resistance-nodulation-cell division (RND super family) also demonstrate the contribution of efflux antibacterial drugs, especially in *Escherichia coli* or *Pseudomonas aeruginosa*.[51] Elimination by efflux pumps, accompanied by outer membrane impermeability, often leads to more resistant Gram-negative bacteria than Gram-positive bacteria against the antibacterial lead compounds.[52,53]

Natural compounds, as valuable sources of chemical structures with diverse biological activities, have great significance for many new drug-screening processes. For example,

Table 1 Most common clinical forms of skin infections, affected location, and treatment protocols

Pathogen name	Common infected form	Category	Treatment protocol	Refs.
Streptococcus pyogenes (Group A beta-hemolytic streptococci)	Perianal cellulitis, nonbullous Impetigo and erysipelas	Gram-positive	*Perianal cellulitis*: Penicillin, a cephalosporin and a macrolide *Nonbullous impetigo*: Topical mupirocin, azithromycin, first-generation cephalosporins and broad-spectrum fluoroquinolones *Erysipelas*: Penicillin, first-generation cephalosporin, glycopeptide antibiotic and macrolides	12–15
Staphylococcus aureus	Bullous impetigo	Gram-positive	*Bullous impetigo*: Topical mupirocin, azithromycin, first-generation cephalosporins and broad-spectrum fluoroquinolones	16–19
Pseudomonas aeruginosa	Hot tub folliculitis	Gram-negative	*Mixed folliculitis*: Clindamycin or erythromycin lotion, topical mupirocin, fluoroquinolones *Pseudomonas aeruginosa folliculitis*: Fluoroquinolones	20–22
Klebsiella, *Enterobacter*, and *Proteus* species	Long-term antibiotic therapy folliculitis	Gram-negative	Isotretinoin, trimethoprim-sulfamethoxazole	23–25
Haemophilus influenza type b (Hib)	Cellulitis, periorbital or orbital cellulitis	Gram-negative	*Patients without underlying disease*: Penicillinase-resistant penicillin, first-generation cephalosporin, amoxicillin-clavulanate, macrolide, or fluoroquinolone *Young children, diabetic and immunosuppressive patients*: Hib vaccine and coverage with a parenteral second- or third-generation cephalosporin *Prophylaxis of recurrent cellulitis*: Erythromycin, penicillin, or clindamycin	26–29
Varicella zoster virus	Herpes zoster, or zona	Varicella zoster virus	Zoster vaccine, acyclovir, valacyclovir or famciclovir	30–32
Candida albicans	Skin candidiasis	Yeast	Topical miconazole, clotrimazole, ketoconazole and nystatin. Oral fluconazole	33,34
Epidermophyton, *Microsporum*, and *Trichophyton* spp.	Superficial and cutaneous mycoses	Dermatophytes	Topicalterbinafine, naftifine and itraconazole	35–37
Human papillomavirus (HPV), *molluscum contagiosum virus* (MCV), *herpes simplex virus*	Raised lesions, warts, genital and nongenital herpes	Virus	Acyclovir, valaciclovir (valacyclovir), famciclovir, and penciclovir	38–41

xenobiotics, which are produced in a competing environment by microorganisms (especially by *Streptomyces*) as a defense mechanism, have been investigated as a source of new antimicrobial compounds.[54,55]

From a chemical structure activity relationship (QSAR) point of view, high structure diversity, in addition to heteroatoms increasing content (mainly oxygen) in structures compared to synthetic compounds, has caused higher average polarity for the natural lead compounds. This causes a better distribution of these compounds in other natural systems, such as the human body. Combined with average polarity (Log *D*), the molecular size (MW) of the lead compounds also plays a significant role in achieving a good level of bacterial cellular permeability.[56,57] Although high polar antibacterial agents such as aminoglycosides can operate either without cellular permeation or lipid membrane penetration, other compounds such as cationic peptides act directly on the cytoplasmic membrane and interact with negatively charged head groups of bacterial phospholipids and then penetrate into the cytoplasmic membrane. This process causes conductance events that lead to the leakage of cell contents and cell death.[58–61]

Considering the need to find new active antibiotics to counteract microbial resistance, and the various factors of QSAR and microbial characters that influence the effectiveness of the natural lead structures as new antimicrobial candidates, understanding the most potential natural remedies, their active chemical classes, and their physicochemical properties is essential.

3. NATURAL REMEDIES IN SKIN INFECTION DISEASES

Application of plant therapeutic remedies for treatment of infectious diseases has a long history in many countries.[62,63] Relatively speaking, a larger number of plants have been used in management of skin infections compared to other organic infections. These remedies comprise a diverse range of hydrophilic or lipophilic extracts and compounds. Folk therapeutic resources, especially plants, have played an important role over a broad spectrum of activities in the management of dermatological conditions[64] (Table 2). The observed antimicrobial, antiinflammatory, and antifungal effects of many of these ethnobotanicals are meaningfully distinctive.[80] Screening of these ethnobotanicals for search of active phytochemical compounds, biological activities, and clinical advantage is a global effort.[81] Many of these plant species are wild growing, endemic, or distinct, which means that drawing attention to these valuable sources to discover and develop new antibiotic compounds is very important.

These traditional usages of medicinal plants have provided an opportunity for developing complementary therapeutics or supplements in many countries. For example, the extract of *Melissa officinalis* L. leaves is a well-known treatment for herpes labialis in Europe.[82] Clinical trials have shown that balm mint cream is effective in the treatment of herpes simplex labialis without any side effects.[82–84]

Table 2 Bioactive extracts of traditionally used plant remedies with antibacterial effects

Plant species (family)	Plant parts used	Indications	Refs.
Acacia zanzibarica S. Moore Taub, *Acacia arabica* (Lam.) Willd. (Mimosaceae)	Root bark, dried fruits	Fungal skin infections	65
Agathisanthemum bojeri Klotzsch., *Geophila repens* (L.) I.M. Johnston (Rubiaceae)	Leaves, whole plant	Fungal skin infections	66
Clutia abyssinica Jaub. & Spach. *Euphorbia drummondii* Boiss. *E. amygdaloides* L. *E. bungei* Boiss. *E. marschalliana* Boiss. (Euphorbiaceae)	Leaves, whole plant	Fungal skin infections, skin sores, warts, dermal infections	67,68
Convolvolus commutatus Boiss. (Convolvolaceae)	Aerial parts	Infected wounds	67,68
Alcea gorganica (Rech. f., Allen & Esfand.) Zohary (Malvaceae)	Flower and root	Infected wounds	67,68
Verbascum gossypinum M. Bieb. (Scrophulariaceae)	Flower and leaves	Dermal infection and inflammations, wart	67,68
Ehretia amoena Klotzsch *Anchusa aegyptiaca* DC. *Cacciniama cranthera* (Banks & Sol.) Brand Var. *macranthera* (Boraginaceae)	Root bark	Fungal skin infections, skin dermatitis and herpes	67–69
Physalis edulis Sims., *Hyoscyamu sreticulatus* L. (Solanaceae)	Leaves	Fungal skin infections, skin problems	69
Dolichosar genteus Willd. (Leguminosae)	Leaves	Fungal skin infections	69
Synaptolepis kirkii Oliver (Thymelaeaceae)	Root	Fungal skin infections	69
Tetracera boiviniana Baill (Dilleniaceae)	Leaves	Fungal skin infections	69
Biscutella didyma L. (Brassicaceae)	Fruit	Skin problems (burns, fungal attacks)	69
Crataegus spp.(Rosaceae)	Leaf, fruit, flower	Skin fungal infections	69
Cajanus cajan (L.) Millsp., *Trigonella foenum-graecum* L. *Astragalus jolderensis* B. Fedtsch. (Fabaceae)	Latex and leaves	Active against fungal infections, inflammations of skin	67,68,70–72
Quercus infectoria Oliv. (Fagaceae)	Bark, leaves, fruits	Skin infections	70–72

Continued

Table 2 Bioactive extracts of traditionally used plant remedies with antibacterial effects—cont'd

Plant species (family)	Plant parts used	Indications	Refs.
Trigonella foenum-graecum L. (Leguminosae)	Seeds, leaves	Skin complaints	70–72
Combretum molle R.Br. ex G.Don, (Combretaceae)	Leaves	Active against fungal infections	73,74
Tabebuia ellanedae Lorentz *ex* Griseb. (Bigoniaceae)	Bark	Active against fungal infections	75
Sarcostemma viminale (L.) R. (Asclepiadaceae)	Stem	Skin sores	76
Polygonum pulchrum Blume, Bijdr. Fl. Ned. *Rumex conglomeratus* Murr. (Polygonaceae)	Roots, leaves	Skin infections, dermal inflammation and infected wounds	67,68,77
Cissus populnea Guill. & Perr. (Vitaceae)	Root	Infected wounds	78
Commiphora habessinica (*O.Berg*) *Engl.* (*Burseraceae*)	Oleo-gum-resin	Fungal skin infections	79

The abdominal gland secretion of *Lytta vesicatoria* (Spanish fly), which is used as a rubifacient and counterirritant, is today used in treating warts, in particular warts caused by human papillomavirus (HPV) and *Molluscum contagiosum virus* (MCV). *Epicauta* genus and some other insects, including *Hycleus lugens* (blister beetle) and *Chauliognathus lugubris* (soldier beetles), have demonstrated high potential action against skin warts.[85,86] The active compound that is isolated from these insects is a terpene called cantharidin (***1***), which can be used topically to treat some benign skin lesions like warts and molluscum.[87–89]

Another example of a traditional remedy, which is used by American Indians, is *Podophyllum peltatum* (mayapple). This plant is used as an emetic and cathartic compound. The roots and rhizomes of this plant contain a lignan known as podophyllotoxin (***2***), which is used as a topical treatment for external genital warts, caused by some types of the HPV.[90,91]

One of the recent fields in the discovery of antibiotics is research on the antimicrobial peptides (AMPs) of arthropods.[92,93] AMPs are categorized into several classes based on their primary structures (amino acid sequences), secondary structures, and functional group similarity, e.g., proline-rich, cysteine-rich peptides purified from *Centruroides limpidus* and *Rapana venosa*.[94–96] Many of these peptides exhibit a broad spectrum of antimicrobial activities against Gram-negative and Gram-positive bacteria and can be considered an interesting research area for the development of new potent antimicrobials.[97] The AMPs of some of the common insects such as *Musca domestica* (houseflies) and

Amblyomma hebraeum (female tick) have shown strong activity against Gram-negative bacteria.[98–101]

Other sources of natural bioactive compounds can be snails, molluscs, and marine arthropods that are exposed to a large number of microbial pathogens in their living environment. The immune system of these creatures has to protect them from different pathogens.[102,103] Many of these specific immunities come from various peptides that show antibacterial properties and help them adapt to the threatening pathogens in the environment[104] (Fig. 1).

Due to the occurrence of bacterial resistance, allergies, and organ complications such as nephrological and hepatological adverse effects, the priority of the administration route of any antibacterial drug is localized to the infection area, especially topical forms for dermal infections.

Despite the huge structural diversity of these antimicrobial classes, several factors limit the application of these isolated compounds as "lead compounds" in future research. One of the most important factors is bioavailability. Many bacterial pathogens, especially *Stratum corneum*, colonize on the epidermis, the skin surface.[105] Other skin structures, particularly sebaceous and sweat glands, could also become infected.[106] In other words, most skin infections are superficial, and thus the property of surface adsorption and skin penetration and permeation is essential for a candidate antibiotic to pass the drug development process.

Fig. 1 Some plants and insects used as traditional and modern antibacterial compounds: (A) *Anchusa aegyptiaca*, (B) *Convolvulus althaeoides*, (C) *Euphorbia amygdaloides*, (D) *Lytta vesicatoria*, (E) *Rapana venosa*, (F) *Chauliognathus lugubris*, (G) *Centruroides limpidus*, (H) *Hycleus lugens*, (I) *Amblyomma hebraeum*, (J) *Musca domestica*, (K) *Epicauta* sp., (L) *Rumex conglomeratus.*

Although most antimicrobials have been isolated from microorganisms, plants are still a good source of new antimicrobial compounds. Plants' first and secondary metabolites, especially fatty acids (FAs), alkaloids, terpenoids and phenolic compounds, which are produced in very diverse structures by all plant species, have offered potent and interesting antimicrobials to researchers. A great deal of research has gone into these compounds, leading to the discovery of many valuable candidates that are now in different stages of clinical trial.

In this chapter, some of these compounds are listed according to their structural class and sources.

4. ALKALOIDS

Alkaloids are a large and structurally diverse group of natural products of microbial, plant, and animal origins. Natural alkaloids and their semisynthetic derivatives as lead compounds have inspired the development of several antimicrobial drugs. For example, purification of quinine resulted in the synthesis of the quinolones and structural changes in similar natural compounds yielding azomycin, metronidazole, and other synthetic drugs.

The nitrogen of alkaloids may occur in the form of a primary amine (RNH2), a secondary amine (R2NH), or a tertiary amine (R3N). In terms of chemical structure, there are two broad divisions of alkaloids:

1. Heterocyclic alkaloids (also known as typical alkaloids) that contain nitrogen in the heterocycle;
2. Nonheterocyclic alkaloids (also known as atypical alkaloids or protoalkaloids) that contain nitrogen in a side chain.

Heterocyclic alkaloids are typically classified according to their ring structure, although the structural complexity of these rings may lead to subdivisions. Nonheterocyclic alkaloids are often classified according to biosynthetic pathways, because their amino acid precursors remain largely intact in the alkaloid structure. Classification of these compounds based on their origin is also possible. Alkaloids are insoluble or sparingly soluble in water, unless reacted with an acid to form a salt. Despite their salts, free alkaloids are soluble in nonpolar solvents such as chloroform.

Several alkaloids from natural sources are reported to possess potent antimicrobial properties and have been considered as candidates to develop new drugs. Subsequent research has been more thorough, and several potently antibacterial alkaloid monomers have been identified in the indole, indolizidine, isoquinoline, piperazine, quinoline, quinolone, and polyamine classes. Alkaloid dimmers with similar levels of activity have been found in the bisindole, indole-quinoline, pyridoacridine, bispyrrole, and pyrrole-imidazole classes (Table 3).

Ramiflorines A (***3***)and B (***4***) which belong to indole alkaloids have been isolated from the stem bark of *Aspidosperma ramiflorum* (Apocynaceae) and have shown significant

Table 3 Natural chemical antimicrobial compounds with their lowest recorded MICs

	Compound name	Microorganisms	MIC (μg/mL)
1	Cantharidin		
2	Podophyllotoxin		
3	Ramiflorines A	*S. aureus*	25
4	Ramiflorines B	*S. aureus*	50
5	Canthin-6-one	*C. glabrata*	12.8
		Trichosporonbeigelii	12.8
		T. cutaneum	12.8
		Multidrug-resistant (MDR) *S. aureus*	8–64
		(MRSA) *S. aureus*	8–64
6	8-Hydroxycanthin-6-one	Multidrug-resistant (MDR) *S. aureus*	8–64
		(MRSA) *S. aureus*	8–64
7	5-Methoxycanthin-6-one	*Trichophyton mentagrophytes* var. *interdigitale*	12.3
		Trichosporon beigelii	22.7
		T. cutaneum	90.8
8	Topsentin		
9	Deoxytopsentin	(MRSA) *S. aureus*	12.5
10	Hamacanthin A	(MRSA) *S. aureus*	3.12
11	Hamacanthin B	(MRSA) *S. aureus*	6.25
12	Keramamine A	*S. aureus*	6.5
13	Keramamine B	*S. aureus*	25
14	5,14-Dimethylbudmunchiamine L1	(MRSA) *S. aureus*	5.5
15	5-Normethylbudmunchiamine K	(MRSA) *S. aureus*	6.5
16	Condaline A	*S. aureus*	25
		S. epidermidis	6.25
17	Scutianine E	*S. aureus*	12.5
		S. epidermidis	3.12
18	6-Nitro-8-methoxy-7, 8-dihydronitidine	*S. aureus*	4
19	Cryptopleurine	*C. albicans*	0.1
20	Julandine	*C. albicans*	12.5
21	Justicidin B	*C. albicans*	4
22	Vitrofolal C	(MRSA) *S. aureus*	≤65
23	Vitrofolal C	(MRSA) *S. aureus*	≤65
24	Detetrahydroconidendrin	(MRSA) *S. aureus*	≤65
25	Heyneanol A	(MRSA) *S. aureus*	2
		Enterococcus faecium	2–4
		S. aureus	2–4
		S. pyogenes	2–4
26	(*E*)-Resveratrol 3-(6″-galloyl)-O-β-D-glucopyranoside	(MRSA) *S. aureus*	125–≥250

Continued

Table 3 Natural chemical antimicrobial compounds with their lowest recorded MICs—cont'd

	Compound name	Microorganisms	MIC (μg/mL)
27	Eupomatenoid-5	*S. aureus*	1.56
		T. rubrum	50
28	Eupomatenoid-6	*S. aureus*	3.12
		M. gypseum	0.5
		T. mentagrophytes	1
29	Conocarpan	*T. rubrum*	6.2
30	Eupomatenoid-3	*T. rubrum*	6.2
31	Mangostanin	Penicillin sensitive *S. aureus*	4
		(MRSA) *S. aureus*	4
32	Xanthone V1	*S. aureus*	1.1
33	Nigrolineaxanthone N	(MRSA) *S. aureus*	4
34	Cochinchinone A	(MRSA) *S. aureus*	9.37
		P. aeruginosa	4.7
35	1,3,7-Trihydroxy-2, 4-diisoprenylxanthone	(MRSA) *S. aureus*	37.5
		Vancomycin-resistant *E. faecalis* (VRE)	75
		P. aeruginosa	4.7
		C. albicans	75
36	Celebixanthonemethylether	*P. aeruginosa*	4.7
37	Cochinchinone L	(MRSA) *S. aureus*	37.5
		P. aeruginosa	4.7
		C. albicans	18.75
38	3-Acetoxy-7-geranyloxy-1-hydroxyxanthone	(MRSA) *S. aureus*	18.75
		P. aeruginosa	4.7
39	1,3-Diacetoxy-7-geranyloxyxanthone	*E. faecalis*	75
		(MRSA) *S. aureus*	37.5
		P. aeruginosa	4.7
		C. albicans	18.75
40	7-Acetoxy-3-geranyloxy-1-hydroxy-xanthone	(MRSA) *S. aureus*	18.75
		P. aeruginosa	4.7
41	1,7-Diacetoxy-3-geranyloxyxanthone	(MRSA) *S. aureus*	37.5
		P. aeruginosa	4.7
42	3-Geranyloxy-1,7-dihydroxyxanthone	(MRSA) *S. aureus*	9.3
		P. aeruginosa	4.7
		C. albicans	37.5
43	7-Geranyloxy-1,3-dihydroxyxanthone	(MRSA) *S. aureus*	18.5
		P. aeruginosa	4.7
		C. albicans	75
44	Kanzonol C	*E. aerogens*	4.9
		C. gabrata	4.9
		C. albicans	4.9

Table 3 Natural chemical antimicrobial compounds with their lowest recorded MICs—cont'd

	Compound name	Microorganisms	MIC (μg/mL)
		E. cloacae	4.9
		M. morganii	4.9
		S. faecalis	4.9
45	Isobavachalcone	*C. albicans*	0.3
		E. aerogens	0.3
		E. cloacae	0.6
		M. morganii	1.2
		Proteus vulgaris	1.2
		P. mirabilis	1.2
		T. rubrum	1.2
46	Stipulin	*S. faecalis*	2.44
		P. aeruginosa	19.53
47	4-Hydroxylonchocarpin	*E. cloacae*	9.8
		M. morganii	9.8
		S. aureus	9.8
		S. faecalis	9.8
		T. rubrum	9.8
48	Pseudopterosin P	*E. faecalis*	3.5
		S. aureus	2
		S. pyogenes	0.8
49	Pseudopterosins A	*E. faecalis*	2.4
		S. aureus	2.1
		S. pyogenes	0.8
50	16α-Hydroxy-cleroda-3,13(14) *Z*-dine-15,16-olide	*S. aureus*	7.5
51	Totarol	Penicillin sensitive *S. aureus*	0.78
		Propionibacterium acnes	3.9
		(MRSA) *S. aureus*	0.78
52	Fusicoserpenol A	*C. albicans*	2[a]
53	Dolabeserpenoic acid A	*C. albicans*	5[a]
54	1,2-Dioxanes,mycaperoxides A	Herpes simplex virus type-1	0.25–1
55	1,2-Dioxanes,mycaperoxides B	Herpes simplex virus type-1	0.25–1
56	Halisulphate	*S. aureus*	6[b]
57	(*E*)-8(17),12-Labadiene-15,16-dial		
58	(*E*)-8β(17)-Epoxylabd-12-ene-15,16-dial		
59	5*R*,8*R*,9*R*,10*R*-Labdane-13(*E*)-ene-8a,15-diol	*C. albicans*	1
60	14α-Hydro-xyisopimara-7,15-dien-1-on	*C. albicans*	0.5[a]
61	7β-Hydroxyisopimara-8(14),15-dien-1-one	*C. albicans*	0.5[a]

[a]In the TLC bioautographic assay.
[b]In dick diffusion assay.

activity against *Staphylococcus aureus* (MIC=25 and 50 μg/mL).[107] Other indole-quinolone alkaloids, canthin-6-one (***5***) and its derivatives such as 8-hydroxycanthin-6-one (***6***) and 5-methoxycanthin-6-one (***7***), which are extracted from *Allium* and *Zanthoxylum* species, exhibit a broad spectrum of activities against *Staphylococcus aureus* and *Candida albicans*.[108–110]

Topsentin (***8***) and hamacanthin (bisindole alkaloids), isolated from the marine sponge *Spongosorite* species, also show suitable activity.

Among hamacanthin compounds, deoxytopsentin (***9***), hamacanthin A (***10***), and hamacanthin B (***11***) exhibit significant antibacterial activity against methicillin-resistant *Staphylococcus aureus* (MIC=12.5, 3.12, and 6.25 μg/mL).[111]

Other alkaloids possessing acceptable activity have macrocyclic structures. This class of compounds has shown good antibacterial properties in many investigations.[112] Keramamine A (***12***) and B (***13***) two β-carboline alkaloid, which belong to this class and are isolated from the Japanese marine sponge *Pellina* sp., act as a potent antibacterial compound against *Staphylococcus aureus* (MIC=6.5, 25 μg/mL). Other macrocyclicspermine alkaloid 5,14-dimethylbudmunchiamine L1 (***14***) and 5-Normethylbudmunchiamine K (***15***), isolated from *Albizia schimperiana* Oliv and *Albizia gummifera*, components of traditional remedies for the treatment of bacterial infections, exhibit significant activity against methicillin-resistant *Staphylococcus aureus* (MIC=5.5, 6.5 μg/mL).[113,114]

The cyclopeptide alkaloids, isolated from the Brazilian plant *Scutia buxifolia* root bark, condaline A (***16***) and scutianine E (***17***), show antimicrobial activity against *Staphylococcus aureus*, *Staphylococcus epidermidis* with MIC=25.0, 12.5, and 6.25, 3.12 μg/mL respectively.[115] Among other classes of alkaloids, the benzophenanthrene alkaloids contain the main portion of antibacterial alkaloids. One interesting example is 6-nitro-8-methoxy-7,8-dihydronitidine (***18***), which is isolated from *Zanthoxylum atchoum* and exhibits potent antibacterial activity against *Staphylococcus aureus* (MIC=4 μg/mL).[116]

Research into the anticandidal activity of *Boehmeria cylindrica* (L.) SW (Urticaceae) has resulted in the isolation of cryptopleurine (***19***) and Julandine (***20***), which are clearly potent against *Candida albicans* (MIC=0.1, 12.5 μg/mL)[117] (Table 3).

5. PHENOLIC COMPOUNDS (LIGNANS, XANTHONES)

Several lignans isolated from plants exhibit potent antimicrobial activities against human skin pathogenic bacteria and fungi such as *Staphylococcus aureus* and *Candida albicans*. Justicidin B (***21***), which is isolated as the major component of *Phyllanthus piscatorum*, inhibited the growth of *Candida albicans* (MIC=4 μg/mL).[118] Vitrofolal C, D (***22***, ***23***) and detetrahydroconidendrin (***24***), which belong to the phenylnaphthalene type lignans from *Vitex rotundifolia*, have been found to act against methicillin-resistant *Staphylococcus aureus*

(MRSA) (MIC > 64 μg/mL).[119] A hydrobenzofuranoide lignan, Heyneanol A (***25***), isolated from the root extract of *Vitis* sp., has showed antibacterial activity against *Staphylococcus aureus*, *Streptococcus pyogenes*, and MRSA (MIC = 2, 4, 2 μg/mL).[120,121]

In other cases, stilbene derivatives also show antimicrobial activity, particularly against Gram-positive pathogens. (*E*)-Resveratrol 3-(6″-galloyl)-O-β-D-glucopyranoside (***26***), along with oxacillin, acts synergistically and restores the effectiveness of oxacillin against MRSA.[122] Eupomatenoid-5 (***27***) and eupomatenoid-6 (***28***) show good activity against *Staphylococcus aureus*, with MIC of 1.56, 3.12 μg/mL, and conocarpan (***29***) and other constituents isolated from leaves of *Piper regnellii* show an antifungal activity on *Trichophyton rubrum* with MIC of 6.2 μg/mL.

Neolignans eupomatenoid-3 (***30***) and eupomatenoid-5 (***27***), isolated from various *Piper* species, show antifungal activity on the dermatophytes with MIC of 50.0, 6.2 μg/mL on *Trichophyton rubrum* and eupomatenoid-6 (***28***) derived from *Piper fulvescens* (Miq.) C. DC. with MIC of 0.5, 1.0 μg/mL on *Microsporum gypseum* and *Trichophyton mentagrophytes*, respectively.[123,124]

Among other phenolic compounds, xanthones and quinones are known for their potent MRSA-inhibitory properties and are therefore considered as lead compounds for synthesizing new antibacterial agents.[125–127] Mangostanin (***31***), a tetraoxygenated xanthone isolated from the fruit of *Garcinia cowa*, is a strong growth inhibitor against two penicillin-sensitive and methicillin-resistant strains of *Staphylococcus aureus* with MIC values of 4.0 μg/mL.[128]

In studies on the compounds of the roots of *Cratoxylum formosum*, several potent antibacterial xanthones have been isolated. Xanthone V1 (***32***) is active against all the tested pathogens including *Staphylococcus aureus* with a MIC of 1.1 μg/mL.[129] Another compound from this chemical class that shows excellent antibacterial activity on MRSA *Staphylococcus aureus* (MIC = 4 μg/mL) is nigrolinea xanthone N (***33***), which is isolated from the leaves of *Garcinia nigrolineata*.[130]

A preliminary screening of xanthones containing natural sources for antibacterial activity specifically against *Pseudomonas aeruginosa* results in the isolation of several xanthones from native Thai plant *Cratoxylum cochinchinense*, which has long been used as a folk remedy for treatment of ulcers, fever, and coughs. Cochinchinone A (***34***), 1,3,7-trihydroxy-2,4-diisoprenylxanthone (***35***), celebixanthone methyl ether (***36***), cochinchinone L (***37***), 3-acetoxy-7-geranyloxy-1-hydroxyxanthone(***38***), 1,3-diacetoxy-7-geranyloxyxanthone (***39***), 7-acetoxy-3-geranyloxy-1-hydroxy-xanthone (***40***), 1,7-diacetoxy-3-geranyloxyxanthone (***41***), 3-geranyloxy-1,7-dihydroxyxanthone (***42***), and 7-geranyloxy-1,3-dihydroxyxanthone (***43***), which are isolated from this plant, demonstrate antibacterial potency with a MIC 4.7 μg/mL against *Pseudomonas aeruginosa*.[131,132]

Kanzonol compounds and derivatives are well known for their antimicrobial potential. Kanzonol C (***44***), isobavachalcone (***45***), stipulin (***46***), and 4-hydroxylonchocarpin (***47***) show broad-spectrum bacteriostatic activities against fungi, Gram-negative and

Gram-positive bacteria. Among these compounds, isobavachalcone (***45***) is highly active against *Trichophyton rubrum*, *Candida albicans*, *Microsporum morganii* (MIC = 1.2, 0.3, and 1.2 μg/mL), *Proteus vulgaris*, *Proteus mirabilis*, *Enterobacter aerogens*, and *Enterobacter cloacae* (MIC = 1.2, 1.2, 0.3, and 0.6 μg/mL).[133,134]

6. DITERPENES

Diterpenes are the most important plant metabolites that are derived from geranyl geranyl pyrophosphate (GGPP) and are classified into several categories, namely phytanes, labdanes, halimane, clerodanes, pimaranes, abietanes, cassanes, rosanes, vouacapanes, podocarpanes, trachlobanes, kauranes, aphidicolanes, stemodanes, stemaranes, bayeranes, atisanes, gibberellanes, taxanes, cembranes, daphnanes, tiglianes, and ingenanes classes. The various biological activities, especially antimicrobial activity, have aroused interest in the identification of novel antibiotic compounds from this class of compounds (Table 3).

Pseudopterosin P (***48***) and pseudopterosins A (***49***), two isolated diterpenes from the Caribbean sea plume *Pseudopterogorgia elisabethae*, show antibacterial activity against *Streptococcus pyogenes*, *Staphylococcus aureus*, and *Enterococcus faecalis* with MIC of 0.8, 2, 3.5 and 0.8, 2.1, 2.4 μg/mL.[135]. A clerodanditerpene, isolated from *Premnaoligotricha* Baker (Verbenaceae) and named 16α-hydroxy-cleroda-3,13(14)*Z*-dine-15,16-olide (***50***), shows antibacterial activity against *Staphylococcus aureus* (MIC = 7.5 μg/mL).[136] Among the diterpenes, totarol and its derivatives have high antibacterial properties noticeably on methicillin-resistant strains of *Staphylococcus aureus* (MRSA) and *Pseudomonas aeruginosa*. The antimicrobial mechanism of totarol has been studied comprehensively.[137,138] Haraguchi et al. have shown that totarol (***51***) can inhibit cells' oxygen consumption, proton translocation in respiratory cycle, and oxidation of NADH in membranes. Based on these observations, several enzymes have been inhibited in bacterial cells by this compound including NADH-DPIP reductase and NADH-CoQ reductase.[139,140]

Two antifungal fusicoccanediterpen, fusicoserpenol A (***52***), dolabeserpenoic acid A (***53***) are isolated from *Hypoestes serpens* (Vahl) R. Br. (Acanthaceae) by bioautographic Thin Layer Chromatography (TLC) assay-guided fractionation and open column chromatography. The minimum amounts of compounds (***52***, ***53***) required for inhibiting *Candida albicans* in the TLC bioautographic assay are 2 μg for (***52***) and 5 μg for (***53***).

Marine organisms are rich sources of bioactive compounds with therapeutic properties. Two nor sesquiterpene 1,2-dioxanes,mycaperoxides A (***54***) and B (***55***) are isolated from Thai sponge genus *Mycale*. These display antiviral activity against herpes simplex virus type-1 with inhibition concentration 0.25–1.0 μg/mL.[141] Halisulphate (***56***), another biologically active sulphated sesquiterpene isolated from the *Halichondriidae* sponge family, shows growth inhibition on *Staphylococcus aureus* at a concentration of 5 μg/dick (Fig. 2).

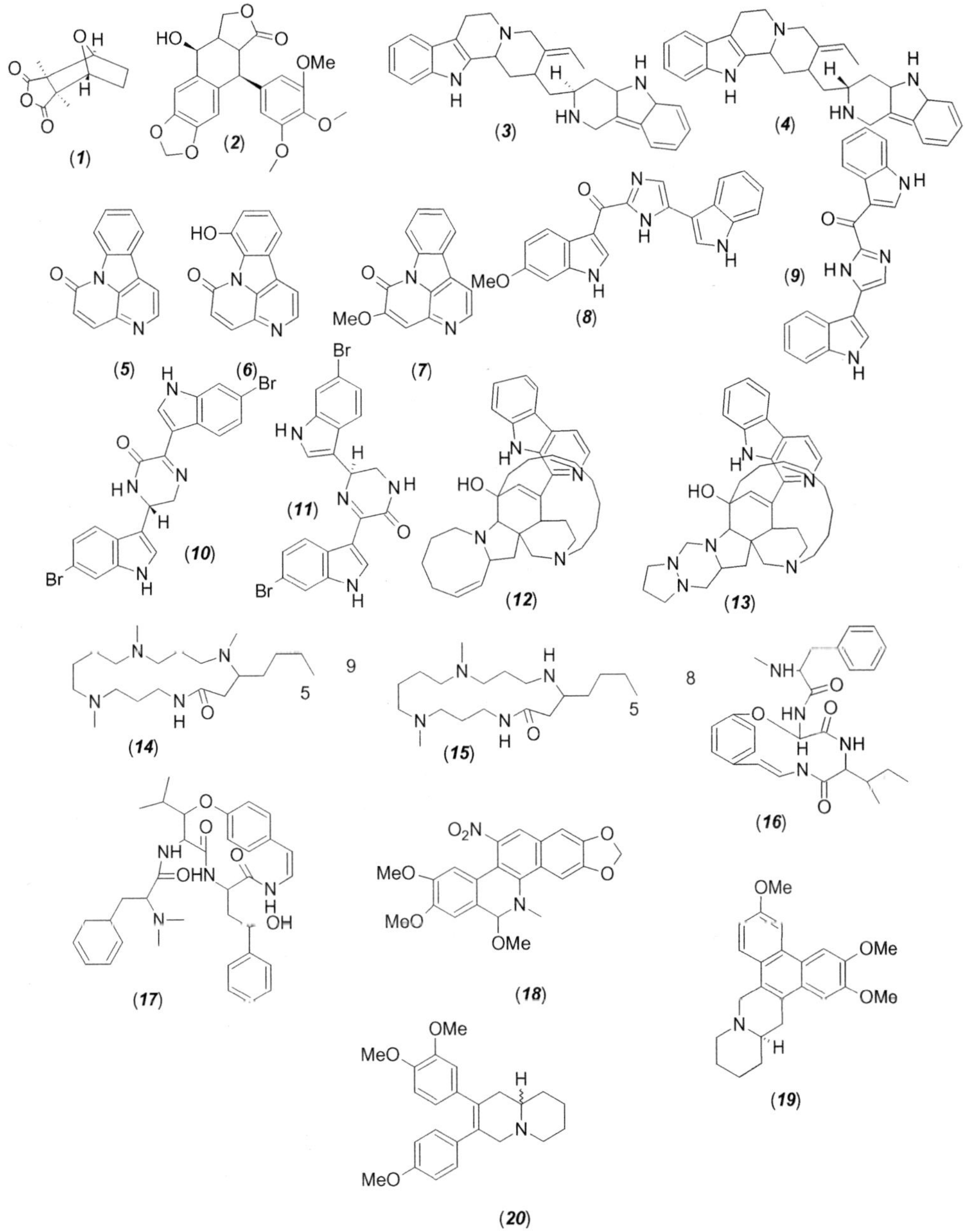

Fig. 2 Structures of some isolated bioactive chemical compounds.

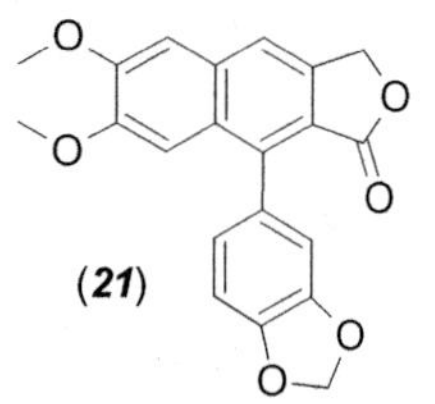
(*21*)

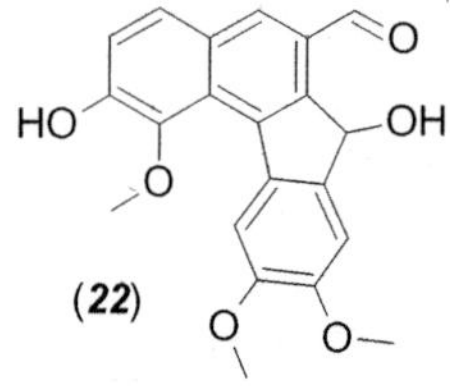
(*22*)

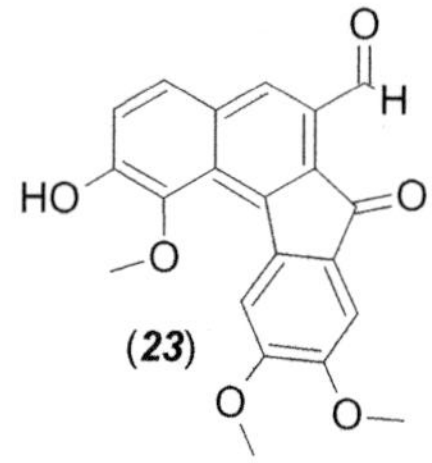
(*23*)

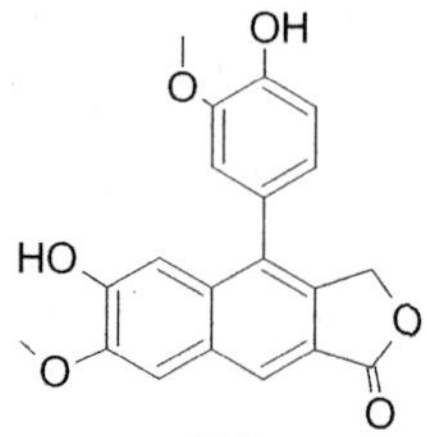
(*24*)

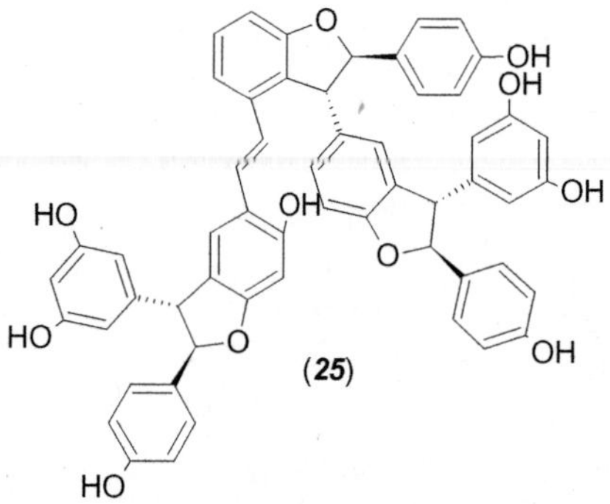
(*25*)

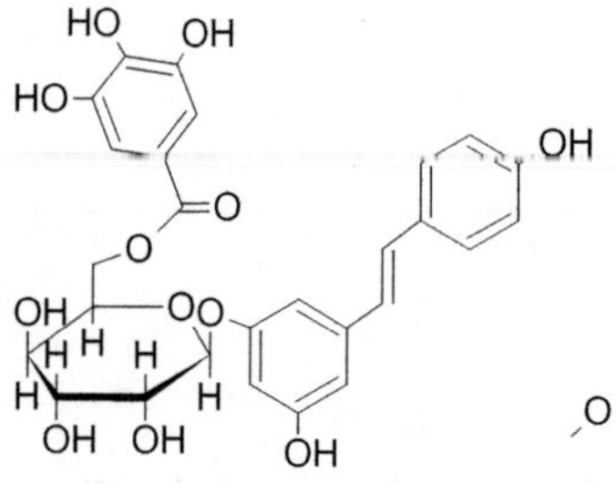
(*26*)

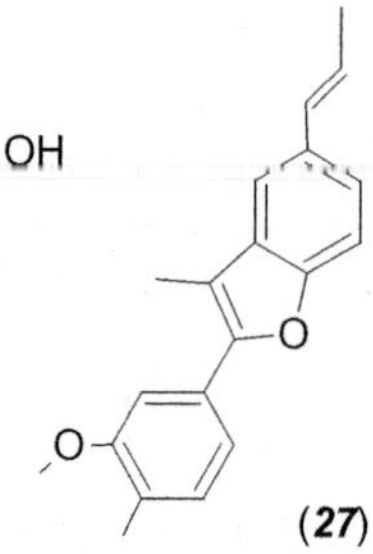
(*27*)

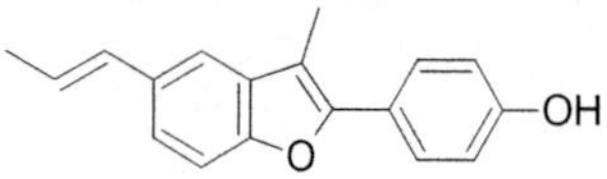
(*28*)

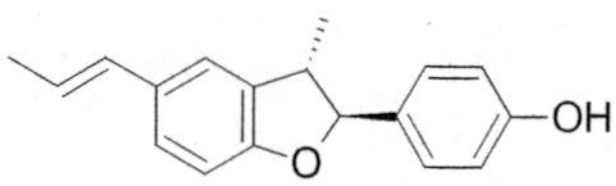
(*29*)

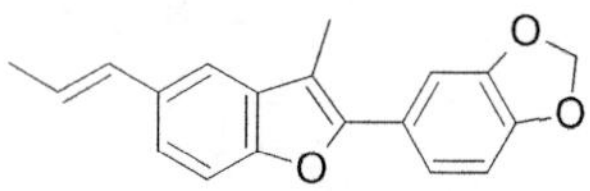
(*30*)

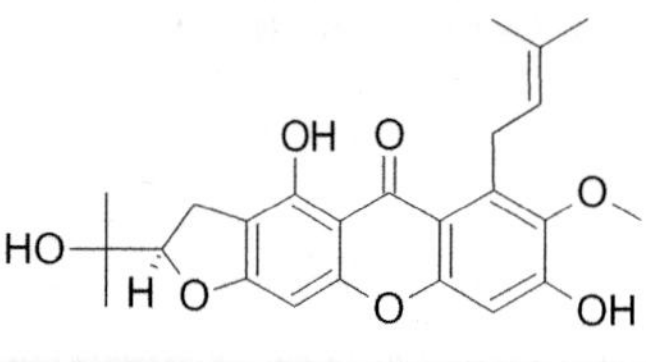
(*31*)

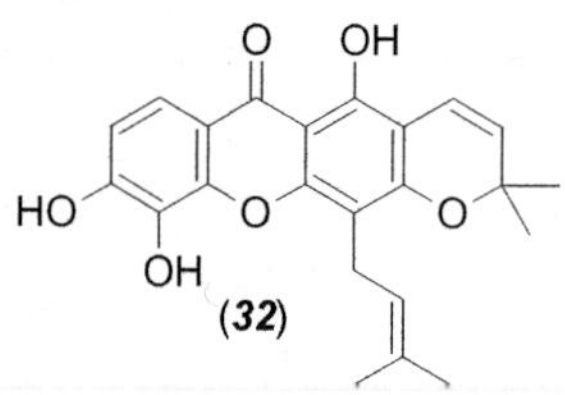
(*32*)

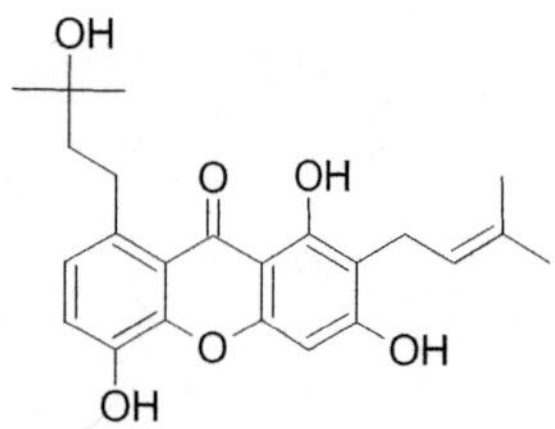
(*33*)

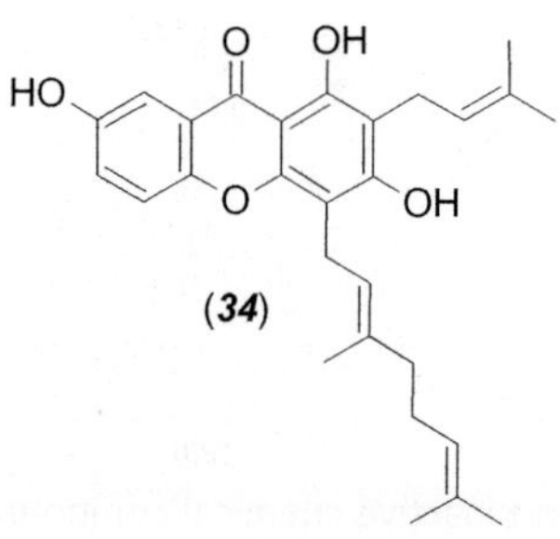
(*34*)

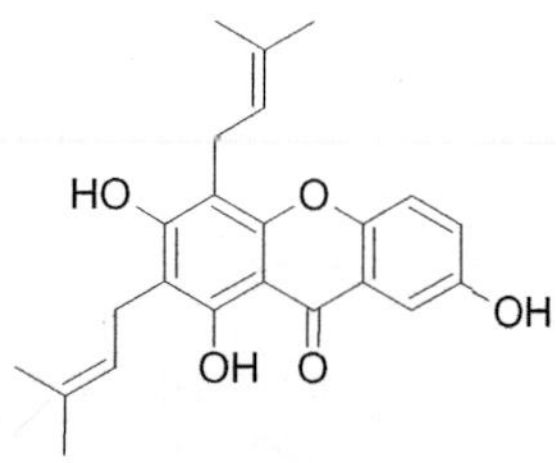
(*35*)

Fig. 2—Cont'd

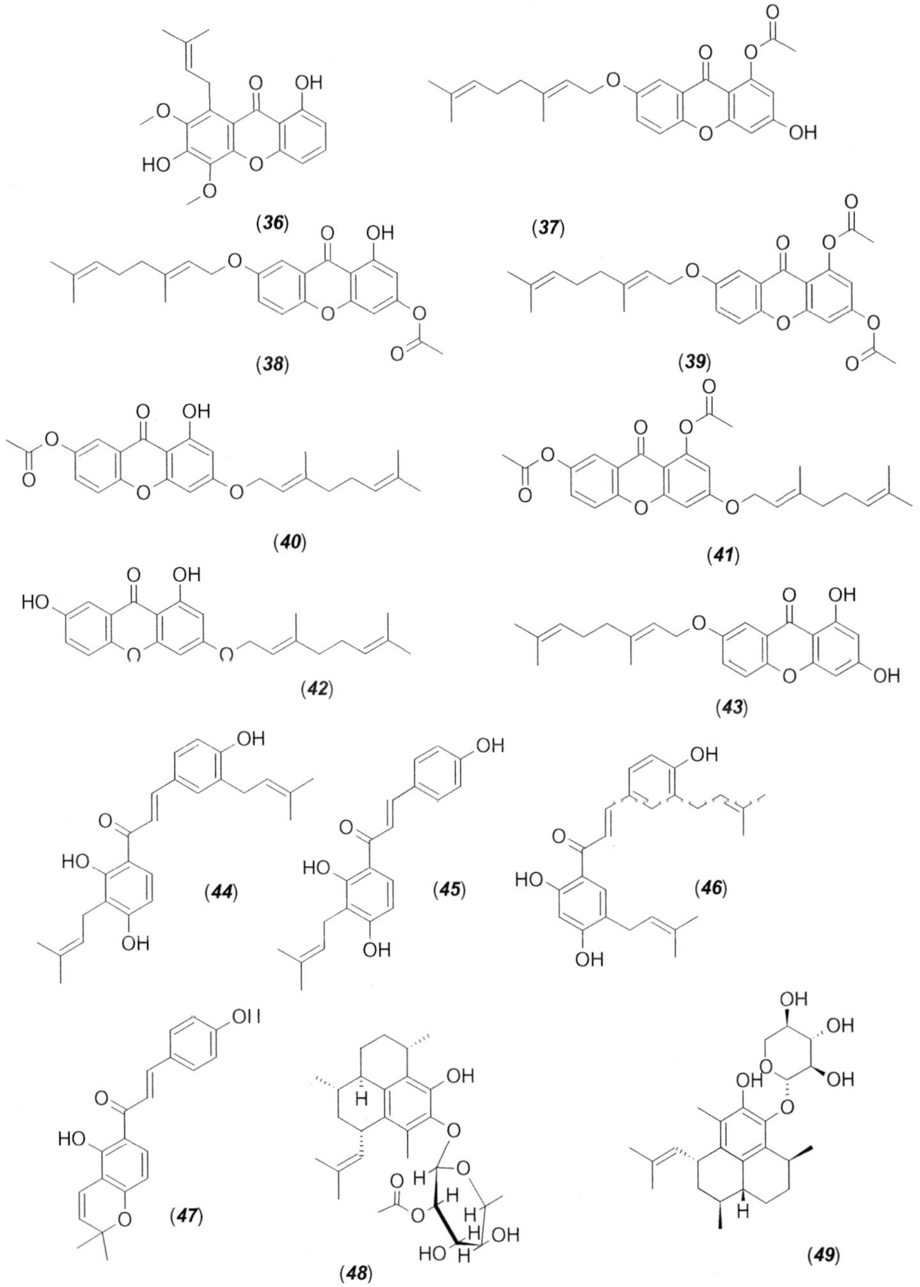

Fig. 2—Cont'd

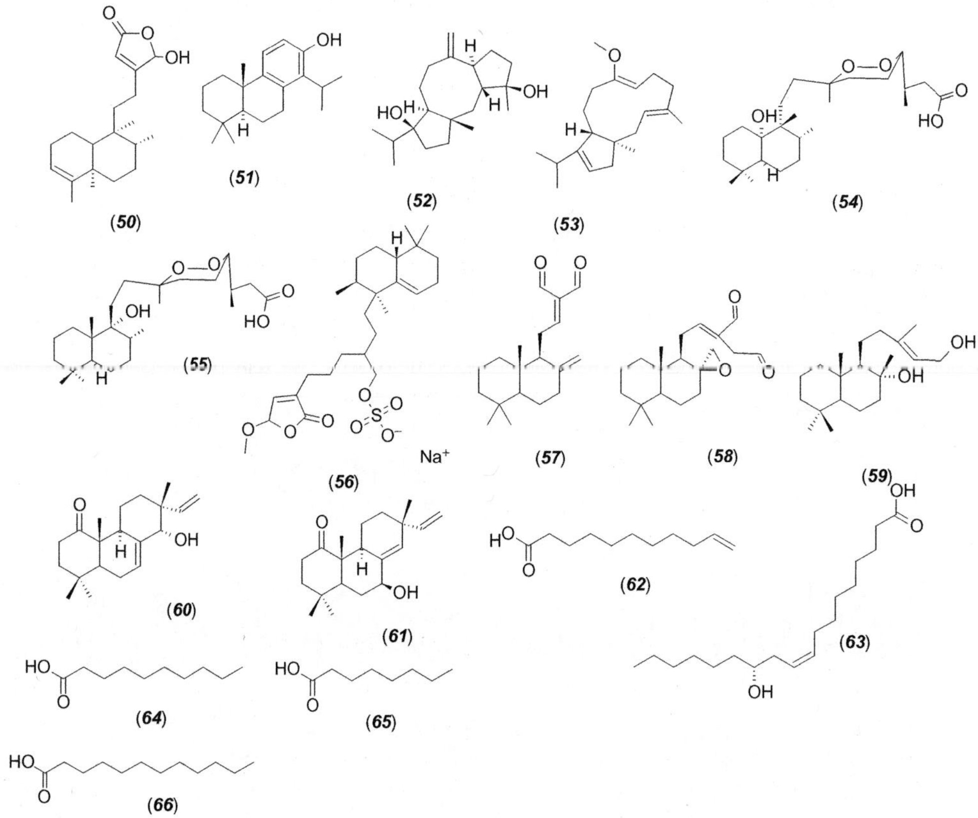

Fig. 2—Cont'd

Some impressive antifungal activities of labdane diterpene have encouraged researchers to focus on isolation and synthesis of this diterpene.[142] In this regard, investigation into two plants of the *Zingiberaceae* family, namely *Alpinia galanga* and *Aframomum daniellii*, have yielded two labedan compounds: (*E*)-8(17),12-labadiene-15,16-dial (***57***), (*E*)-8β(17)-epoxylabd-12-ene-15,16-dial (***58***), which exhibit strong antifungal activity against *Candida albicans*.[143] Another labdane compound, which is isolated from the leaves of *Cistus incanus*, 5*R*,8*R*,9*R*,10*R*-labdane-13(*E*)-ene-8a,15-diol (***59***), is active against *Candida albicans* at concentration of 1.0 μg/mL.[144] Other two labdane class diterpenes, 14α-hydro-xyisopimara-7,15-dien-1-on (***60***), 7β-hydroxyisopimara-8(14),15-dien-1-one (***61***), show more antifungal potency as compared to the reference compound, miconazole. The minimum amount of these compounds required for inhibition of *Candida albicans* in the TLC bioautographic assay is 0.5 μg.[145]

Although numerous natural compounds have shown excellent antifungal activities, there is insufficient research on human dermatophytic fungi compared to other fungi.[146–148] It seems more research is needed on the antifungal potential of diterpenes on human dermatophyte is for the development of new antifungal drugs.[149]

7. FAs AND LIPIDS

Different biological activities have been discovered in natural oils and lipids obtained from biological sources, including algae, animals, and plants. Oils and lipids have been used for different skin conditions since ancient times. They have been used as a vehicle to prepare topical formulations or as solely as active drugs (e.g., sesame oil) for burns, wound healings, and skin infections (e.g., chaulmoogra oil for the treatment of leprosy).[150–152]

FAs, which are the major constituents of natural lipids and oils, are typically bound to other compounds such as glycerol, sugars, or phosphate groups. Free fatty acids (FFAs) can be obtained through enzymatic action or acid hydrolysis.[153] Although FAs may not be as effective as other antimicrobials, they cause less environmental contamination. They are biodegradable and are accepted as food additives, and in many cases their application is generally safe. In many cases, occurrence of resistance is unlikely, and pathogenic fungi are less likely to become resistant to antifungal FAs.[154] FFAs also exhibit a definite role in host defenses against potential pathogenic or opportunistic microorganisms. They can exhibit microbial growth inhibition or can directly kill microbes.

The significance of FFAs in innate human immune system, particularly in the defense of skin and mucosal surfaces, has already been established.[155–157] FFAs and some of their polar derivatives (i.e., phytosphingosine, sphingosine, and dihydrosphingosine) from epidermal *Stratum corneum* exhibit antibacterial activity against a range of Gram-positive and some Gram-negative bacteria such as *Staphylococcus aureus*, *Streptococcus pyogenes*, *Streptococcus epidermidis*, and *Micrococcus* sp.[156] C16:1*n*-10 is reportedly the most important antibacterial FFA in the human skin exudate. It has been suggested that treating skin with C16:1*n*-10 would enhance its protection against colonization of opportunistic pathogen, particularly *Staphylococcus aureus*.[158,159] These compounds have exhibited effectiveness against various drug-resistant bacteria, including methicillin-resistant *Staphylococcus aureus* (MRS).[156]

The possible mechanisms of antibacterial action of FFAs are as follows:

1. Interfering with the cell membrane or direct lysis of bacterial cells.
2. Disruption of the electron transport chain and oxidative phosphorylation.
3. Interfering with cellular energy production.
4. Inhibition of enzyme activity.
5. Impairment of nutrient uptake.
6. Generation of peroxidation and auto-oxidation degradation products.

FA salts can be used as saponifiable agents in soap with or without other antibacterial. Jang et al. have investigated different types of FA salts for their antibacterial activity against

Staphylococcus spp. when used in ultrapure soft water. They also evaluated the therapeutic effects of FFA salts on a spontaneous canine model of dermatitis. They reported that oleic acid salt and linoleic acid salt reduced the number of these bacteria. Several recent studies have focused on preparing nano-formulations (nano-emulsions, liposomes, solid lipid nanoparticles, and controlled release hydrogels) of FAs and other lipids (with or without conjunctions of other antimicrobials) for treatment of skin infections and dermatitis.[160,161]

The antifungal and antibacterial activities of FAs have been reviewed by Pohl et al.[154] Some commercial forms of these antifungal FAs now have a widespread clinical application. For example, undecylenic acid (***62***) is a semisynthetic FA prepared by pyrolysis of ricinoleic acid (***63***) (a FA derived from castor oil). Undecylenic acid (***62***) is used as an active antifungal (especially against *Candida albicans*) in topical formulations for skin infections such as diaper rash, athlete's foot, ringworm, and tinea cruris.[158,159]

Some FAs such as capric acids (***64***), caprilic acid (***65***) and lauric acid (***66***) of coconut oil and their derivatives have been used to treat pediculosis.[162,163]

The antimicrobial activity of FFAs is determined by their length of carbon chain and the presence, number, position, and orientation of double bonds (*cis* orientation in natural FFAs). Moreover, it appears that the —OH in the carboxyl group is essential to their antibacterial activity. Although some esterified FFAs have been reported to possess antimicrobial properties, these effects are reduced compared to FFAs and many of them have no activity.[164–167] In comparison with saturated FFAs, medium- and long-chain unsaturated FFAs exhibit more antibacterial activities against Gram-positive bacteria than Gram-negatives.[168] Besides, with the same length of carbon, unsaturated FFAs have been reported to be more potent than saturated FFAs.[169,170] FFAs with *cis*-oriented double bounds are more potent than the *trans*-FFAs. Within monounsaturated FFAs, the most potent molecules usually have 14 or 16 carbons.[168]

However, there are some exceptions, and some FFAs with 14, 16, or 18 carbon atoms have been reported to have more activity against certain species of bacteria (12). But the most active saturated FFAs have been reported to contain 10 or 12 carbons in the chain.[171,172]

Since different studies use different bacterial organisms and experimental methodologies, it is difficult to compare the potency of these FFAs. It seems that introducing some standard protocols for performing experiments and reporting results is essential to have a better prospect of developing drugs from this class of compounds.

8. CARBOHYDRATES

Natural polysaccharides have long been used for wound healing in the folk medicine of different societies, and several studies have shown the effectiveness of some of them such as alginates,[173] fucoidans,[174] chitin and chitosan,[175] pectins,[173] hyaluronic acid, polysaccharides of *Aloe vera, Opuntia ficus-indica, Malva sylvestris*, sum mushrooms and alga.[176–180]

They increase cell proliferation and work as humectant agents to provide a wet dressing to prevent the wounded are from drying. Effects of natural beta glucan on skin health have been reviewed by Du et al.[181] These effects include antioxidant, antiwrinkle, antiviral, antiultraviolet light, wound healing, and moisturizing effects.

On the other hand, polysaccharides from plant,[182,183] mushrooms[184–187] and natural manna[188,189] have been reported to boost immune system in vitro, in vivo, and in clinical trials on healthy or immune patients.[190] Many of these polysaccharides can stimulate IL-12, which can protect against lethal group A streptococcal skin infection by depositing interferon γ in the infection site.[191–193]

These immunomodulatory polysaccharides might protect against skin infections indirectly, but more studies are needed to evaluate this hypothesis. In a study by Koray et al., the beneficial effects of natural beta glucan (β-(1→3)glucan) were reported on patients suffering from recurrent aphthous stomatitis. The authors believed that the observed effect was mediated by immune potentiating properties of the beta glucan.[194]

Moreover, polysaccharide and oligosaccharide of microbial cell wall are good candidates to be considered as lead compounds for developing vaccines against different bacterial pathogens, and recently some successes have been achieved by researchers for bacteria, protozoa, helminths, viruses, and fungi.[195]

Plant polysaccharides and oligosaccharides such as β D [1→4] polymaltose, β D [2→1] poly (fructo-furanosyl) α-D-glucose, β-D-[1→3]-glucan have been used as adjuvants and/or delivery vehicles or backbones for the synthesis of conjugated vaccines.[196]

Unfortunately, the available knowledge about direct antibacterial properties of many bioactive polysaccharides and natural carbohydrates is negligible. One of the well studied antibacterial polysaccharides is chitosan and its derivatives. Chitosan is the deacetylated derivative of chitin (the second-most abundant polysaccharide and the main component in the shells of crustaceans, such as shrimp, crab and lobster, and in the cell walls of some fungi). Chitosan is found in some fungi, but its quantity is so limited that its commercial form is produced by alkaline deacetylation of chitin. The antimicrobial activity of chitosan against various bacteria and fungi has been well established.[197] Its biological activities, apart from its biodegradability, nontoxicity, and polycationic nature, make it a good antimicrobial candidate to be used in a variety of medicines, pharmaceutics, and textiles.

Several possible mechanisms for this antimicrobial effect have been suggested. The most accepted mechanism for chitosan antimicrobial effect is the interaction of its positively charged groups of chitosan with the negatively charged residues at the cell surface of many fungi and bacteria. This interaction causes extensive cell surface disruption and alters cell permeability, thereby triggering leakage of intracellular substances and inhibition of the normal metabolism of microorganisms, leading finally to cell death.[198,199] Another mechanism for lower MW chitosan derivatives is the interaction of the positively charged chitosan with the DNA of some fungi and bacteria, causing inhibition of microbial RNA and protein synthesis.[200]

N-Carboxybutylchitosans (NCBC), semisynthetic derivatives of chitosan, have been introduced as polymers with potential to be used in the cosmetic industry. The NCBC has a definite antimicrobial property against *Candida albicans* and Gram-positive bacteria, and it has been suggested that NCBC has a high potential application as wound dressings.[201] Another semisynthetic derivative of chitosan, chitosan-iodine complex (both in dry state and in solution) exhibits a significant bactericidal action against numerous types of the microorganisms that commonly infect the skin and wounds.[202]

Recently, several natural polysaccharides have been reported for their antimicrobial properties. For example, a novel polysaccharide from *Streptomyces virginia* H03 has shown antimicrobial activity against *Staphylococcus aureus* (MIC = 15.6 μg/mL).[203] Polysaccharide extracted from the mushroom *Lintinus edodes* has been reported to inhibit *Staphylococcus aureus* with the MIC of 25 μg/mL.[204]

Additionally, some natural polysaccharides such as those extracted from *Sclerotium glucanicum, Ganoderma lucidum, Grifola frondosa, Pleurotus tuber-regium* have been reported to be effective against herpes simplex virus.[176,205–207] Many of these polysaccharides can be used in relatively large doses with very few side effects. Their administration in oral, topical, or parenteral form is usually considered safe in several clinical trials.

9. CONCLUSION

Given the great structural diversity of natural compounds, attention to their bioactivity, especially antimicrobial activity, is a predictable option in medicinal chemistry. Based on the physiological skin anatomy, attention to the physiochemical properties of the new lead compounds is a necessity. Research into such parameters as permeability, distribution, accumulation, and extraction of the new lead compounds with basically consideration of antimicrobial potencies is essential. Researches on primary and secondary metabolites showed that some classes of these compounds such as alkaloids, phenolic compounds, diterpenes, FAs, and some carbohydrates can be considered as a potent source for screening of new antibacterial structures. Therefore, knowledge of the chemical classes of the most potential natural antimicrobials is a necessary requirement for microbiologists, pharmacists, and clinicians.

REFERENCES

1. Chuong C, Nickoloff B, Elias P, Goldsmith L, Macher E, Maderson P, et al. What is the 'true' function of skin? *Exp Dermatol* 2002;**11**(2):159–87.
2. Freinkel RK, Woodley DT. *The biology of the skin*. New York, NY: CRC Press; 2001. p. 255–75.
3. Bos JD, Zonneveld I, Das PK, Krieg SR, van der Loos CM, Kapsenberg ML. The skin immune system (SIS): distribution and immunophenotype of lymphocyte subpopulations in normal human skin. *J Invest Dermatol* 1987;**88**(5):569–73.

4. Grice EA, Kong HH, Conlan S, Deming CB, Davis J, Young AC, et al. Topographical and temporal diversity of the human skin microbiome. *Science* 2009;**324**(5931):1190–2.
5. Kloos WE, Musselwhite MS. Distribution and persistence of *Staphylococcus* and *Micrococcus* species and other aerobic bacteria on human skin. *Appl Microbiol* 1975;**30**(3):381–95.
6. Roth RR, James WD. Microbiology of the skin: resident flora, ecology, infection. *J Am Acad Dermatol* 1989;**20**(3):367–90.
7. Segre JA. Epidermal barrier formation and recovery in skin disorders. *J Clin Invest* 2006;**116**(5): 1150–8.
8. Frykberg RG, Zgonis T, Armstrong DG, Driver VR, Giurini JM, Kravitz SR, et al. Diabetic foot disorders: a clinical practice guideline (2006 revision). *J Foot Ankle Surg* 2006;**45**(5):S1–S66.
9. Wheat LJ. Infection and diabetes mellitus. *Diabetes Care* 1980;**3**(1):187–97.
10. Kotilainen P, Nikoskelainen J, Huovinen P. Emergence of ciprofloxacin-resistant coagulase-negative staphylococcal skin flora in immunocompromised patients receiving ciprofloxacin. *J Infect Dis* 1990;**161**(1):41–4.
11. Stevens DL, Bisno AL, Chambers HF, Everett ED, Dellinger P, Goldstein EJ, et al. Practice guidelines for the diagnosis and management of skin and soft-tissue infections. *Clin Infect Dis* 2005; **41**(10):1373–406.
12. Barzilai A, Choen H-A. Isolation of group A streptococci from children with perianal cellulitis and from their siblings. *Pediatr Infect Dis J* 1998;**17**(4):358–60.
13. Mogielnicki NP, Schwartzman JD, Elliott JA. Perineal group A streptococcal disease in a pediatric practice. *Pediatrics* 2000;**106**(2):276–81.
14. Herbst RA. Perineal streptococcal dermatitis/disease. *Am J Clin Dermatol* 2003;**4**(8):555–60.
15. Vayalumkal JV, Jadavji T. Children hospitalized with skin and soft tissue infections. *Pediatr Drugs* 2006; **8**(2):99–111.
16. Amagai M, Matsuyoshi N, Wang Z, Andl C, Stanley J. Toxin in bullous impetigo and staphylococcal scalded-skin syndrome targets desmoglein 1. *Nat Med* 2000;**6**(11):1275–7.
17. Stanley JR, Amagai M. Pemphigus, bullous impetigo, and the staphylococcal scalded-skin syndrome. *N Engl J Med* 2006;**355**(17):1800–10.
18. Johnston GA. Treatment of bullous impetigo and the staphylococcal scalded skin syndrome in infants. *Expert Rev Anti Infect Ther* 2004;**2**(3):439–46.
19. Eells LD, Mertz PM, Piovanetti Y, Pekoe GM, Eaglstein WH. Topical antibiotic treatment of impetigo with mupirocin. *Arch Dermatol* 1986;**122**(11):1273–6.
20. Yu Y, Cheng AS, Wang L, Dunne WM, Bayliss SJ. Hot tub folliculitis or hot hand–foot syndrome caused by *Pseudomonas aeruginosa*. *J Am Acad Dermatol* 2007;**57**(4):596–600.
21. Chandrasekar PH, Rolston KV, Kannangara DW, LeFrock JL, Binnick SA. Hot tub-associated dermatitis due to *Pseudomonas aeruginosa*: case report and review of the literature. *Arch Dermatol* 1984, **120**(10):1337–40.
22. Stulberg DL, Penrod MA, Blatny RA. Common bacterial skin infections. *Am Fam Physician* 2002; **66**(1):119–28.
23. Böni R, Nehrhoff B. Treatment of gram-negative folliculitis in patients with acne. *Am J Clin Dermatol* 2003;**4**(4):273–6.
24. Blankenship ML. Gram-negative folliculitis: follow-up observations in 20 patients. *Arch Dermatol* 1984;**120**(10):1301–3.
25. Plewig G, Nikolowski J, Wolff HH. Action of isotretinoin in acne rosacea and gram-negative folliculitis. *J Am Acad Dermatol* 1982;**6**(4):766–85.
26. Stanley T, Jogose M. *Haemophilus influenzae* type b cellulitis. *N Z Med J* 1991;**104**(917):334–6.
27. Ambati BK, Ambati J, Azar N, Stratton L, Schmidt EV. Periorbital and orbital cellulitis before and after the advent of *Haemophilus influenzae* type B vaccination. *Ophthalmology* 2000;**107**(8):1450–3.
28. Shapiro ED, Wald ER, Brozanski BA. Periorbital cellulitis and paranasal sinusitis: a reappraisal. *Pediatr Infect Dis J* 1982;**1**(2):91–4.
29. Hauser A, Fogarasi S. Periorbital and orbital cellulitis. *Pediatr Rev* 2010;**31**(6):242–9.
30. Balfour Jr. H. *Varicella zoster virus* infections in immunocompromised hosts. A review of the natural history and management. *Am J Med* 1988;**85**(2A):68–73.

31. Degreef H, FHZCS Group. Famciclovir, a new oral antiherpes drug: results of the first controlled clinical study demonstrating its efficacy and safety in the treatment of uncomplicated *herpes zoster* in immunocompetent patients. *Int J Antimicrob Agents* 1994;**4**(4):241–6.
32. Beutner KR, Friedman DJ, Forszpaniak C, Andersen PL, Wood MJ. Valaciclovir compared with acyclovir for improved therapy for *herpes zoster* in immunocompetent adults. *Antimicrob Agents Chemother* 1995;**39**(7):1546–53.
33. Pappas PG, Rex JH, Sobel JD, Filler SG, Dismukes WE, Walsh TJ, et al. Guidelines for treatment of candidiasis. *Clin Infect Dis* 2004;**38**(2):161–89.
34. Rex JH, Walsh TJ, Sobel JD, Filler SG, Pappas PG, Dismukes WE, et al. Practice guidelines for the treatment of candidiasis. *Clin Infect Dis* 2000;**30**(4):662–78.
35. Van Cutsem J, Van Gerven F, Janssen P. Activity of orally, topically, and parenterally administered itraconazole in the treatment of superficial and deep mycoses: animal models. *Rev Infect Dis* 1987;**9**(Suppl. 1):S15–32.
36. Hector RF. An overview of antifungal drugs and their use for treatment of deep and superficial mycoses in animals. *Clin Tech Small Anim Pract* 2005;**20**(4):240–9.
37. Elewski B. Cutaneous mycoses in children. *Br J Dermatol* 1996;**134**(s46):7–11.
38. Neefe JR, Goldstone SE, Winnett MT, Siegel M, Boux LJ. Human papilloma virus treatment. Google Patents; 2004.
39. Meadows KP, Tyring SK, Pavia AT, Rallis TM. Resolution of recalcitrant molluscum contagiosum virus lesions in human immunodeficiency virus-infected patients treated with cidofovir. *Arch Dermatol* 1997;**133**(8):987–90.
40. Hanson D, Diven DG. Molluscum contagiosum. *Dermatol Online J* 2003;**9**(2).
41. Whitley RJ, Roizman B. *Herpes simplex virus* infections. *Lancet* 2001;**357**(9267):1513–8.
42. Neu HC. The crisis in antibiotic resistance. *Science* 1992;**257**(5073):1064–73.
43. Sunagawa M, Matsumura H, Inoue T, Fukasawa M, Kato M. A novel carbapenem antibiotic, SM-7338 structure-activity relationships. *J Antibiot* 1990;**43**(5):519–32.
44. Powers J-PS, Hancock RE. The relationship between peptide structure and antibacterial activity. *Peptides* 2003;**24**(11):1681–91.
45. Biava M, Porretta GC, Poce G, Deidda D, Pompei R, Tafi A, et al. Antimycobacterial compounds. Optimization of the BM 212 structure, the lead compound for a new pyrrole derivative class. *Bioorg Med Chem* 2005;**13**(4):1221–30.
46. Abraham MH, Ibrahim A, Zissimos AM, Zhao YH, Comer J, Reynolds DP. Application of hydrogen bonding calculations in property based drug design. *Drug Discov Today* 2002;**7**(20):1056–63.
47. Lipinski CA. Drug-like properties and the causes of poor solubility and poor permeability. *J Pharmacol Toxicol Methods* 2000;**44**(1):235–49.
48. Kubinyi H. *3D QSAR in drug design: volume 1: theory methods and applications*. Dordrecht, The Netherland: Kluwer Academic Publisher; 1994. p. 155–369.
49. Bugg T, Walsh C. Intracellular steps of bacterial cell wall peptidoglycan biosynthesis: enzymology, antibiotics, and antibiotic resistance. *Nat Prod Rep* 1992;**9**(3):199–215.
50. Costerton J, Cheng K-J. The role of the bacterial cell envelope in antibiotic resistance. *J Antimicrob Chemother* 1975;**1**(4):363–77.
51. Delcour AH. Outer membrane permeability and antibiotic resistance. *Biochim Biophys Acta* 2009;**1794**(5):808–16.
52. Webber M, Piddock L. The importance of efflux pumps in bacterial antibiotic resistance. *J Antimicrob Chemother* 2003;**51**(1):9–11.
53. Okusu H, Ma D, Nikaido H. AcrAB efflux pump plays a major role in the antibiotic resistance phenotype of *Escherichia coli* multiple-antibiotic-resistance (Mar) mutants. *J Bacteriol* 1996;**178**(1):306–8.
54. Li X, Schuler MA, Berenbaum MR. Molecular mechanisms of metabolic resistance to synthetic and natural xenobiotics. *Annu Rev Entomol* 2007;**52**:231–53.
55. Nishino K, Yamaguchi A. Role of xenobiotic transporters in bacterial drug resistance and virulence. *IUBMB Life* 2008;**60**(9):569–74.

56. Epand RM, Epand RF. Lipid domains in bacterial membranes and the action of antimicrobial agents. *Biochim Biophys Acta* 2009;**1788**(1):289–94.
57. Wu M, Maier E, Benz R, Hancock RE. Mechanism of interaction of different classes of cationic antimicrobial peptides with planar bilayers and with the cytoplasmic membrane of *Escherichia coli*. *Biochemistry* 1999;**38**(22):7235–42.
58. Kagan BL, Selsted ME, Ganz T, Lehrer RI. Antimicrobial defensin peptides form voltage-dependent ion-permeable channels in planar lipid bilayer membranes. *Proc Natl Acad Sci U S A* 1990;**87**(1):210–4.
59. Silvestro L, Gupta K, Weiser JN, Axelsen PH. The concentration-dependent membrane activity of cecropin A. *Biochemistry* 1997;**36**(38):11452–60.
60. Taber HW, Mueller J, Miller P, Arrow A. Bacterial uptake of aminoglycoside antibiotics. *Microbiol Rev* 1987;**51**(4):439.
61. Hancock RE. Alterations in outer membrane permeability. *Annu Rev Microbiol* 1984;**38**(1):237–64.
62. Pirzada A, Shaikh W, Usmanghani K, Mohiuddin E. Antifungal activity of *Dodonaea viscosa* Jacq extract on pathogenic fungi isolated from superficial skin infection. *Pak J Pharm Sci* 2010;**23**(3):337–40.
63. Duraipandiyan V, Ignacimuthu S. Antibacterial and antifungal activity of *Cassia fistula* L.: an ethnomedicinal plant. *J Ethnopharmacol* 2007;**112**(3):590–4.
64. Lipipun V, Kurokawa M, Suttisri R, Taweechotipatr P, Pramyothin P, Hattori M, et al. Efficacy of Thai medicinal plant extracts against herpes simplex virus type 1 infection in vitro and in vivo. *Antiviral Res* 2003;**60**(3):175–80.
65. El Nabi OMA, Reisinger EC, Reinthaler FF, Still F, Eibel U, Krejs GJ. Antimicrobial activity of *Acacia nilotica* (L.) Willd. ex Del. var. nilotica (Mimosaceae). *J Ethnopharmacol* 1992;**37**(1):77–9.
66. Portillo A, Vila R, Freixa B, Adzet T, Cañigueral S. Antifungal activity of Paraguayan plants used in traditional medicine. *J Ethnopharmacol* 2001;**76**(1):93–8.
67. Palombo EA, Semple SJ. Antibacterial activity of traditional Australian medicinal plants. *J Ethnopharmacol* 2001;**77**(2):151–7.
68. Ghorbani A. Studies on pharmaceutical ethnobotany in the region of Turkmen Sahra, north of Iran: (part 1): general results. *J Ethnopharmacol* 2005;**102**(1):58–68.
69. Mosaddegh M, Naghibi F, Moazzeni H, Pirani A, Esmaeili S. Ethnobotanical survey of herbal remedies traditionally used in Kohghiluyeh va Boyer Ahmad province of Iran. *J Ethnopharmacol* 2012;**141**(1):80–95.
70. Boily Y, Van Puyvelde L. Screening of medicinal plants of Rwanda (Central Africa) for antimicrobial activity. *J Ethnopharmacol* 1986;**16**(1):1–13.
71. Weniger B, Rouzier M, Daguilh R, Henrys D, Henrys J, Anton R. Traditional medicine in the Central Plateau of Haiti. 2. Ethnopharmacologic inventory. *J Ethnopharmacol* 1986;**17**(1):13–30.
72. Miraldi E, Ferri S, Mostaghimi V. Botanical drugs and preparations in the traditional medicine of West Azerbaijan (Iran). *J Ethnopharmacol* 2001;**75**(2):77–87.
73. Baba-Moussa F, Akpagana K, Bouchet P. Antifungal activities of seven West African *Combretaceae* used in traditional medicine. *J Ethnopharmacol* 1999;**66**(3):335–8.
74. Khan MN, Ngassapa O, Matee MI. Antimicrobial activity of Tanzanian chewing sticks against oral pathogenic microbes. *Pharm Biol* 2000;**38**(3):235–40.
75. Guiraud P, Steiman R, Campos-Takaki G-M, Seigle-Murandi F, de Buochberg MS. Comparison of antibacterial and antifungal activities of lapachol and β-lapachone. *Planta Med* 1994;**60**(04):373–4.
76. Latz PK. *Bushfires and Bushtucker. Aboriginal Plant Use in Central Australia.* Alice Springs: Indiana University, IAD Press; 1995.
77. Vlietinck A, Van Hoof L, Totte J, Lasure A, Berghe DV, Rwangabo P, et al. Screening of hundred Rwandese medicinal plants for antimicrobial and antiviral properties. *J Ethnopharmacol* 1995;**46**(1):31–47.
78. Kone W, Atindehou KK, Terreaux C, Hostettmann K, Traore D, Dosso M. Traditional medicine in North Côte-d'Ivoire: screening of 50 medicinal plants for antibacterial activity. *J Ethnopharmacol* 2004;**93**(1):43–9.
79. Hamedi A, Zomorodian K, Safari F. Antimicrobial activity of four medicinal plants widely used in Persian folk medicine. *Res J Pharmacogn* 2015;**2**(1):25–33.

80. Abbasi AM, Khan M, Ahmad M, Zafar M, Jahan S, Sultana S. Ethnopharmacological application of medicinal plants to cure skin diseases and in folk cosmetics among the tribal communities of North-West Frontier Province, Pakistan. *J Ethnopharmacol* 2010;**128**(2):322–35.
81. Ahmad I, Beg AZ. Antimicrobial and phytochemical studies on 45 Indian medicinal plants against multi-drug resistant human pathogens. *J Ethnopharmacol* 2001;**74**(2):113–23.
82. Schnitzler P, Schuhmacher A, Astani A, Reichling J. *Melissa officinalis* oil affects infectivity of enveloped herpesviruses. *Phytomedicine* 2008;**15**(9):734–40.
83. Wölbling R, Leonhardt K. Local therapy of herpes simplex with dried extract from *Melissa officinalis*. *Phytomedicine* 1994;**1**(1):25–31.
84. Koytchev R, Alken R, Dundarov S. Balm mint extract (Lo-701) for topical treatment of recurring herpes labialis. *Phytomedicine* 1999;**6**(4):225–30.
85. Dossey AT. Insects and their chemical weaponry: new potential for drug discovery. *Nat Prod Rep* 2010;**27**(12):1737–57.
86. Cherniack EP. Bugs as drugs, Part 1: insects: the "new" alternative medicine for the 21st century. *Altern Med Rev* 2010;**15**(2):124–35.
87. Moed L, Shwayder TA, Chang MW. Cantharidin revisited: a blistering defense of an ancient medicine. *Arch Dermatol* 2001;**137**(10):1357–60.
88. Kartal Durmazlar SP, Atacan D, Eskioglu F. Cantharidin treatment for recalcitrant facial flat warts: a preliminary study. *J Dermatolog Treat* 2009;**20**(2):114–9.
89. Gibbs S, Altman DG, Harvey I, Sterling J, Stark R. Papers local treatments for cutaneous warts: systematic review commentary: systematic reviewers face challenges from varied study designs. *Br Med J* 2002;**325**(7362):461.
90. Von Krogh G. Topical treatment of penile condylomata acuminata with podophyllin, podophyllotoxin and colchicine. A comparative study. *Acta Derm Venereol* 1977;**58**(2):163–8.
91. Beutner K. Podophyllotoxin in the treatment of genital warts. *Sexually transmitted diseases: advances in diagnosis and treatment*. Basel: Karger Publishers; 1996. p. 227–32.
92. Reddy K, Yedery R, Aranha C. Antimicrobial peptides: premises and promises. *Int J Antimicrob Agents* 2004;**24**(6):536–47.
93. Schauber J, Gallo RL. Antimicrobial peptides and the skin immune defense system. *J Allergy Clin Immunol* 2008;**122**(2):261–6.
94. Brogden KA. Antimicrobial peptides: pore formers or metabolic inhibitors in bacteria? *Nat Rev Microbiol* 2005;**3**(3):238–50.
95. De La Vega RR, Garcia B, D'Ambrosio C, Diego-Garcia E, Scaloni A, Possani L. Antimicrobial peptide induction in the haemolymph of the Mexican scorpion *Centruroides limpidus limpidus* in response to septic injury. *Cell Mol Life Sci* 2004;**61**(12):1507–19.
96. Dolashka P, Moshtanska V, Borisova V, Dolashki A, Stevanovic S, Dimanov T, et al. Antimicrobial proline-rich peptides from the hemolymph of marine snail *Rapana venosa*. *Peptides* 2011;**32**(7):1477–83.
97. Shai Y. Mode of action of membrane active antimicrobial peptides. *Pept Sci* 2002;**66**(4):236–48.
98. Liang Y, Wang J-X, Zhao X-F, Du X-J, Xue J-F. Molecular cloning and characterization of cecropin from the housefly (*Musca domestica*), and its expression in *Escherichia coli*. *Dev Comp Immunol* 2006;**30**(3):249–57.
99. Wang J-X, Zhao X-F, Liang Y-L, Li L, Zhang W, Ren Q, et al. Molecular characterization and expression of the antimicrobial peptide defensin from the housefly (*Musca domestica*). *Cell Mol Life Sci* 2006;**63**(24):3072–82.
100. Lai R, Takeuchi H, Lomas LO, Jonczy J, Rigden DJ, Rees HH, et al. A new type of antimicrobial protein with multiple histidines from the hard tick, *Amblyomma hebraeum*. *FASEB J* 2004;**18**(12):1447–9.
101. Yu D, Sheng Z, Xu X, Li J, Yang H, Liu Z, et al. A novel antimicrobial peptide from salivary glands of the hard tick, *Ixodes sinensis*. *Peptides* 2006;**27**(1):31–5.
102. Tincu JA, Taylor SW. Antimicrobial peptides from marine invertebrates. *Antimicrob Agents Chemother* 2004;**48**(10):3645–54.
103. Mitta G, Vandenbulcke F, Roch P. Original involvement of antimicrobial peptides in mussel innate immunity. *FEBS Lett* 2000;**486**(3):185–90.

104. Charlet M, Chernysh S, Philippe H, Hetru C, Hoffmann JA, Bulet P. Innate immunity Isolation of several cysteine-rich antimicrobial peptides from the blood of a mollusc, *Mytilus edulis*. *J Biol Chem* 1996;**271**(36):21808–13.
105. Elias PM. The skin barrier as an innate immune element. *Semin Immunopathol* 2007;**29**:3–14.
106. Elias PM, Schmuth M. Abnormal skin barrier in the etiopathogenesis of atopic dermatitis. *Curr Allergy Asthma Rep* 2009;**9**(4):265–72.
107. Tanaka J, Da Silva C, De Oliveira A, Nakamura C, Dias Filho B. Antibacterial activity of indole alkaloids from *Aspidosperma ramiflorum*. *Braz J Med Biol Res* 2006;**39**(3):387–91.
108. Odebiyi O, Sofowora E. Antimicrobial alkaloids from a Nigerian chewing stick (*Fagara zanthoxyloides*). *Planta Med* 1979;**36**(07):204–7.
109. O'Donnell G, Gibbons S. Antibacterial activity of two canthin-6-one alkaloids from *Allium neapolitanum*. *Phytother Res* 2007;**21**(7):653–7.
110. Thouvenel C, Gantier JC, Duret P, Fourneau C, Hocquemiller R, Ferreira ME, et al. Antifungal compounds from *Zanthoxylum chiloperone* var. *angustifolium*. *Phytother Res* 2003;**17**(6):678–80.
111. Oh K-B, Mar W, Kim S, Kim J-Y, Lee T-H, Kim J-G, et al. Antimicrobial activity and cytotoxicity of bis (indole) alkaloids from the sponge *Spongosorites* sp. *Biol Pharm Bull* 2006;**29**(3):570–3.
112. Monks NR, Lerner C, Henriques AT, Farias FM, Schapoval EE, Suyenaga ES, et al. Anticancer, antichemotactic and antimicrobial activities of marine sponges collected off the coast of Santa Catarina, southern Brazil. *J Exp Mar Bio Ecol* 2002;**281**(1):1–12.
113. Samoylenko V, Jacob MR, Khan SI, Zhao J, Tekwani BL, Midiwo JO, et al. Antimicrobial, antiparasitic and cytotoxic spermine alkaloids from *Albizia schimperiana*. *Nat Prod Commun* 2009;**4**(6):791.
114. Rukunga GM, Waterman PG. New macrocyclic spermine (budmunchiamine) alkaloids from *Albizia gummifera*: with some observations on the structure-activity relationships of the budmunchiamines. *J Nat Prod* 1996;**59**(9):850–3.
115. Morel AF, Maldaner G, Ilha V, Missau F, Silva UF, Dalcol II. Cyclopeptide alkaloids from *Scutia buxifolia* Reiss and their antimicrobial activity. *Phytochemistry* 2005;**66**(21):2571–6.
116. Yao-Kouassi PA, Caron C, Ramiarantsoa H, Prost É, Harakat D, Le Magrex-Debar É, et al. New nitro-benzo [c] phenanthridine and indolopyridoquinazoline alkaloids from *Zanthoxylum atchoum*. *C R Chim* 2015;**18**(8):891–7.
117. Macfoy C, Danosus D, Sandit R, Jones TH, Garraffo HM, Spande TF, et al. Alkaloids of anuran skin: antimicrobial function? *Z Naturforsch C* 2005;**60**(11–12):932–7.
118. Gertsch J, Tobler RT, Brun R, Sticher O, Heilmann J. Antifungal, antiprotozoal, cytotoxic and piscicidal properties of Justicidin B and a new arylnaphthalide lignan from *Phyllanthus piscatorum*. *Planta Med* 2003;**69**(05):420–4.
119. Kawazoe K, Yutani A, Tamemoto K, Yuasa S, Shibata H, Higuti T, et al. Phenylnaphthalene compounds from the subterranean part of *Vitex rotundifolia* and their antibacterial activity against methicillin-resistant *Staphylococcus aureus*. *J Nat Prod* 2001;**64**(5):588–91.
120. Peng SC, Cheng CY, Sheu F, Su CH. The antimicrobial activity of heyneanol A extracted from the root of Taiwanese wild grape. *J Appl Microbiol* 2008;**105**(2):485–91.
121. Famakin J, Katerere D. Isolation of an antibacterial stilbene from *Combretum woodii* (Combretaceae) leaves. *Afr J Biotechnol* 2005;**4**(10):1167–71.
122. Okasaka M, Takaishi Y, Kogure K, Fukuzawa K, Shibata H, Higuti T, et al. New Stilbene derivatives from *Calligonum l eucocladum*. *J Nat Prod* 2004;**67**(6):1044–6.
123. Xu W-H, Li X-C. Antifungal compounds from *Piper* species. *Curr Bioact Compd* 2011;**7**(4):262–7.
124. Freixa B, Vila R, Ferro EA, Adzet T, Cañigueral S. Antifungal principles from *Piper fulvescens*. *Planta Med* 2001;**67**(09):873–5.
125. Townsend D, Ashdown N, Bolton S, Bradley J, Duckworth G, Moorhouse E, et al. The international spread of methicillin-resistant *Staphylococcus aureus*. *J Hosp Infect* 1987;**9**(1):60–71.
126. Omura S, Iwai Y, Hinotozawa K, Tanaka H, Takahashi Y, Nakagawa A. OM-704A, a new antibiotic active against gram-positive bacteria produced by *Streptomyces* sp. *J Antibiot* 1982;**35**(11):1425–9.
127. Malkina ND, Dudnik YV, Lysenkova LN, Lazhko EI, Galatenko O, Katrukha GS. 167-A, a new antibiotic produced by a mutant of an inactive wild strain of *Amycolata autotrophica*. *J Antibiot* 1994;**47**(3):342–8.

128. Panthong K, Pongcharoen W, Phongpaichit S, Taylor WC. Tetraoxygenated xanthones from the fruits of *Garcinia cowa*. *Phytochemistry* 2006;**67**(10):999–1004.
129. Boonsri S, Karalai C, Ponglimanont C, Kanjana-Opas A, Chantrapromma K. Antibacterial and cytotoxic xanthones from the roots of *Cratoxylum formosum*. *Phytochemistry* 2006;**67**(7):723–7.
130. Rukachaisirikul V, Kamkaew M, Sukavisit D, Phongpaichit S, Sawangchote P, Taylor WC. Antibacterial xanthones from the leaves of *Garcinia nigrolineata*. *J Nat Prod* 2003;**66**(12):1531–5.
131. Boonnak N, Karalai C, Chantrapromma S, Ponglimanont C, Fun H-K, Kanjana-Opas A, et al. Anti-*Pseudomonas aeruginosa* xanthones from the resin and green fruits of *Cratoxylum cochinchinense*. *Tetrahedron* 2009;**65**(15):3003–13.
132. Boonnak N, Karalai C, Chantrapromma S, Ponglimanont C, Kanjana-Opas A, Chantrapromma K, et al. Chromene and prenylated xanthones from the roots of *Cratoxylum formosum ssp. pruniflorum*. *Chem Pharm Bull (Tokyo)* 2010;**58**(3):386–9.
133. Mbaveng AT, Ngameni B, Kuete V, Simo IK, Ambassa P, Roy R, et al. Antimicrobial activity of the crude extracts and five flavonoids from the twigs of *Dorstenia barteri* (Moraceae). *J Ethnopharmacol* 2008;**116**(3):483–9.
134. Kuete V, Simo IK, Ngameni B, Bigoga JD, Watchueng J, Kapguep RN, et al. Antimicrobial activity of the methanolic extract, fractions and four flavonoids from the twigs of *Dorstenia angusticornis* Engl. (Moraceae). *J Ethnopharmacol* 2007;**112**(2):271–7.
135. Ata A, Win HY, Holt D, Holloway P, Segstro EP, Jayatilake GS. New antibacterial diterpenes from *Pseudopterogorgia elisabethae*. *Helv Chim Acta* 2004;**87**(5):1090–8.
136. Habtemariam S, Gray AI, Waterman PG. Antibacterial diterpenes from the aerial parts of *Premna oligotricha*. *Planta Med* 1992;**58**(01):109–10.
137. Nicolson K, Evans G, O'Toole PW. Potentiation of methicillin activity against methicillin-resistant *Staphylococcus aureus* by diterpenes. *FEMS Microbiol Lett* 1999;**179**(2):233–9.
138. Kubo I, Muroi H, Himejima M. Antibacterial activity of totarol and its potentiation. *J Nat Prod* 1992; **55**(10):1436–40.
139. Haraguchi H, Oike S, Muroi H, Kubo I. Mode of antibacterial action of totarol, a diterpene from *Podocarpus nagi*. *Planta Med* 1996;**62**(02):122–5.
140. Muroi H, Kubo I. Bactericidal effects of anacardic acid and totarol on methicillin-resistant *Staphylococcus aureus* (MRSA). *Biosci Biotechnol Biochem* 1994;**58**(10):1925–6.
141. Tanaka J, Higa T, Suwanborirux K, Kokpol U, Bernardinelli G, Jefford CW. Bioactive norsesterterpene 1, 2-dioxanes from a Thai sponge, *Mycale* sp. *J Org Chem* 1993;**58**(11):2999–3002.
142. Herlem D, Kervagoret J, Khuong-Huu F. Total synthesis of (±)-galanolactone: NMR in CDCl3, at 400 MHz for 1H, TMS as reference. *Tetrahedron Lett* 1989;**30**(5):553–6.
143. Kimbu SF, Ngadjui B, Sondengam LB, Njimi TK, Connolly JD, Fakunle CO. A new labdane diterpenoid from the seeds of *Aframomum daniellii*. *J Nat Prod* 1987;**50**(2):230–1.
144. Chinou I, Demetzos C, Harvala C, Roussakis C, Verbist J. Cytotoxic and antibacterial labdane-type diterpenes from the aerial parts of *Cistus incanus subsp. creticus*. *Planta Med* 1994;**60**(01):34–6.
145. Rasoamiaranjanahary L, Guilet D, Marston A, Randimbivololona F, Hostettmann K. Antifungal isopimaranes from *Hypoestes serpens*. *Phytochemistry* 2003;**64**(2):543–8.
146. Kato H, Kodama O, Akatsuka T. Oryzalexin E, a diterpene phytoalexin from UV-irradiated rice leaves. *Phytochemistry* 1993;**33**(1):79–81.
147. Salah MA, Bedir E, Toyang NJ, Khan IA, Harries MD, Wedge DE. Antifungal clerodane diterpenes from *Macaranga monandra* (L) Muell. et Arg. (Euphorbiaceae). *J Agric Food Chem* 2003;**51**(26):7607–10.
148. Kusumoto N, Ashitani T, Murayama T, Ogiyama K, Takahashi K. Antifungal abietane-type diterpenes from the cones of *Taxodium distichum* Rich. *J Chem Ecol* 2010;**36**(12):1381–6.
149. Odds FC, Brown AJ, Gow NA. Antifungal agents: mechanisms of action. *Trends Microbiol* 2003;**11**(6):272–9.
150. Hamedi A, Zarshenas MM, Sohrabpour M, Zargaran A. Herbal medicinal oils in traditional Persian medicine. *Pharm Biol* 2013;**51**(9):1208–18.
151. Parascandola J. Chaulmoogra oil and the treatment of leprosy. *Pharm Hist* 2003;**45**(2):47–57.
152. Kiran K, Asad M. Wound healing activity of *Sesamum indicum* L seed and oil in rats. *Indian J Exp Biol* 2008;**46**(11):777.

153. Hamedi A, Ghanbari A, Razavipour R, Saeidi V, Zarshenas MM, Sohrabpour M, et al. *Alyssum homolocarpum* seeds: phytochemical analysis and effects of the seed oil on neural stem cell proliferation and differentiation. *J Nat Med* 2015;**69**(3):387–96.
154. Pohl CH, Kock JL, Thibane VS. Antifungal free fatty acids: a review. *Science against microbial pathogens: current research and technological advances*, vol. 1; 2011. p. 61–71.
155. Thormar H, Hilmarsson H. The role of microbicidal lipids in host defense against pathogens and their potential as therapeutic agents. *Chem Phys Lipids* 2007;**150**(1):1–11.
156. Drake DR, Brogden KA, Dawson DV, Wertz PW. Thematic review series: skin lipids. Antimicrobial lipids at the skin surface. *J Lipid Res* 2008;**49**(1):4–11.
157. Nguyen MT, Hanzelmann D, Härtner T, Peschel A, Götz F. Skin-specific unsaturated fatty acids boost the *Staphylococcus aureus* innate immune response. *Infect Immun* 2016;**84**(1):205–15.
158. Shi D, Zhao Y, Yan H, Fu H, Shen Y, Lu G, et al. Antifungal effects of undecylenic acid on the biofilm formation of *Candida albicans*. *Int J Clin Pharmacol Ther* 2016;**54**(5):343–53.
159. Van der Steen M, Stevens CV. Undecylenic acid: a valuable and physiologically active renewable building block from castor oil. *ChemSusChem* 2009;**2**(8):692–713.
160. Jackman JA, Yoon BK, Li D, Cho N-J. Nanotechnology formulations for antibacterial free fatty acids and monoglycerides. *Molecules* 2016;**21**(3):305.
161. Tamm I, Heinämäki J, Laidmäe I, Rammo L, Paaver U, Ingebrigtsen SG, et al. Development of suberin fatty acids and chloramphenicol-loaded antimicrobial electrospun nanofibrous mats intended for wound therapy. *J Pharm Sci* 2016;**105**(3):1239–47.
162. Crammer B, Ikan R, Mumicuoglu YK, Weinstein V. Licidal compositions containing carboxylic acids. Google Patents; 1991.
163. Feldmeier H. Pediculosis capitis: new insights into epidemiology, diagnosis and treatment. *Eur J Clin Microbiol Infect Dis* 2012;**31**(9):2105–10.
164. Huang CB, George B, Ebersole JL. Antimicrobial activity of n-6, n-7 and n-9 fatty acids and their esters for oral microorganisms. *Arch Oral Biol* 2010;**55**(8):555–60.
165. Agoramoorthy G, Chandrasekaran M, Venkatesalu V, Hsu M. Antibacterial and antifungal activities of fatty acid methyl esters of the blind-your-eye mangrove from India. *Braz J Microbiol* 2007;**38**(4):739–42.
166. Chandrasekaran M, Kannathasan K, Venkatesalu V. Antimicrobial activity of fatty acid methyl esters of some members of Chenopodiaceae. *Z Naturforsch C* 2008;**63**(5–6):331–6.
167. Mourshid SS, Badr JM, Risinger AL, Mooberry SL, Youssef DT. Penicilloitins A and B, new antimicrobial fatty acid esters from a marine endophytic Penicillium species. *Z Naturforsch C* 2016;**71**(11–12):387–92.
168. Desbois AP, Smith VJ. Antibacterial free fatty acids: activities, mechanisms of action and biotechnological potential. *Appl Microbiol Biotechnol* 2010;**85**(6):1629–42.
169. Desbois AP, Lawlor KC. Antibacterial activity of long-chain polyunsaturated fatty acids against *Propionibacterium acnes* and *Staphylococcus aureus*. *Mar Drugs* 2013;**11**(11):4544–57.
170. Zheng CJ, Yoo J-S, Lee T-G, Cho H-Y, Kim Y-H, Kim W-G. Fatty acid synthesis is a target for antibacterial activity of unsaturated fatty acids. *FEBS Lett* 2005;**579**(23):5157–62.
171. Wille J, Kydonieus A. Palmitoleic acid isomer (C16: 1Δ6) in human skin sebum is effective against gram-positive bacteria. *Skin Pharmacol Physiol* 2003;**16**(3):176–87.
172. Sun CQ, O'Connor CJ, Roberton AM. Antibacterial actions of fatty acids and monoglycerides against *Helicobacter pylori*. *FEMS Immunol Med Microbiol* 2003;**36**(1–2):9–17.
173. Munarin F, Tanzi M, Petrini P. Advances in biomedical applications of pectin gels. *Int J Biol Macromol* 2012;**51**(4):681–9.
174. Murakami K, Aoki H, Nakamura S, Nakamura S-I, Takikawa M, Hanzawa M, et al. Hydrogel blends of chitin/chitosan, fucoidan and alginate as healing-impaired wound dressings. *Biomaterials* 2010;**31**(1):83–90.
175. Francesko A, Tzanov T. Chitin, chitosan and derivatives for wound healing and tissue engineering. *Biofunctionalization of polymers and their applications*. Springer-Verlag Berlin Heidelberg: Springer; 2010. p. 1–27.

176. Zhang M, Cheung PC, Ooi VE, Zhang L. Evaluation of sulfated fungal β-glucans from the sclerotium of *Pleurotus tuber-regium* as a potential water-soluble anti-viral agent. *Carbohydr Res* 2004;**339**(13):2297–301.
177. Minagawa T, Okamura Y, Shigemasa Y, Minami S, Okamoto Y. Effects of molecular weight and deacetylation degree of chitin/chitosan on wound healing. *Carbohydr Polym* 2007;**67**(4):640–4.
178. Trombetta D, Puglia C, Perri D, Licata A, Pergolizzi S, Lauriano E, et al. Effect of polysaccharides from *Opuntia ficus-indica* (L.) cladodes on the healing of dermal wounds in the rat. *Phytomedicine* 2006;**13**(5):352–8.
179. Ghosh PK, Gaba A. Phyto-extracts in wound healing. *J Pharm Pharm Sci* 2013;**16**(5):760–820.
180. Hong E-G, Kim M-G. Effect of Polysaccharide from *Schizophyllum commune* on burn and wound healing. *Korean Chem Eng Res* 2006;**44**(1):87–91.
181. Du B, Bian Z, Xu B. Skin health promotion effects of natural beta-glucan derived from cereals and microorganisms: a review. *Phytother Res* 2014;**28**(2):159–66.
182. Schepetkin IA, Quinn MT. Botanical polysaccharides: macrophage immunomodulation and therapeutic potential. *Int Immunopharmacol* 2006;**6**(3):317–33.
183. Kumar S, Gupta P, Sharma S, Kumar D. A review on immunostimulatory plants. *Chin J Integr Med* 2011;**9**(2):117–28.
184. Ruthes AC, Smiderle FR, Iacomini M. Mushroom heteropolysaccharides: a review on their sources, structure and biological effects. *Carbohydr Polym* 2016;**136**:358–75.
185. Meng X, Liang H, Luo L. Antitumor polysaccharides from mushrooms: a review on the structural characteristics, antitumor mechanisms and immunomodulating activities. *Carbohydr Res* 2016;**424**:30–41.
186. Singdevsachan SK, Auroshree P, Mishra J, Baliyarsingh B, Tayung K, Thatoi H. Mushroom polysaccharides as potential prebiotics with their antitumor and immunomodulating properties: a review. *Bioact Carbohydr Diet Fibre* 2016;**7**(1):1–14.
187. Kristiansen B. Production of fungal extracellular immune stimulating compounds. Google Patents; 2016.
188. Hamedi A, Farjadian S, Karami MR. Immunomodulatory properties of *Trehala manna* decoction and its isolated carbohydrate macromolecules. *J Ethnopharmacol* 2015;**162**:121–6.
189. Hamedi A, Farjadian S, Karami MR. Immunomodulatory properties of taranjebin (camel's thorn) manna and its isolated carbohydrate macromolecules. *J Evid Based Complement Altern Med* 2015;**20**(4):269–74.
190. Ramberg JE, Nelson ED, Sinnott RA. Immunomodulatory dietary polysaccharides: a systematic review of the literature. *Nutr J* 2010;**9**(1):1.
191. Raeder RH, Barker-Merrill L, Lester T, Boyle MD, Metzger DW. A pivotal role for interferon-γ in protection against group A streptococcal skin infection. *J Infect Dis* 2000;**181**(2):639–45.
192. Masihi KN. Fighting infection using immunomodulatory agents. *Expert Opin Biol Ther* 2001;**1**(4):641–53.
193. Rice PJ, Adams EL, Ozment-Skelton T, Gonzalez AJ, Goldman MP, Lockhart BE, et al. Oral delivery and gastrointestinal absorption of soluble glucans stimulate increased resistance to infectious challenge. *J Pharmacol Exp Ther* 2005;**314**(3):1079–86.
194. Koray M, Ak G, Kürklü E, Tanyeri H, Aydin F, Oguz FS, et al. The effect of β-glucan on recurrent *Aphthous Stomatitis*. *J Evid Based Complementary Altern Med* 2009;**15**(2):111–2.
195. Astronomo RD, Burton DR. Carbohydrate vaccines: developing sweet solutions to sticky situations? *Nat Rev Drug Discov* 2010;**9**(4):308–24.
196. Rosales-Mendoza S, Salazar-González JA, Decker EL, Reski R. Implications of plant glycans in the development of innovative vaccines. *Expert Rev Vaccines* 2016;**15**:915–25.
197. Lim S-H, Hudson SM. Review of chitosan and its derivatives as antimicrobial agents and their uses as textile chemicals. *Polym Rev (Phila Pa)* 2003;**43**(2):223–69.
198. Helander I, Nurmiaho-Lassila E-L, Ahvenainen R, Rhoades J, Roller S. Chitosan disrupts the barrier properties of the outer membrane of Gram-negative bacteria. *Int J Food Microbiol* 2001;**71**(2):235–44.
199. Simoncic B, Tomsic B. Structures of novel antimicrobial agents for textiles—a review. *Text Res J* 2010; **80**(16):1721–37.

200. Hadwiger L, Kendra D, Fristensky B, Wagoner W. Chitosan both activates genes in plants and inhibits RNA synthesis in fungi. *Chitin in nature and technology*. Plenum Press, New York: Springer; 1986. p. 209–14.
201. Muzzarelli R, Weckx M, Filippini O, Lough C. Characteristic properties of N-carboxybutyl chitosan. *Carbohydr Polym* 1989;**11**(4):307–20.
202. De Rosa A, Rossi A, Affaitati P. Process for the preparation of iodinated biopolymers having disinfectant and cicatrizing activity, and the iodinated biopolymers obtainable thereby. Google Patents; 1996.
203. He F, Yang Y, Yang G, Yu L. Studies on antibacterial activity and antibacterial mechanism of a novel polysaccharide from *Streptomyces virginia* $H0_3$. *Food Control* 2010;**21**(9):1257–62.
204. Zhu H, Sheng K, Yan E, Qiao J, Lv F. Extraction, purification and antibacterial activities of a polysaccharide from spent mushroom substrate. *Int J Biol Macromol* 2012;**50**(3):840–3.
205. Marchetti M, Pisani S, Pietropaolo V, Seganti L, Nicoletti R, Degener A, et al. Antiviral effect of a polysaccharide from *Sclerotium glucanicum* towards *herpes simplex* virus type 1 infection. *Planta Med* 1996;**62**(04):303–7.
206. Eo S-K, Kim Y-S, Lee C-K, Han S-S. Possible mode of antiviral activity of acidic protein bound polysaccharide isolated from *Ganoderma lucidum* on herpes simplex viruses. *J Ethnopharmacol* 2000;**72**(3):475–81.
207. Gu C-Q, Li J-W, Chao F, Jin M, Wang X-W, Shen Z-Q. Isolation, identification and function of a novel anti-HSV-1 protein from *Grifola frondosa*. *Antiviral Res* 2007;**75**(3):250–7.

CHAPTER 15

Role of Medicinal Plants and Bioactive Compounds Against Skin Disease–Causing Microbes, With Special Emphasis on Their Mechanisms of Action

M.J. Kaneria*, K.D. Rakholiya†, S.V. Chanda*
*Saurashtra University, Rajkot, Gujarat, India
†Main Dry Farming Research Station, Junagadh Agricultural University, Targhadia, Gujarat, India

1. INTRODUCTION

The skin is the human body's largest organ and covers the other organs, tissues, bones, and muscles of the body, providing thermoregulation, protection, percutaneous absorption, and a remarkably good barrier against microbial infections, as well as having secretory and sensory activities.[1,2] Skin infections can be caused by various microbes, including bacteria, fungi, viruses and protozoa. Skin infections can occur in various sizes and locations, ranging from a spot to covering the entire body surface, as well as entering into the bloodstream. Many skin disorders are transmitted via person-to-person contact and cause serious life-threatening disease and mortality and morbidity throughout the world[3,4]. According to a survey made by Patel and Patel[5] focusing on the north Gujarat region, a major skin problem was itching (80%) with redness (18%), swelling and inflammation (17%), and the majority of the public used antihistaminic drugs to combat itching.

Some people are at a particular risk of developing skin infections. For example, people with diabetes can have poor blood flow, especially in the hands and feet, and the excessive levels of sugar in their blood decrease the ability of white blood cells to combat infections. Those who are older, who are infected by human immunodeficiency virus or other immune issues, and who are undergoing chemotherapy or treatments with other drugs that suppress the immune system, are at higher risk as well because they have a weakened immune system. Skin that is inflamed or damaged by sunburn, scratching, or other trauma is more likely to become infected. In fact, any break in the skin predisposes a person to infection.

The Microbiology of Skin, Soft Tissue, Bone and Joint Infections
ISSN 2451-9006, http://dx.doi.org/10.1016/B978-0-12-811079-9.00015-X

The main purpose of this chapter is to present information on combating skin disease through use of medicinal plants and bioactive compounds. Skin infections can occur over the whole life of human beings, from newborn till death, and many ointments and synthetic drugs are available in the market to treat these ailments, but they have drawbacks. This chapter discusses the alternative use of herbal drugs, and the secondary metabolites present in herbal drugs, obtained first from the knowledge of indigenous people and subsequently proven by the scientific method to be effective.

2. TYPES OF SKIN INFECTIONS

Skin infections are caused by: (1) bacterial infection, and (2) fungal infection (Fig. 1).

2.1 Bacterial Skin Infections

Many types of bacteria can infect the skin. The most common are *Staphylococcus* and *Streptococcus*. Skin infections caused by less-common bacteria may develop in people while hospitalized or living in a nursing home, while gardening, or while swimming in a pond, lake, or ocean.

2.1.1 Impetigo

Impetigo is a highly contagious and common bacterial surface skin infection caused by strains of both *staph* and *strep* bacteria. Impetigo is more common in children than in adults.

2.1.2 Scarlet Fever

Scarlet fever is caused by *streptococci*. It affects especially children, and causes fever and scarlet rash.

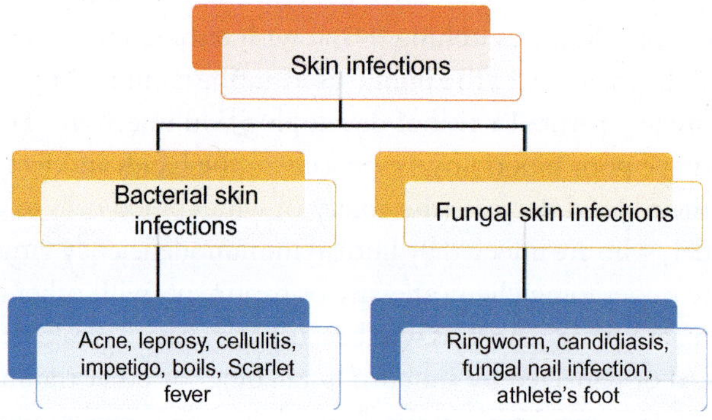

Fig. 1 Types of skin infections.

2.1.3 *Leprosy (Hansen's Disease)*

Leprosy is caused by a slow-growing type of bacteria called *Mycobacterium leprae.*

2.1.4 *Methicillin-Resistant* Staphylococcus aureus *(MRSA) Skin Infections*

Certain strains of bacteria, such as methicillin-resistant *Staphylococcus aureus* (MRSA), are resistant to common antibiotics and can be harder to treat. *Staphylococci* are widely regarded as the most important colonizers of the human skin, both in terms of frequency and sources for infection.[6]

2.1.5 *Acne*

Propionibacterium acnes has been identified as one of the main causative agents in the pathogenesis of pimples. This normally begins at puberty and varies in severity from a few spots on the face, back and chest to more extensive eruptions, mostly occurring in the younger years. Due to the facial pimples it can become a more serious problem that can be embarrassing, sap self-confidence and also cause scarring. At present, there is no "therapy" for acne, although the available remedies may be very effective in preventing the formation of new spots. Saising and Voravuthikunchai[7] used rhodomyrtone, an isolated compound from *Rhodomyrtus tomentosa* (Aiton), while Hammer[8] used tea tree oil for the treatment of *Propionibacterium acnes* and both got promising results.

2.2 Fungal Skin Infections

Superficial fungal infections are caused by many fungi, which can invade various parts of the human body.[9]

2.2.1 *Candidiasis (Yeast Infection)*

Candidiasis is an infection caused by a group of yeast organisms. There are more than 20 species of *Candida*, the most common being *Candida albicans*.

2.2.2 *Ringworm (Tinea) and Athlete's Foot (Tinea Pedis)*

Ringworm is caused by group of fungi collectively known as dermatophytes, which consist of three genera viz. *Epidermophyton*, *Microsporum* and *Trichophyton*. They are the prevailing causes of fungal infection of the skin, hair, and nails due to their ability to utilize keratin.[10] They are also responsible for Athlete's foot. This infection occurs in any part of the foot, but most often grows between the toes. Many researchers working on dermatophytes got positive results against these spp. (Table 1). Sobrinho et al.[32] worked on *Baccharis trinervis* and got a promising result against the *Trichophyton* species. *Eucalyptus citriodora* extract was able to inhibit both genus *Microsporum* and *Trichophyton*.[34] According to our review, only *Artemisia sieberi* was able to kill *Epidermophyton* spp.[33] More research is needed to search out new medicinal drugs that inhibit the *Epidermophyton* spp.

Table 1 Medicinal plants used for skin diseases

No.	Botanical name/family	Method used	Positive responds microorganisms	Country	References
Different skin diseases					
1	*Pongamia pinnata* L. (Fabaceae), *Naregamia alata* Wight & Arn. (Meliaceae), *Randia dumetorum* Lam. (Rubiaceae), *Girardinia diversifolia* (Link.) Friis.(Urticaceae)	Indigenous knowledge	–	India	11
2	*Senna alata* L. (Fabaceae), *Areca catechu* L. (Arecaceae), *Knema globularia* (Lamk.) Warb. (Myristicaceae), *Tamarindus indica* L. (Fabaceae)			Thailand	12
3	*Senecio chrysanthemoides* DC. (Asteraceae), *Verbascum thapsus* L. (Scrophulariaceae)			Pakistan	Ahmed and Akhtar[13]
4	25 different medicinal plants			Kenya	14
5	*Aloe ferox* Mill. (Xanthorrhoeaceae)			South Africa	Afolayan et al.[15]
6	*Curcuma longa* L. (Zingiberaceae)			India	Rath and Padhy[16]
7	*Ricinus communis* L. (Euphorbiaceae), *Tridax procumbens* L. (Asteraceae), *Azadirachta indica* A. Juss. (Meliaceae), *Ageratum conyzoides* L. (Asteraceae), *Allium cepa* L. (Amaryllidaceae)			India	Sharma et al.[17]

Wound healing property					
8	61 different plants belonging to 36 families	MIC, MBC	–	South Africa	18
9	*Eucalyptus camaldulensis* Dehnh. (Myrtaceae)	GC-MS, MIC, Bioautographic, checkerboard method, time kill curve	*Acinetobacter baumannii*	Serbia	Knezevic et al.[19]
Antiacne					
10	Rhodomyrtone from *Rhodomyrtus tomentosa* (Aiton) Hassk. (Myrtaceae)	MIC	*Propionibacterium acnes*	Thailand	Saising and Voravuthikunchai[7]
11	*Thea sinensis* L. (Theaceae)	–	*Propionibacterium acnes*	Australia	Hammer[8]
12	*Rosa hybrida* L. (Rosaceae)	MIC	*Propionibacterium acnes*	South Korea	20
Scabies					
13	*Saraca asoca* (Roxb.) Willd. (Fabaceae)	Indigenous knowledge	–	India	Kingston et al.[21]
Antibacterial					
14	*Ilex paraguariensis* A. St.-Hil. (Aquifoliaceae)	–	MRSA	United States	22
15	*Ilex paraguariensis* (Aquifoliaceae) A. St. Hil. small Molecules	Disc diffusion method, GC-MS	MRSA	United States	23
16	*Curcuma longa* L. (Zingiberaceae)	Disc diffusion method, SEM	*Staphylococcus aureus*	India	24
17	*Vangueriam adagascariensis* J.F. Gmel. (Rubiaceae) + Chloramphenicol	MIC	*Streptococcus* group A	Mauritius	25
18	*Passiflora cincinnata* Mast. (Passifloraceae)	MIC	*Staphylococcus aureus*	Brazil	26

Continued

Table 1 Medicinal plants used for skin diseases—cont'd

No.	Botanical name/family	Method used	Positive responds microorganisms	Country	References
19	*Alchornea cordifolia* Mull. Arg. (Euphorbiaceae)	MIC, Bioautography, Mass spectroscopy, Fourier transformed infrared spectrometry (FTIR), nuclear magnetic resonance (NMR)	*Staphylococcus aureus*, *Klebsiella pneumoniae*	South Africa	27
20	*Plukenetia volubilis* L. (Euphorbiaceae)	Cytotoxic effects, Adherence assays	*Staphylococcus aureus*	France	28
21	*Thea sinensis* L. (Theaceae)	Time-kill assays, MIC	*Staphylococcus aureus*	China	29
22	*Azadirachta indica* A. Juss (Meliaceae)	MIC, AFM, FT-MS, HPLC-FT-MS	MRSA	Brazil	30
23	*Citrus limon* L. (Rutaceae)	Sandwich-type immunoassays and genotyping	*Enterococcus faecalis*, *Bacillus subtilis*, *Salmonella typhimurium*, *Shigella sonnei*	South Africa	31
Antifungal					
24	*Baccharis trinervis* (Lam.) Pers. (Asteraceae)	MIC, MFC, checkerboard assay	*Trichophyton rubrum*, *Candida albicans*	Brazil	32
25	*Artemisia sieberi* L. (Asteraceae)	MIC, MFC	*Trichophyton mentagrophytes*, *T. rubrum*, *Epidermophyton floccosum*, *Microsporum gypseum*, *M. canis*	Iran	33
26	*Eucalyptus citriodora* (Hook.) K.D. Hill & L.A.S. Johnson (Myrtaceae)	Disc diffusion method, disc volatilization method, agar dilution method	*M.canis*, *M. gypseum*, *Trichophyton mentagrophytes*, *Trichophyton rubrum*	Algeria	34
27	*Thymus vulgaris* L. (Lamiaceae) *Cinnamomum cassia* L. (Lauraceae)	MIC, MFC	*Aspergillus flavus*	Serbia	Pekmezovic et al.[35]
28	Fluconazole + Red propolis	MIC, checkerboard assay	*Candida parapsilosis*, *C. glabrata*	Brazil	36
29	*Daucus carota* L. (Apiaceae)	MIC	*Cryptococcus neoformans*	Portugal	37
30	Propolis	–	*Candida* spp., *S. cerevisiae*, *Trichophyton* spp.	Brazil	38

31	*Citrus aurantifolia* (Christm.) Swingle (Rutaceae) + fluconazole	Photoactivation	*Candida albicans, Tricophyton rubrum*	Iran	39
32	*Cassia fistula* L.(Fabaceae)	Disc diffusion assay	*Candida albicans*	India	40
33	Chitooligosaccharides	Disc diffusion assay, MFC	*Trichophyton rubrum*	China	41
34	*Manihot multifida* L. Crantz (Euphorbiaceae)	Minimal inhibitory concentration (MIC)	*Candida albicans*	Brasil	42
35	*Cassia fistula* L. (Fabaceae)	Agar well diffusion assay	*Aspergillus niger*	India	43
Antimicrobial					
36	*Etlingera fimbriobracteata* (K.Schum.) R.M.Sm. (Zingiberaceae)	Disc diffusion assay, MIC	*Bacillus subtilis, Bacillus spizizenii, Staphylococcus aureus, Escherichia coli, Pseudomonas aeruginosa, Candida albicans, Saccharomyces cerevisiae*	Turkey	44
37	Essential oil of *Pimenta dioica* L. Merr. (Myrtaceae), *Psiadia arguta* Voigt (Asteraceae), *Piper betle* L. (Piperaceae) + gentamicin	Disc diffusion method, MIC, MBC, MFC, GC-MS	*Staphylococcus epidermidis Propionibacterium acnes, Candida albicans, Candida tropicalis, Aspergillus niger*	Mauritius	Aumeeruddy-Elalfia et al.[45]
38	*Curcuma longa* L., *Curcuma amada* L. (Zingiberaceae)	Agar well diffusion method	*Staphylococcus aureus, Bacillus subtilis, Pseudomonas aeruginosa, Escherichia coli, Candida tropicalis, Candida albicans, Candida neoformans, Cryptococcus leuteolus*	India	46
39	Selected essential oils	MIC	*Bacillus cereus, Staphylococcus aureus, Enterococcus faecalis, Escherichia coli, Pseudomonas aeruginosa, Candida albicans, Cryptococcus neoformans*	South Africa, United Kingdom	47

MRSA, Methicillin resistant *Staphylococcus aureus; GC-MS*, Gas chromatography-mass spectrometry; *FT-MS*, Fourier transform mass spectrometry; *HPLC-FT-MS*, liquid chromatography coupled to ultra-high-resolution mass spectrometry; *AFM*, Atomic force microscopy; *NMR*, Nuclear magnetic resonance.

2.2.3 Fungal Nail Infections (Tinea Unguium)

A fungal nail infection occurs when a fungus attacks a fingernail, a toenail, or the skin under the nail, called the nail bed. Nail infections are more common in people who live in hot or humid climates.

3. CAUSES OF ANTIBIOTIC RESISTANCE

The main cause of resistance to antibiotics is their repeated and improper use, which has led to the increase in drug-resistant bacteria.[48–51] There are three main types of misuse:

(1) Incomplete use: Not finishing the entire antibiotic prescription (often because people start to feel better and stop taking the medication).

(2) Inappropriate use: Taking antibiotics for a viral infection (antibiotics do not treat viral infections, such as colds or the flu).

(3) Unnecessary use: Taking antibiotics "just in case" (for example, taking them prior to a vacation or special event to try to prevent illness).

Bacteria develop antibiotic resistance in several ways, as explained in our previous publications.[52,53]

There is a dire need to search for new antibacterial agents, mainly because of the increase in persisting infectious diseases and resistance to modern antibiotics.[54–56] Moreover, bacterial skin infection treatment becomes complicated especially in immunocompromised patients due to weakened immunity, for example in cancer, AIDS, transplant, etc. In addition to this problem, antibiotics are sometimes associated with adverse effects on the host, which include hypersensitivity, depletion of beneficial gut and mucosal microorganisms, immune suppression and allergic reactions.[57,58] All these factors are associated with synthetic drugs; therefore, there is a need to search for safe, economically affordable by common users, and alternative natural drugs. Therefore, one of the best approaches is the use of local medicinal plants from indigenous resources as therapeutic agents.

Looking at the numerous pharmaceutical skin formulations available, it can be observed that microbes are now more resistant to the existing drugs; in addition, these drugs often have undesired adverse effects, such as burning sensations, stinging, photohypersensitivity, skin irritation and, more severe, anaphylactic shock.

4. INDIGENOUS KNOWLEDGE

Indigenous knowledge of the use of medicinal plants to cure various ailments dates back to the beginnings of human civilization.[59,60] According to the World Health Organization (WHO) about 70%–95% of the world's population in developing countries relies mainly on indigenous medicinal plants for their primary health care. Traditional medicine has not only gained in popularity and approval, but it is sometimes the only system available in many rural areas. Furthermore, the use of medicinal plants to treat skin infections is very common in many rural areas all over the world.

The term *ethnobotany* was first coined by an American botanist, John Harshburger, in 1896, in an attempt to study the plants used by primitive and aboriginal people.[61] Since then, it has been defined as the traditional knowledge of indigenous communities about surrounding plant diversity and how various people make use of indigenous plants found in their localities. As shown in Table 1, many scientists have gathered information about use of plants by indigenous people to treat skin diseases. These include Kingston et al.[21], Sharma et al.[17] and Rath and Padhy[16] from India, who reported use of *Saraca asoca* (Fabaceae), *Ricinus communis* (Euphorbiaceae), *Tridax procumbens*, *Azadirachta indica*, *Ageratum conyzoides* (Asteraceae), *Allium cepa* (Amaryllidaceae) and *Curcuma longa* to treat scabies and fungal diseases. Apart from India, people from other countries also treat skin diseases on the basis of traditional knowledge; for example, *Aloe ferox* used by Afolayan et al.[15] was reported from South Africa, and *Senecio chrysanthemoides* is used for skin diseases in Pakistan.[13]

5. SCIENTIFIC KNOWLEDGE APPLIED TO USE OF MEDICINAL PLANTS

It is the interface between indigenous humans and their use of the flora around them that has become a significant facet of biological diversity considerations. Herbs are also known as "miracle elixirs of life" and their secondary metabolites are recommended in the rem edies of various skin conditions, including lupus, varicose ulcers, leprosy, psoriasis, and eczema, along with their wound-healing properties.[62] Knezevic et al.[19] reported the wound-healing property of *Eucalyptus camaldulensis* oil.

From traditional knowledge, many scientists have proved the efficacy of plants and discovered the potential bioactive constitutes present in them experimentally and scientifically. Moreover, traditional medicine has even more benefits than other health care approaches, as it is inexpensive, readily available and could cure diseases permanently.[63–66] On the basis of our review, we can say that reports by numerous scientists throughout the world are varied in the extraction techniques, plant parts (stem, root, leaves, peel, etc.), essential oils and microorganisms to treat skin diseases (Table 1). It was found that some of the plants in the survey have been cited in the ethnobotanical survey of other foreign countries, especially in the treatment of skin diseases, e.g., *Eucalyptus camaldulensis* showed wound-healing properties.[19] Pekmezovic et al.[35] included *Thymus vulgaris*, *Cinnamomum cassia* in their work. The information showed that these plants have potential for curing skin diseases. Essential oils of medicinal plants exhibit potentialities toward different skin disease–causing microorganisms. According to Kon and Rai[67] essential oils are used against multidrug-resistant strains. Many components of essential oils have been screened for antibacterial and antifungal activities, which are shown in Table 1. Zouari et al.[68] reported essential oils of *Artemisia saharae* contain 75 compounds that inhibit bacterial strains.

6. PLANTS AS A RICH SOURCE OF BIOACTIVE COMPOUNDS (SECONDARY METABOLITES)

Plants synthesize and accumulate a diverse array of secondary metabolites, also known as natural products, which have biological activity and provide a rich source for modern pharmacy.[69–71] There are two types of plant metabolism: (1) Primary metabolism, and (2) secondary metabolism.

Plant secondary metabolites can be divided into three major, chemically distinct, groups: terpenoids, flavonoids and allied phenolic and polyphenolic compounds; nitrogen-containing alkaloids; and sulfur-containing compounds. Secondary metabolites show different antimicrobial activity because of the differences in the bioactive constituents present in them. According to reported data, tannins, flavonoids, alkaloids and other phenolic compounds are responsible for the active antimicrobial compounds[72] and some of the structures are shown in Fig. 2.

7. COMBINED PREPARATIONS USED TO TREAT SKIN INFECTIONS

Another aspect in the investigation of plants used to treat skin infections is the investigation of pharmacological interactions between plants used in combination.[73] A number of plant combinations are used to treat various bacterial and fungal skin infections (Table 1). Aumeeruddy-Elalfia et al.[45] reported *Pimenta dioica*, *Psiadia arguta* and *Piper betle* against *Staphylococcus epidermidis* when combined with gentamicin. Therefore, the development of new products (where special attention has been given to natural products with dermatological relevance) can lead to potentially effective agents, which may be additionally less expensive, relatively safe for human use and be environmentally friendly, and also affordable to a majority of economically underprivileged communities.[74–79]

8. CONCLUSIONS AND FUTURE PERSPECTIVES

Natural drugs have great potential to cure various kinds of skin diseases caused by bacterial and fungal strains based on this review, which screened many indigenous people depending on medicinal plants all over the country. On the basis of traditional knowledge with its holistic and systemic approach, various scientists have worked on different parts of particular medicinal plants (leaf, stem, flower, root, bark, fruit) using different methods to cure skin-related problems. This experimental base can serve as a

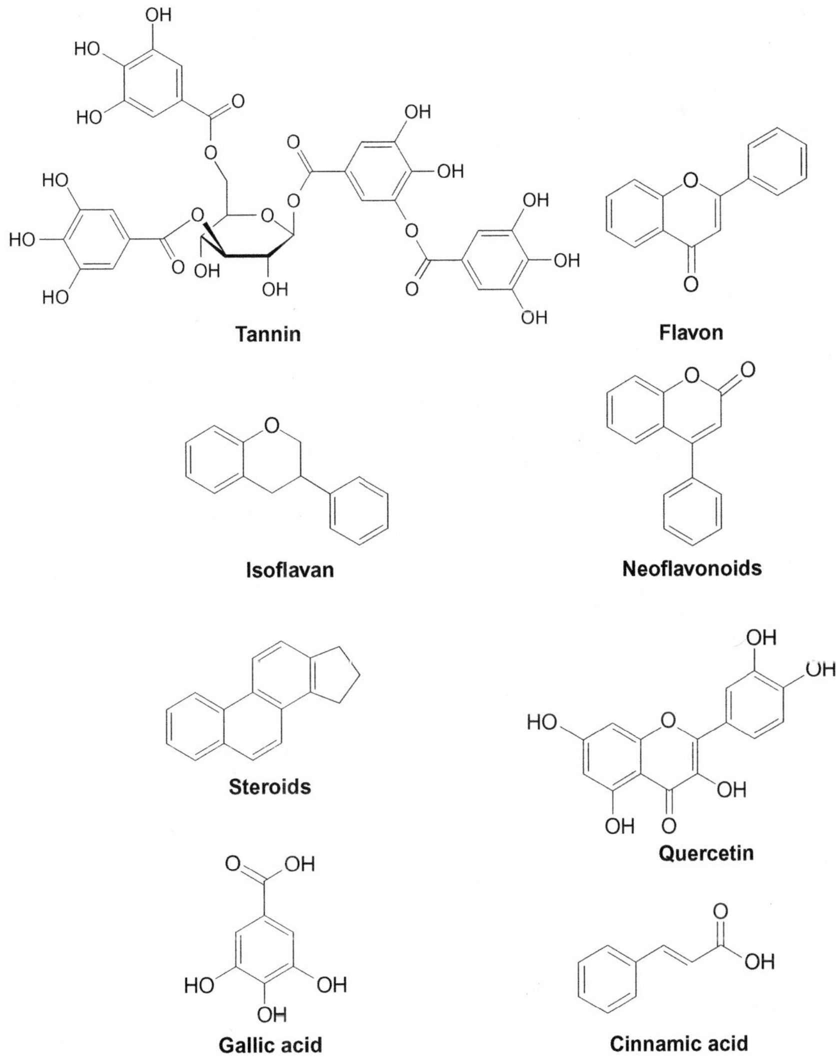

Fig. 2 Structure of secondary metabolites.

pioneering and powerful discovery engine for easily accessible sources of newer, safer and most affordable medicines.

Several promising medicinal plants used in various countries, alone or in combination with synthetic drugs, might be of future relevance to modern medicine not only in treating skin diseases but also as potential sources for maintaining proper health. The potent plants showed biological activity due to the phytoconstituents present in them. Further research is warranted to isolate the compounds responsible for the observed biological activity.

REFERENCES

1. Chanda S, Baravalia Y. Screening of some plant extracts against some skin diseases caused by oxidative stress and microorganisms. *Afr J Biotechnol* 2010;**9**:3210–7.
2. Otto M. Physical stress and bacterial colonization. *FEMS Microbiol Rev* 2014;**38**:1250–70.
3. Fair RJ, Tor Y. Antibiotics and bacterial resistance in the 21st century. *Perspect Med Chem* 2014;**6**:25–64.
4. Leprince C, Desroches M, Emirian A, Coutureau C, Anais L, Fihman V, et al. Distribution and antimicrobial susceptibility of bacteria from adults with community-acquired pneumonia or complicated skin and soft tissue infections in France: the nationwide French PREMIUM study. *Diagn Microbiol Infect Dis* 2015;**83**:175–82.
5. Patel G, Patel NJ. Epidemiological study of skin (dermatological) diseases and its treatment in North Gujarat. *Asian J Pharm Clin Res* 2010;**3**:40–2.
6. Otto M. *Staphylococcus* colonization of the skin and antimicrobial peptides. *Expert Rev Dermatol* 2010;**5**:183–95.
7. Saising J, Voravuthikunchai SP. Anti-*Propionibacterium acnes* activity of rhodomyrtone, an effective compound from *Rhodomyrtu stomentosa* (Aiton) Hassk leaves. *Anaerobe* 2012;**18**:400–4.
8. Hammer KA. Treatment of acne with tea tree oil (melaleuca) products: a review of efficacy, tolerability and potential modes of action. *Int J Antimicrob Agents* 2015;**45**:106–10.
9. Kaushik N, Pujalte GFA, Reese ST. Superficial fungal infections. *Prim Care* 2015;**42**:501–16.
10. Adams BB. Tinea corporis gladiatorum. *J Am Acad Dermatol* 2002;**47**:286.
11. Bhat P, Hegde GR, Hegde G, Mulgund GS. Ethnomedicinal plants to cure skin diseases—an account of the traditional knowledge in the coastal parts of Central Western Ghats, Karnataka, India. *J Ethnopharmacol* 2014;**151**:493–502.
12. Neamsuvan O, Kama A, Salaemae A, Leesen S, Waedueramae N. A survey of herbal formulas for skin diseases from Thailand's three southern border provinces. *J Herb Med* 2015;**5**:190–8.
13. Ahmed MJ, Akhtar T. Indigenous knowledge of the use of medicinal plants in Bheri, Muzaffarabad, Azad Kashmir, Pakistan. *Eur J Integr Med* 2016. http://dx.doi.org/10.1016/j.eujim.2016.01.006.
14. Omwenga EO, Hensel A, Shitandi A, Goycoolea FM. Ethnobotanical survey of traditionally used medicinal plants for infections of skin, gastrointestinal tract, urinary tract and the oral cavity in Borabu sub-county, Nyamira county, Kenya. *J Ethnopharmacol* 2015;**176**:508–14.
15. Afolayan AJ, Grierson DS, Mbeng WO. Ethnobotanical survey of medicinal plants used in the management of skin disorders among the Xhosa communities of the Amathole District, Eastern Cape, South Africa. *J Ethnopharmacol* 2014;**153**:220–32.
16. Rath S, Padhy RN. Monitoring in vitro antibacterial efficacy of 26 Indian spices against multidrug resistant urinary tract infecting bacteria. *Integr Med Res* 2016. http://dx.doi.org/10.1016/j.imr.2014.04.004.
17. Sharma J, Gairola S, Sharma YP, Gaur RD. Ethnomedicinal plants used to treat skin diseases by Tharu community of district Udham Singh Nagar, Uttarakhand, India. *J Ethnopharmacol* 2014;**158**:140–206.
18. Agyare C, Boakye YD, Bekoe EO, Hensel A, Dapaah SO, Appiah T. Review: African medicinal plants with wound healing properties. *J Ethnopharmacol* 2016;**177**:85–100.
19. Knezevic P, Aleksic V, Simin N, Svircev E, Petrovic A, Mimica-Dukic N. Antimicrobial activity of *Eucalyptus camaldulensis* essential oils and their interactions with conventional antimicrobial agents against multi-drug resistant *Acinetobacter baumannii*. *J Ethnopharmacol* 2016;**178**:125–36.
20. Park D, Shin K, Choi Y, Guo H, Cha Y, Kim S, et al. Antimicrobial activities of ethanol and butanol fractions of white rose petal extract. *Regul Toxicol Pharm* 2016;**76**:57–62.
21. Kingston C, Jeeva S, Jeeva GM, Kiruba S, Mishra BP, Kannan D. Indigenous knowledge of using medicinal plants in treating skin diseases in Kanyakumari district, Southern India. *Indian J Tradit Knowl* 2009;**8**:196–200.
22. Burris KP, Higginbotham KL, Stewart Jr CN. Aqueous extracts of yerba mate as bactericidal agents against methicillin-resistant *Staphylococcus aureus* in a microbiological medium and ground beef mixtures. *Food Control* 2015;**50**:748–53.
23. Rempe CS, Burris KP, Woo HL, Goodrich B, Gosnell DK, Tschaplinski TJ, et al. Computational ranking of yerba mate small molecules based on their predicted contribution to antibacterial activity against methicillin-resistant *Staphylococcus aureus*. *PLoS ONE* 2015;**10**:e0123925.

24. Gupta A, Mahajan S, Sharma R. Evaluation of antimicrobial activity of *Curcuma longa* rhizome extract against *Staphylococcus aureus*. *Biotechnol Rep* 2015;**6**:51–5.
25. Mahomoodally MF, Dilmohamed S. Antibacterial antibiotic potentiating activity of Vangueria madagascariensis leaves and ripe fruit pericarp against human pathogenic clinical bacterial isolates. *J Tradit Complement Med* 2015. http://dx.doi.org/10.1016/j.jtcme.2015.09.002.
26. Siebra ALA, Oliveira LR, Martins OBPB, Siebra DC, Albuquerque RS, Lemos ICS, et al. Potentiation of antibiotic activity by *Passiflora cincinnata* Mast. front of strains *Staphylococcus aureus* and *Escherichia coli*. *Saudi J Biol Sci* 2016. http://dx.doi.org/10.1016/j.sjbs.2016.01.019.
27. Noundou XS, Krause RWM, van Vuuren SF, Ndinteh DT, Olivier DK. Antibacterial effects of *Alchornea cordifolia* (Schumach. and Thonn.) Müll. Arg extracts and compounds on gastrointestinal, skin, respiratory and urinary tract pathogens. *J Ethnopharmacol* 2016;**179**:76–82.
28. Gonzalez-Aspajo G, Belkhelfa H, Haddioui-Hbabi L, Bourdy G, Deharo E. Sacha Inchi oil (*Plukenetia volubilis* L.), effect on adherence of *Staphylococus aureus* to human skin explant and keratinocytes *in vitro*. *J Ethnopharmacol* 2015;**171**:330–4.
29. Shi C, Zhao X, Yan H, Meng R, Zhang Y, Li W, et al. Effect of tea tree oil on *Staphylococcus aureus* growth and enterotoxin production. *Food Control* 2016;**62**:257–63.
30. Quelemes PV, Perfeito MLG, Guimarães MA, dos Santos RC, Lima DF, Nascimento C, et al. Effect of neem (*Azadirachta indica* A. Juss) leaf extract on resistant *Staphylococcus aureus* biofilm formation and *Schistosoma mansoni*worms. *J Ethnopharmacol* 2015;**175**:287–94.
31. Otang WM, Afolayan AJ. Antimicrobial and antioxidant efficacy of *Citrus limon* L. peel extracts used for skin diseases by Xhosa tribe of Amathole District, Eastern Cape, South Africa. *S Afr J Bot* 2016;**102**:46–9.
32. Sobrinho ACN, de Souza EB, Rocha MFG, Albuquerque MJR, Bandeira PN, dos Santos HS, et al. Chemical composition, antioxidant, antifungal and hemolytic activities of essential oil from *Baccharis trinervis* (Lam.) Pers. (Asteraceae). *Ind Crop Prod* 2016;**84**:108–15.
33. Khosravi RL, Shokri H, Farahnejat Z, Chalangari R, Katalin M. Antimycotic efficacy of Iranian medicinal plants towards dermatophytes obtained from patients with dermatophytosis. *Chin J Nat Med* 2013;**11**:43–8.
34. Tolba H, Moghrani H, Benelmouffok A, Kellou D, Maachia R. Essential oil of Algerian *Eucalyptus citriodora*: chemical composition, antifungal activity. *J Med Mycol* 2015;**25**:128–33.
35. Pekmezovic M, Rajkovic K, Barac A, Senerovic L, Arsenijevic VA. Development of kinetic model for testing antifungal effect of *Thymus vulgaris* L. and *Cinnamomum cassia* L. essential oils on *Aspergillus flavus* spores and application for optimization of synergistic effect. *Biochem Eng J* 2015;**99**:131–7.
36. Pippi B, Lana AJ, Moraes RC, Guez CM, Machado M, de Oliveira LF, et al. In vitro evaluation of the acquisition of resistance, antifungal activity and synergism of Brazilian red propolis with antifungal drugs on *Candida* spp. *J Appl Microbiol* 2015;**118**:839–50.
37. Valente J, Zuzarte M, Resende R, Gonçalves MJ, Cavaleiro C, Pereira CF, et al. *Daucus carota* subsp. gummifer essential oil as a natural source of antifungal and anti-inflammatory drugs. *Ind Crop Prod* 2015;**65**:361–6.
38. Freires IA, de Alencar SM, Rosalen PL. A pharmacological perspective on the use of Brazilian Red Propolis and its isolated compounds against human diseases. *Eur J Med Chem* 2016;**110**:267–79.
39. Fekrazad R, Mir DDS, Barghi VG, Shams-Ghahfarokhi M. Eradication of *C. albicans* and *T. rubrum* with photoactivated indocyanine green, *Citrus aurantifolia* essential oil and fluconazole. *Photodiagnosis Photodyn Ther* 2015;**12**:289–97.
40. Bhalodia NR, Shukla VJ. Antibacterial and antifungal activities from leaf extracts of *Cassia fistula* L.: An ethnomedicinal plant. *J Adv Pharm Technol Res* 2011;**2**:104–9.
41. Mei Y, Dai X, Yang W, Xu X, Liang Y. Antifungal activity of chitooligosaccharides against the dermatophytes *Trichophyton rubrum*. *Int J Biol Macromol* 2015;**77**:330–5.
42. Fabri RL, De Sa DS, Pereira APO, Scio E, Pimenta DS, Chedier LM. Antimicrobial, antioxidant and cytotoxicity potential of *Manihot multifida* (L.) Crantz (Euphorbiaceae). *An Acad Bras Cienc* 2015;**87**:303–11.
43. Kumar S, Jena PK, Sabnam S, Kumari MK, Tripathy PK. Study of plants used against the skin diseases with special reference to *Cassia fistula* L. among the king (Dongaria Kandha) of Niyamgiri: a primitive tribe of Odisha, India. *Int J Drug Dev Res* 2012;**4**:256–64.

44. Ud-Daula AFMS, Demirci F, Salim KA, Demirci B, Lim LBL, Baser KHC, et al. Chemical composition, antioxidant and antimicrobial activities of essential oils from leaves, aerial stems, basal stems, and rhizomes of *Etlingera fimbriobracteata* (K.Schum.) R.M.Sm. *Ind Crop Prod* 2016;**84**:189–98.
45. Aumeeruddy-Elalfia Z, Gurib-Fakim A, Mahomoodally F. Antimicrobial, antibiotic potentiating activity and phytochemical profile of essential oils from exotic and endemic medicinal plants of Mauritius. *Ind Crop Prod* 2015;**71**:197–204.
46. Chanda S, Baravalia Y. Novel leads from herbal drugs for infectious skin diseases. In: Mendez-Vilas A, editor. *Current research, technology and education topics in applied microbiology and microbial biotechnology*, vol. 2. Badajoz, Spain: Formatex; 2010. p. 451–6.
47. Maree J, Kamatou G, Gibbons S, Viljoen A, Vuuren SV. The application of GC-MS combined with chemometrics for the identification of antimicrobial compounds from selected commercial essential oils. *Chemometr Intell Lab* 2014;**130**:172–81.
48. Levy SB. Factors impacting on the problem of antibiotic resistance. *J Antimicrob Chemother* 2002;**49**:25–30.
49. Chanda S, Baravalia Y, Kaneria M, Rakholiya K. Fruit and vegetable peels—strong natural source of antimicrobics. In: Mendez-Vilas A, editor. *Current research, technology and education topics in applied microbiology and microbial biotechnology*, vol. 2. Badajoz, Spain: Formatex; 2010. p. 444–50.
50. Rakholiya K, Kaneria M, Chanda S. Inhibition of microbial pathogens using fruit and vegetable peel extracts. *Int J Food Sci Nutr* 2014;**65**:733–9.
51. Rakholiya K, Kaneria M, Chanda S. *In vitro* assessment of novel antimicrobial from methanol extracts of matured seed kernel and leaf of *Mangifera indica* L. (Kesar Mango) for inhibition of *Pseudomonas* spp. and their synergistic potential. *Am J Drug Discovery Dev* 2015;**5**:13–23.
52. Chanda S, Rakholiya K. Indian Combination therapy: synergism between natural plant extracts and antibiotics against infectious diseases. In: Mendez-Vilas A, editor. *Science against microbial pathogens: communicating current research and technological advances*, vol. 1. Badajoz, Spain: Formatex Research Center; 2011. p. 520–9.
53. Rakholiya KD, Kaneria MJ, Chanda SV. Medicinal plants as alternative sources of therapeutics against multidrug-resistant pathogenic microorganisms based on their antimicrobial potential and synergistic properties. In: Kon KV, Rai MK, editors. *Fighting multidrug resistance with herbal extracts, essential oils and their components*. New York, NY: Elsevier; 2013. p. 165–79.
54. Chanda S, Kaneria M, Vaghasiya Y. Evaluation of antimicrobial potential of some Indian medicinal plants against some pathogenic microbes. *Indian J Nat Prod Res* 2011;**2**:225–8.
55. Rakholiya K, Vaghela P, Rathod T, Chanda S. Comparative study of hydroalcoholic extracts of *Momordica charantia* L. against food borne pathogens. *Indian J Pharm Sci* 2014;**76**:148–56.
56. Assiri AMA, Elbanna K, Al-Thubiani A, Ramadan MF. Cold-pressed oregano (*Origanum vulgare*) oil: a rich source of bioactive lipids with novel antioxidant and antimicrobial properties. *Eur Food Res Technol* 2015. http://dx.doi.org/10.1007/s0217-015-2607-7.
57. Chanda S, Rakholiya K, Nair R. Antimicrobial activity of *Terminalia catappa* L. leaf extracts against some clinically important pathogenic microbial strains. *Chinese Med* 2011;**2**:171–7.
58. Chanda S, Rakholiya K, Parekh J. Indian medicinal herb: Antimicrobial efficacy of *Mesua ferrea* L. seed extracted in different solvents against infection causing pathogenic strains. *J Acute Dis* 2013;**2**:277–81.
59. Malla B, Gauchan DP, Chhetri RB. An ethnobotanical study of medicinal plants used by ethnic people in Parbat district of western Nepal. *J Ethnopharmacol* 2015;**165**:103–17.
60. Sonagara J, Rakholiya K, Padalia H, Chanda S, Kaneria M. In vitro antimicrobial activity: *Salvadora* species. In: Meghwal M, Goyal MR, Kaneria M, editors. *Emerging technologies and innovations in food engineering*. Harrington Park, NJ: Apple Academic Press; 2016.
61. Shosan LO, Fawibe OO, Ajiboye AA, Abeegunrin TA, Agboola DA. Ethnobotanical survey of medicinal plants used in curing some diseases in infants in Abeokuta South local government area of Ogun State, Nigeria. *Am J Plant Sci* 2014;**5**:3258–68.
62. Dhama K, Tiwari R, Chakraborty S, Saminathan M, Kumar A, Karthik K, et al. Evidence based antibacterial potentials of medicinal plants and herbs countering bacterial pathogens especially in the era of emerging drug resistance: an integrated update. *Int J Pharmacol* 2014;**10**:1–43.
63. Tabassum N, Hamdani M. Plants used to treat skin diseases. *Pharmacogn Rev* 2014;**8**:52–60.

64. Chanda S, Dudhatra S, Kaneria M. Antioxidative and antibacterial effects of seeds and fruit rind of nutraceutical plants belonging to the Fabaceae family. *Food Funct* 2010;**1**:308–15.
65. Kaneria M, Chanda S. Evaluation of antioxidant and antimicrobial capacity of *Syzygium cumini* L. leaves extracted sequentially in different solvents. *J Food Biochem* 2013;**37**:168–76.
66. Rakholiya K, Kaneria M, Chanda S. Mango pulp: a potential source of natural antioxidant and antimicrobial agents. In: Gupta VK, editor. *Medicinal plants: phytochemistry, pharmacology and therapeutics*. vol. 3. New Delhi: Daya Publication House; 2014. p. 253–84.
67. Kon KV, Rai MK. Combining essential oils with antimicrobial agents to overcome multidrug-resistant bacteria. In: Kon KV, Rai MK, editors. *Fighting multidrug resistance with herbal extracts, essential oils and their components*, vol. 1. Atlanta, GA: Elsevier; 2013. p. 149–64.
68. Zouari S, Ayadi I, Fakhfakh N, Jdir H, Aloui L, Kossentini M, et al. Essential oil variation in wild populations of *Artemisia saharae* (Asteraceae) from Tunisia: chemical composition, antibacterial and antioxidant properties. *Bot Stud* 2014;**55**:76.
69. Bandaranayake WM. Bioactivities, bioactive compounds and chemical constituents of mangrove plants. *Wetl Ecol Manag* 2002;**10**:421–52.
70. Rakholiya K, Kaneria M, Trivedi N, Chanda S. Antimicrobial efficiency and phenolic content of acetone extract and its fractions from Mango peels (ripe and unripe). In: Gupta VK, editor. *Medicinal plants: phytochemistry, pharmacology and therapeutics*, vol. 4. New Delhi: Daya Publication House; 2016.
71. Yang L, Yang C, Li C, Zhao Q, Liu L, Fang X, et al. Recent advances in biosynthesis of bioactive compounds in traditional Chinese medicinal plants. *Sci Bull* 2016;**61**:3–17.
72. Rakholiya KD, Kaneria MJ, Chanda SV. Isolation, validation and characterization of major bioactive constituents by various techniques from mango ripe seed. In: Meghwal M, Goyal MR, Kaneria M, editors. *Emerging technologies and innovations in food engineering*. Waretown, NJ: Apple Academic Press; 2016.
73. Mabona U, Van Vuuren SF. Southern African medicinal plants used to treat skin diseases. *S Afr J Bot* 2013;**87**:175–93.
74. Njoroge GN, Bussmann RW. Ethnotherapeautic management of skin diseases among the Kikuyus of Central Kenya. *J Ethnopharmacol* 2007;**111**:303–7.
75. Chanda S, Kaneria M. Indian nutraceutical plant leaves as a potential source of natural antimicrobial agents. In: Mendez-Vilas A, editor. *Science against microbial pathogens: communicating current research and technological advances*. vol. 2. Badajoz, Spain: Formatex Research Center; 2011. p. 1251–9.
76. Rakholiya K, Chanda S. *In vitro* interaction of certain antimicrobial agents in combination with plant extracts against some pathogenic bacterial strains. *Asian Pac J Trop Biomed* 2012;**2**:S876–80.
77. Rakholiya K, Kaneria M, Desai D, Chanda S. Antimicrobial activity of decoction extracts of residual parts (seed and peels) of *Mangifera indica* L. var. Kesar against pathogenic and food spoilage microorganism. In: Mendez-Vilas, editor. *Microbial pathogens and strategies for combating them: science, technology and education*, vol. 2. Badajoz, Spain: Formatex Research Center, 2013. p. 850–6.
78. Chanda S, Rakholiya K, Dholakia K, Baravalia Y. Antimicrobial, antioxidant and synergistic property of two nutraceutical plants: *Terminalia catappa* L. and *Colocasia esculenta* L. *Turk J Biol* 2013;**37**:81–91.
79. Kaneria M, Rakholiya K, Chanda S. *In vitro* antimicrobial and antioxidant potency of two nutraceutical plants of Cucurbitaceae. In: Gupta VK, editor. *Traditional and folk herbal medicine: recent researches*, vol. 2. New Delhi: Daya Publication House; 2014. p. 221–48.

CHAPTER 16

Antimicrobial Strategies in Novel Drug Delivery Systems: Applications in the Treatment of Skin and Soft Tissue Infections

J. Bueno*, F. Demirci[†,‡], K. Husnu Can Baser[§]

[*]Research Center of Bioprospecting and Biotechnology for Biodiversity Foundation (BIOLABB), Colombia
[†]Faculty of Health Sciences, Anadolu University, Eskişehir, Turkey
[‡]Faculty of Pharmacy, Anadolu University, Eskişehir, Turkey
[§]Faculty of Pharmacy, Near East University, Nicosia, N. Cyprus

The problem of chemotherapy of bacterial infections could be solved neither by the experimental medical research worker nor by the chemist alone, but only by the two together working in very close cooperation over many years.

Gerhard Domagk (1895–1964)

1. INTRODUCTION

1.1 Acute Bacterial Skin and Skin Structure Infections

Skin and soft tissue infections (SSTIs) are defined as skin and subjacent soft tissue damage caused by microbial attack with variable severity and clinical presentation, ranging from cellulitis to necrotizing infections.[1] Around 7–10% of patients under medical care can be affected by SSTIs,[2] and their increase has been reported as a cause of hospitalization in about 90% in men and 75% in women[3]. Among SSTIs, the most complicated correspond to acute bacterial skin and skin structure infections (ABSSSI) caused by *Staphylococcus aureus* and *S. pyogenes*, as well as *Enterococcus faecalis* and Gram-negative bacteria such as *Acinetobacter baumanii*. ABSSSIs include cellulitis, erysipelas, wound infection, and major cutaneous abscess with a minimum lesion surface area of 75 cm^2.[4–6] In addition, surgical site infections (SSIs) are a frequent complication of hospitalization and occur in about 2–5% of surgical patients in the United States; they are the second most common form of health care-associated infection (HAI).[7] The etiological agent most associated with this infection is *S. aureus* with a presence in 20% of the cases, with methicillin-resistant *S. aureus* (MRSA) having the worst prognosis and clinical response.[8]

The Microbiology of Skin, Soft Tissue, Bone and Joint Infections
ISSN 2451-9006, http://dx.doi.org/10.1016/B978-0-12-811079-9.00016-1

Under this particular topic, clinical complications of ABSSSIs are necrotizing soft tissue infections (NSTIs) which represent the most difficult challenge of treatment for physicians and surgeons, with a mortality ranging between 25–35%. The usual treatment consists of broad-spectrum antibiotics and aggressive surgical debridement.[9,10]

The treatment for ABSSSIs before the emergence of MRSA consisted of the administration of beta-lactam antibiotics; with the subsequent spread of MRSA strains, the current approaches to skin infections comprise three procedures: Surgical debridement and drainage, wound culture with antimicrobial susceptibility testing profile and prescription of empiric antibiotic therapy.[11] The treatments of choice for ABSSSIs due to MRSA currently are vancomycin, linezolid, tigecycline, daptomycin, ceftaroline, and telavancin.[12–14] In 2002, the Infectious Diseases Society of America (IDSA) issued an alert concerning the lack of advancement in developing novel antimicrobial drugs to treat multidrug-resistant (MDR) infections.[15] Since then, new antibiotics such as oritavancin have shown clinical efficacy with a single dose, in comparison to 7–10 days of vancomycin treatment.[12,16]

Another interesting approach is the development of novel drug delivery systems. The explosion of interest and research in regenerative medicine represents an important opportunity to obtain new methods of antimicrobial drug delivery in skin and soft tissues.[17]

1.2 Skin Microbiome

The role of the skin microbiome on host predisposition to microbial agents has not been extensively explored; this microbial diversity can modulate the growth of invasive pathogens.[18] Also, the cutaneous microenvironment (dry, moist, and sebaceous) can be an important determinant of skin infection by influencing local microbiota.[19] Four dominant phyla of bacteria reside on the skin: The Actinobacteria, Proteobacteria, Firmicutes, and Bacteroidetes.[20] These phyla of bacteria form biofilms of indigenous microbiota in the skin, participating actively in the prevention of skin infection. This protection is a mechanism of defense to prevent the attachment of exogenous bacteria and fungi to the skin surface.[21] Experiments seeking to demonstrate the effects of *Haemophilus ducreyi* infection on the skin microbiome were performed, showing that sites with pustules and skin injury had a high quantity of *Proteobacteria*, *Bacteroidetes*, *Micrococcus*, *Corynebacterium*, *Paracoccus*, and *Staphylococcus* species, but the sites showing resolution of infection presented higher levels of *Actinobacteria* and *Propionibacterium* species.[18] For that reason, the design and development of new antimicrobial strategies for skin and soft tissue treatment needs to integrate new antimicrobial compound discovery and delivery strategies, with the purpose of obtaining combinative antiinfective drugs with the ability of resolving the infection, preventing emergence of resistance and protecting the natural microbiome.[22]

1.3 Antimicrobial Drug Discovery and Bioavailability in Skin

The stratum corneum and viable epidermis offer a barrier to drug penetration across the skin in transdermal delivery.[23] The current methods used to alter the barrier properties of skin can be classified as both chemical and physical methods.[24] Various drug delivery systems have been developed that use different skin penetration enhancers to overcome this barrier.[25] Interest in the development of new skin permeabilization agents for the purpose of delivering bioactive compounds is rising. In this effort, 22 drugs have been approved by the FDA for transdermal delivery.[26]

In seeking for an increase in the bioavailability of antimicrobial agents, an important approach is the use of nanoparticle formulations. These formulations include solid lipid nanoparticles, nanoemulsions, liposomes, ethosomes, and nanotubes.[27] Nanoparticles have the ability to be directed to specific sites of infection, reducing the exposure of normal tissues and microbiome to the antiinfective agent.[28] Also, nanoparticle modification has shown to improve antimicrobial delivery, decreasing systemic toxicity.[29] In addition, the nanoparticle approach has advantages in preventing the emergence of resistance and decreasing side effects of conventional antibiotic formulations; their inclusion in medical devices avoids the primary microbial adhesion. Finally, using nanoparticles as vaccine adjuvants is a promising application to evoke a more efficient immune response.[30]

The aim of this chapter is to present information on the integration of different disciplines such as microbiology, pharmacology, nanotechnology, and physics for the design and development of new antimicrobial applications that can reduce the incidence of ABSSSI as well as their complications, increasing survival and quality of life of patients.

2. DRUG DELIVERY AND THE SKIN

Skin has a large surface area, which makes it an important route for new drug delivery systems. For this purpose, transdermal drug delivery is a good choice for the administration of systemic and local drugs using the skin as an entry point to the body.[31]

In the development of drug delivery systems in skin, it is necessary to take into account the following factors[24]:

- Penetration: Defined as the introduction of a specific compound or formulation into the different strata of the skin.
- Partitioning and diffusion across the stratum corneum into epidermis and upper dermis.
- Permeation: Defined as the penetration of compounds between different layers of skin.
- Absorption: Defined as the uptake of a compound into the systemic circulation.

In this order of ideas for topical and transdermal drug delivery, drug diffusion across the stratum corneum is determined by Fick's second law of diffusion, which predicts the change of concentration of a molecule over time using the following formula[32,33]:

$$J = DCP/L$$

In this model, J is defined as the flux, D is the drug diffusion coefficient, C is the drug concentration in the carrier, P is the coefficient of drug partition, and L is the thickness of the stratum corneum.[34] Lipophilic and hydrophilic small molecules have the ability of penetrating the stratum corneum without carriers, but larger compounds (peptides, proteins, siRNA, or DNA) need skin permeabilization strategies to traverse into the dermis.[35]

Skin permeabilization enhancers are classified as chemical and physical enhancers. Chemical enhancers have the ability to increase drug partitioning into the stratum corneum, to modify the fluidity of lipid layers that comprise the stratum corneum, and to produce disruption of intercellular lipids.[36] The most-investigated chemical penetration enhancers are alcohols, sulfoxides, ozone, pyrrolidones, essential oils (EOs), terpenes, fatty acids, water, and urea[37] (Table 1).

On the other hand, physical enhancers use various devices that apply ultrasound, electric fields, ballistic methods, and micro needles, with the ability of producing a disruption of the skin structure for the delivery of macromolecular drugs.[38]

Another factor to be considered is the penetration of the drug after diffusing across the stratum corneum. This process can occur through three routes: intercellular, intracellular, and transappendageal. In the intercellular route, the drug diffuses across lipid bilayers among corneocytes; the intracellular route is composed of aqueous areas surrounded by polar lipids that create microchannels; the transappendageal route employs hair follicles and other appendages for delivering drugs into deeper skin layers.[39]

Controlling these transdermal drug delivery factors is key to the development of effective antimicrobial therapy with which to prevent and treat ABSSSIs as well as their complications.

2.1 Chemical Enhancers

Fatty acids, surfactants, esters, alcohols, polyalcohols, pyrrolidones, amines, amides, sulfoxides, terpenes, alkanes, and phospholipids are the most-used chemical enhancers to increase topical drug penetration[40] (Table 1). Chemical enhancers facilitate the absorption of drugs through the following mechanisms: Disruption of the stratum corneum, solubilizing, and extracting the keratin and lipids that comprise the stratum corneum, and fluidizing the stratum corneum by dissolving the lipids and improving the

Table 1 Techniques for enhancement of skin permeabilization: Transdermal drug delivery techniques

Chemical enhancers	Physical enhancers	Nanotechnology
Surfactants	Sonophoresis	Liposomes
Alcohols	Laser and thermal ablation	Nanoparticles
Fatty acids	Electroporation	Ethosomes
Esters	Microneedle arrays	Nanofibers
Sulfoxides	Nonthermal plasma	Wafers
Terpenes	Antimicrobial photodynamic therapy	Biodegradable polymers
Phospholipids	Radiofrequency	Nanotubes

partitioning of the drug.[41,42] The disadvantage of these substances is the tendency to cause skin irritation due to their disruptive effect on the lipid structures of the skin. Also, it is necessary to use high concentrations of the enhancers in order to achieve proper absorption of the drug; hence, their use is limited.[43] Surfactants produce the highest degree of lipid disorder in the stratum corneum, followed by terpenes (*e.g.*, nerol). Unfortunately, these findings do not correlate with an increase in absorption, so it is necessary to find new compounds having more activity at lower concentrations and producing less irritation.[41]

Terpenes and EOs have shown an increase in the delivery of small drug compounds into the skin, due to the fact that they have the ability to interact with the intercellular lipids, producing extraction, fluidization, disorder, and separation.[44] EOs have been considered safer than their individual components as chemical enhancers, since they are easily excreted and it has been demonstrated that whole EOs have more enhancement permeation properties than individual terpenes.[45,46] EOs from *Eucalyptus globulus*, *Chenopodium ambrosioides*, and *Cananga odorata* have been found to increase the penetration of the anticancer drug 5-flouorouracil in human skin.[37,47] Also EO from *Ocimum basilicum* has been shown to increase skin permeation *in vitro* and *in vivo* in a model of permeation of indomethacine.[48,49] Among the terpenes that have been reported for skin permeabilization activity are limonene, menthol, 1,8-cineole, carvone, and geraniol.[49]

Finally, in the search for new chemical penetration enhancers, it is very important to take the following properties of an ideal skin permeabilization agent into account: pharmacologically inactive, nonirritating, nontoxic, nonallergenic, compatible with a wide variety of drugs, odorless, tasteless, colorless, inexpensive, with easy excretion and removal from the skin.[46]

2.2 Physical Enhancers

Physical enhancers use physical methods to temporarily cause an alteration of the integrity of the skin as well as its barrier function for increased skin permeabilization. Devices developed for this purpose apply sonophoresis, laser and thermal ablation, electroporation, radiofrequency treatment, iontophoresis, magnetophoresis, high pressure jets, and microneedle arrays (Table 1). The disadvantage of this approach is that physical enhancers are invasive, may cause damage to the skin's barrier properties, and may produce irritation and burns.[50]

2.2.1 Antimicrobial Photodynamic Therapy

Antimicrobial photodynamic therapy (APDT) is a treatment that utilizes a combination of visible light, a photosensitizer (PS) that is capable of being activated by light, and oxygen, for obtaining a cytotoxic effect. Visible light excites the PS to a singlet state, producing cytotoxic effects by reactive oxygen species (ROS).[51] APDT has the potential to kill MDR pathogenic microorganisms with low-resistance induction, so it is a promising approach for treating microbial infections that are refractory to antiinfectives.[52,53] The efficacy of APDT resides in the interaction of PS and light. An ideal PS should comply with the following parameters: high phototoxicity, low dark toxicity, high quantum yield of $^1O_2$ and free radicals, specific interaction with bacteria and proper pharmacokinetics.[54]

The most commonly used PS from natural products is hypericin, an anthraquinone from Saint John's wort (*Hypericum perforatum*). Hypericin under light is excited at 593 nm, presenting anticancer, antiviral, and antibacterial properties.[55] Other PSs from natural sources are the polymers, such as chitosan and cellulose, for use in producing nontoxic nanomedicines.[56] Also, carbon nanostructures such as fullerenes, nanotubes, and graphene can be photoactivated to kill microbes.[53,57]

An interesting new APDT approach, antimicrobial blue light (aBL), has *in vitro* antimicrobial activity without the use of PSs.[58] The mechanism of action of aBL is by excitation of endogenous photosensitizing chromophores with the consequent production of ROS that can kill MDR strains and biofilms.[59] However, the use of blue light for the treatment of wound infections has not been widely studied. Since there are only a few reports on wound infections produced by *Pseudomonas aeruginosa* and *Acinetobacter baumannii* in a mouse model, it is necessary to determinate the utility of this nonantibiotic procedure in clinical trials.[60] Recently, for the treatment of serious infections caused by MRSA, a combination of silver nanoparticles (AgNPs), blue light, and antibiotics has been proposed.[61]

APDT can synergize visible light and photosensitizing drugs with activity in localized SSTIs. Equally, it has been shown that a potential immunomodulatory activity can increase wound healing and repair, which is an added value of this promising therapy.[29]

2.2.2 Pulsed Electric Fields

Currently, pulsed electric fields (PEFs) have been tested as a new physical burn wound disinfection method. With this method, it is possible to create nonthermal and permanent damage in cell membranes by electroporation. PEF technology may represent an important alternative to antibiotics in the treatment and prevention of bacterial contamination of wounds, particularly those contaminated with MDR bacteria. However, PEF technology has yet to be evaluated for wound disinfection in clinical trials.[62,63] It is very important to validate and evaluate the application of PEF technology for treating ABSSSI in the health care setting and reduce the spread of MDR strains in hospital environments.

2.2.3 Nonthermal Plasma

Nonthermal plasma (NTP) consists of partially ionized gas with a low temperature. It has been demonstrated that NTP has antimicrobial activity by inducing membrane and genomic damage as well as by modification of proteins in MRSA, *P. aeruginosa*, and *Candida albicans*.[64] NTP has efficient disinfection and sterilization activity with little effect on the nearby healthy tissue, inactivating Gram-negative and Gram-positive bacteria, fungi, virus, spores, and parasites.[65] The advantage of this method is that its gaseous form penetrates surfaces homogeneously, as well as cavities and fissures. Equally, it offers the possibility of drug delivery noninvasively.[66] NTP is a promising alternative to antibiotics and biocides to control superficial wound and skin infections. Also, in wound healing NTP argon plasma shows high activity in eliminating biofilms in contaminated wounds.[67]

2.3 Nanotechnology

An important alternative is the use of nanotechnology for the prevention and treatment of ABSSSIs, specifically nanoparticles and nanofibers as drug delivery systems for topical and transdermal applications. This approach increases the drug concentration by modification of drug flux.[33,68]

Among these systems, liposomal formulations have been used successfully in skin drug delivery, since liposomes increase skin permeability for encapsulated drugs and reduce the toxic effects. Encapsulating antimicrobial agents in liposomes has the following advantages: protection of encapsulated drugs from enzymatic hydrolysis due to the lipid nature of these formulations they can easily penetrate the microbial cell membrane and increase cellular concentration.[69] Liposomal formulations are the key to the design and development of new effective antibiotic forms with subsequent therapeutic success.[70]

In addition, ethosomes deserve a particular mention because they respond to the need for obtaining an efficient delivery of drugs by topical application on the skin. Ethosomes are composed principally of phospholipids, ethanol, and water in unique vesicular carriers, with the characteristics of possessing a vesicle's bilayer fluidity, increased permeabilization effect, easy mode of preparation and absence of side effects.[71] Previous reports have shown that ethosomal systems can enhance skin penetration of a broad range of compounds. In a model of penetration with Rhodamine Red, it was possible to examine skin penetration and the fluorescence intensity in different skin layers, indicating that ethosomal systems penetrate deeper into the skin.[72] Ethosomes mediate the disruption of the lipid barrier by which small hydrophobic molecules permeate; in topical application there is a combined action mechanism between ethanol, vesicles, and skin lipids enhancing its lipid fluidity.[73] Ethosomal carriers have shown an increase in the skin absorption of erythromycin in a model of skin infection with *S. aureus*, which is promising for the design of new topical antibiotic therapy using ethosomal carriers.[74] One interesting approach to ethosomal systems is the transethosome, which combines ethanol and a penetration enhancer that increases the lipid perturbation in stratum corneum and has superior skin permeation properties.[75] Also, in this delivery nanosystem it has been possible to introduce natural products such as curcumin and quercetin to increase skin absorption.[76,77]

Another application of nanotechnology in skin drug delivery is biodegradable polymers, such as polyesters and polyanhydrides. Gliadel, a polyanhydride wafer approved in 1996, has shown efficacy in the delivery of carmustine to patients with recurrent malignant glioma, receiving approval from the FDA in 2003 as adjunct to surgery. This opens the gates for the use of new biodegradable drug delivery systems.[78]

The major limitation for penetration enhancers is the occurrence of skin irritation, which decreases the efficacy of these substances in drug delivery. For that reason, new variations of semisolid vehicles such as proniosomes and microemulsion gels have been introduced. Proniosomes are nonionic-based surfactant vesicles, which require hydration before drug release in the skin. Proniosomal gels have been shown to be very useful as penetration enhancers that intensify the drug permeation.[79] Posthydration proniosomes are transformed into niosomes with the ability of diffusing through the stratum corneum and adhering to the cell surface, facilitating the penetration of lipophilic drugs across the skin.[24] Niosome technology is an interesting, less expensive, and more stable alternative to liposomes, but further research is needed to assess its effectiveness depending on chemical composition.[79]

2.4 Dermatopharmacokinetics

An important topic to be considered in the development of innovative topical formulations with new drug delivery systems is dermatopharmacokinetics, which describes the drug bioavailability in the stratum corneum in correlation with pharmacodynamic effects.[80] With these parameters, it is possible to establish concentration-time curves for topical compounds. The most-used method for determining skin drug concentration is microdialysis which, although invasive, provides the possibility of obtaining information of high accuracy and reproducibility in the skin layer where the microdialysis fiber is positioned. Also, with dermatopharmacokinetic methods, it is possible to obtain relevant data about release of the drug, penetration of the drug across the stratum corneum and time to watch the pharmacological effect.[27] Determining topical bioavailability of novel formulations is a very important subject in the validation and evaluation of a program for the discovery of new topical antimicrobial drugs for ABSSSI treatment.

3. APPLICATION MODELS

3.1 Antimicrobial Coating

Antimicrobial coatings are materials developed to mitigate the propagation of microbes in different places.[81] In this scheme, various products for skin infection and wound healing, such as an ethosome gel film-coating agent composed of tea polyphenol, asiaticoside, and tea tree oil, have been tested.[82] In addition, silver-coated chitosan films have been evaluated in a rat model of wound healing with good results.[83] This model of application is an important approach in the reduction of HAI because it eliminates the microbial reservoirs in health care settings.[84]

3.2 Antimicrobial Peptides

Antimicrobial peptides (AMPs) are the first line of defense in many multicellular organisms, and are a promising source of new antiinfective agents. The microbial cell membrane is the principal target of these compounds, with efficacy and selectivity.[85] AMPs present broad-spectrum activity, including MDR bacteria. The major disadvantage of this approach is the lability of natural AMPs, due to susceptibility to protease and pH change. Also, AMPs have potential toxicity for oral application. For that reason, it is possible to use AMPs for topical treatment of skin infectious diseases in combination with drug delivery systems, such as liposome encapsulation, with the end of improving stability and reducing toxicity.[86] Equally, natural polymers in a formulation with impregnated freeze-dried wafers have shown activity in the treatment of wound infections.[87]

3.3 Wound Dressing

Wound dressings have been used to clean, cover, and protect the wound from the external environment. A wound dressing must provide a moist environment, remove the excess of exudate, avoid maceration, protect the wound from infection and maintain an adequate exchange of gases.[88] Some dressings also act as drug delivery systems and can be classified as traditional and modern dressings. They are also available as thin films and gels, and hence can be classified as hydrocolloid dressings, alginate dressings, and nonalginate dressings[89] (Table 2).

Nanofibers are another source for developing innovative wound dressings, allowing the possibility of incorporating antimicrobial agents and growth factors into nanofibers, which can help to avoid the formation of microbial biofilms and reduce the potential SSI.[90] Equally, they can be applied in burns and diabetes ulcers, because local delivery of drugs from dressings have shown to be an efficient therapy as an alternative to antibiotic

Table 2 Types of wound care dressings

Wound dressings			
Nonmedicated			Medicated
Based on function	Based on material	Based on physical form	
Debridement dressings Absorbance Adherence Antibacterial Occlusive	Hydrocolloid Alginates Collagen	Film Ointment Foam Gel	Antimicrobial Growth factor Supplements

therapeutics. A wide variety of medicated dressings are available, such as Cutisorb, which is an antibiotic-loaded dressing used for delivery to wounds. Also, foams (Contreet, Coloplast), hydrocolloids (Urgotul SSD, Urgo), alginates (Silvercel, Johnson & Johnson), and Hydrofiber (Aquacel, ConvaTec) are commercially accessible.[91]

On the other hand, for the treatment of exuding wound surfaces lyophilized wafers, which are produced by freeze-drying polymer solutions and gels, have been developed. Lyophilized wafers provide an interesting potential delivery system for pharmacological agents into wound surfaces, since they have the ability to incorporate soluble and insoluble antimicrobial molecules, maintaining a moist environment without damaging newly formed tissue.[92]

3.4 Healing Efficacy in Skin Infections

The wound healing process is classically defined as a series of continuous processes that involve hemostasis, inflammation, proliferation, epithelization, maturation, and remodeling of the scar tissue. Silver-based nanoparticles loaded into nanofibers for improving the promotion of wound healing and obtaining antimicrobial protection have been shown to reduce the time of wound healing. In the search for increasing the *in vivo* efficacy of growth factors, they have been incorporated into polymer nanocarriers and nanoparticles with success in skin wounds.[93] Equally, nitric oxide—releasing nanoparticles have been utilized for the treatment of complex cutaneous infections and wounds, due to NO having the ability to maintain skin homeostasis by regulating circulation, as well as maintaining the protective barrier against microorganisms and modulating immune responses, so it has great activity in wound healing.[4]

Finally, microneedles are very promising physical enhancers in transdermal drug delivery, due to advancements in microfabrication techniques developed in the last few years. Microneedle technology can perforate into the skin layers without exciting nerve endings and offer skin permeabilization of antimicrobial drugs, growth factors and antiinflammatory compounds, making this technique a breakthrough for regenerative medicine and antibiotic treatment.[35]

3.5 Burn Wound Infections

Burns are highly susceptible to infection because burn injury causes damage to the skin, destroying the cutaneous barrier and causing immunosuppression. Also, the disturbance of the skin combined with the denaturation of proteins and lipids facilitates microbial invasion. The compromise of blood circulation makes it difficult to carry out systemic antiinfective treatment, so the choice is topical treatment.[94]

Nanofibers composed of natural and synthetic polymers have shown a potential use for burn treatment. Chitin and chitosan bandages have proven be effective as an antimicrobial wound dressing in infection models with *P. aeruginosa* and *Proteus mirabilis*.[95] Also, the development of nanoemulsions, which are mixtures of oil-in-water stabilized by surfactants and alcohol with droplets ranging from 200 to 600 nm in size, provides an alternative in antimicrobial topical therapy. Nanoemulsion particles are able to fuse with cell membranes, destabilizing the microorganism lipid membrane and producing cell lysis and death.[96] Terpenes such as limonene, 1,8-cineole, α-pinene oxide, and geraniol have been used as nanoemulsions in a burn wound antimicrobial therapy model increasing the skin permeation to silver sulphadiazine (SSD), showing great activity as penetration enhancers.[97] These enhancing effects of terpenes on the skin have the advantage of being reversible and the permeability of the skin remains intact after the treatment.[98,99]

3.6 Wound Healing Research Model

The selection of the ideal *in vivo* wound healing model is necessary to evaluate activity and determine dermatopharmacokinetics of the different formulations. Among the animal models available, the most valuable is the pig animal model for skin permeation studies and microdialysis studies, since it is comparable to human skin. Also, porcine skin could serve as an *in vitro* model. In this model, it is possible to estimate transdermal drug delivery through human skin combining *in vivo* and *in vitro* data, using the following formula:

$$k_{\text{Percutaneous absorption human}} = k_{\text{Percutaneous absorption animal}} \times \left(J_{\text{human}}/J_{\text{animal}}\right)$$

where J is the percutaneous permeation flux.[100]

4. SKIN INFECTIONS, NATURAL PRODUCTS, AND NANOTECHNOLOGY

Natural products have a large potential in the treatment of wounds and skin infections. These natural agents can stimulate healing and regeneration of skin by multiple mechanisms. For that reason, we urge discovery of new antimicrobial agents, chemical enhancers and nanomaterials from this important source.[101] Natural antioxidants have the potential to prevent cell damage caused by free radicals and facilitate wound healing.[102] Biodegradable polymers can also be developed as carriers of pharmaceutical bioactive compounds. For example, alginate which is a low-cost and biodegradable polymer from algal polysaccharides is widely used as wound dressing; equally calcium alginate beads have been used for encapsulation of drugs for slow-release from the polymer matrix to the wound bed.[103]

Various natural product—derived antimicrobial agents such as propolis, EOs and terpenes can be incorporated into nanofibers by electrospinning with chitosan as a stabilizer, for the development of wound bandages and burn treatments. In addition, antimicrobial activity of these nanofibers against MRSA, *Escherichia coli* and *A. baumannii* has been demonstrated.[104,105]

5. CONCLUSIONS AND PERSPECTIVES

ABSSSIs are currently a threat to public health and a source of antimicrobial resistance. In this infectious disease, the skin microbiome plays an important role in maintaining homeostasis and regulating the immune system. For that reason, in the design and development of new antiinfective drugs useful for skin infections, it is necessary to evaluate the microbiome effect for obtaining an effective curing process.[106] Also, it is necessary to take into account the physical methods, such as APDT and NTP, which do not use antibiotics and have the ability to avoid resistance, which can be effective antimicrobial adjuvants of therapy.[56] Another important approach to consider is the development of new antimicrobial surfaces coated with antiadhesive, antibiofilm, and antimicrobial agents in combination with nanomaterials, for obtaining innovative wound dressings and preventing incidence of HAIs in health care settings. In addition, the use of natural products, such as EOs and AMPs for their antioxidant activity useful in wound healing as well as their antimicrobial activity, which can offer chemical diversity in the search for new drugs to solve the antibiotic crisis, is a very interesting approach. Finally, integration of antimicrobial drug discovery with antimicrobial drug delivery development for increasing absorption as well as efficacy and reduction in toxicity is urged.

ACKNOWLEDGMENTS

The authors acknowledge the TUBITAK program 2221 Fellowship program for Visiting Scientists and Scientists on Sabbatical Leave for the support of this research.

REFERENCES

1. Atzori L, Manunza F, Pau M. New trends in cellulitis. *EMJ Dermatol* 2013;**1**:64–76.
2. Ki V, Rotstein C. Bacterial skin and soft tissue infections in adults: a review of their epidemiology, pathogenesis, diagnosis, treatment and site of care. *Can J Infect Dis Med Microbiol* 2008;**19**:173–84.
3. Miller LG, Eisenberg DF, Liu H, Chang CL, Wang Y, Luthra R, et al. Incidence of skin and soft tissue infections in ambulatory and inpatient settings, 2005–2010. *BMC Infect Dis* 2015;**15**:362.
4. Mihu MR, Martinez LR. Novel therapies for treatment of multi-drug resistant *Acinetobacter baumannii* skin infections. *Virulence* 2011;**2**:97–102.
5. Moran GJ, Abrahamian FM, Lovecchio F, Talan DA. Acute bacterial skin infections: developments since the 2005 Infectious Diseases Society of America (IDSA) guidelines. *J Emerg Med* 2013;**44**:e397–412.
6. Pollack Jr CV, Amin A, Ford Jr WT, Finley R, Kaye KS, Nguyen HH, et al. Acute bacterial skin and skin structure infections (ABSSSI): practice guidelines for management and care transitions in the emergency department and hospital. *J Emerg Med* 2015;**48**:508–19.
7. Tsai DM, Caterson EJ. Current preventive measures for health-care associated surgical site infections: a review. *Patient Saf Surg* 2014;**8**:42.
8. Anderson DJ. Surgical site infections. *Infect Dis Clin N Am* 2011;**25**:135–53.
9. Bulger EM, Maier RV, Sperry J, Joshi M, Henry S, Moore FA, et al. A novel drug for treatment of necrotizing soft-tissue infections: a randomized clinical trial. *JAMA Surg* 2014;**149**:528–36.

10. Hakkarainen TW, Kopari NM, Pham TN, Evans HL. Necrotizing soft tissue infections: review and current concepts in treatment, systems of care, and outcomes. *Curr Probl Surg* 2014;**51**:344–62.
11. Russo A, Concia E, Cristini F, de Rosa FG, Esposito S, Menichetti F, et al. Current and future trends in antibiotic therapy of acute bacterial skin and skin-structure infections. *Clin Microbiol Infect* 2016; **22**(Suppl. 2):S27–36.
12. Corey R, Naderer OJ, O'Riordan WD, Dumont E, Jones LS, Kurtinecz M, et al. Safety, tolerability, and efficacy of GSK1322322 in the treatment of acute bacterial skin and skin structure infections. *Antimicrob Agents Chemother* 2014;**58**:6518–27.
13. Joseph WS, Culshaw D, Anuskiewicz S, de Anda C, Prokocimer P. Tedizolid and linezolid for treatment of acute bacterial skin and skin structure infections of the lower extremity versus non-lower extremity: pooled analysis of two phase 3 trials. *J Am Podiatr Med Assoc* 2016; in press.
14. Boucher HW, Talbot GH, Benjamin DK, Bradley J, Guidos RJ, Jones RN, et al. Infectious Diseases Society of America. 10 × '20 progress—development of new drugs active against Gram-negative bacilli: an update from the Infectious Diseases Society of America. *Clin Infect Dis* 2013;**56**:1685–94.
15. Corey GR, Kabler H, Mehra P, Gupta S, Overcash JS, Porwal A, et al. Single-dose oritavancin in the treatment of acute bacterial skin infections. *N Engl J Med* 2014;**370**:2180–90.
16. Deck DH, Jordan JM, Holland TL, Fan W, Wikler MA, Sulham KA, et al. Single-dose oritavancin treatment of acute bacterial skin and skin structure infections: SOLO trial efficacy by Eron severity and management setting. *Infect Dis Ther* 2016;**5**:353–61.
17. Gainza G, Villullas S, Pedraz JL, Hernandez RM, Igartua M. Advances in drug delivery systems (DDSs) to release growth factors for wound healing and skin regeneration. *Nanomedicine* 2015;**11**:1551–73.
18. Van Rensburg JJ, Lin H, Gao X, Toh E, Fortney KR, Ellinger S, et al. The human skin microbiome associates with the outcome of and is influenced by bacterial infection. *MBio* 2015;**6**:e01315-15.
19. Horton JM, Gao Z, Sullivan DM, Shopsin B, Perez-Perez GI, Blaser MJ. The cutaneous microbiome in outpatients presenting with acute skin abscesses. *J Infect Dis* 2015;**211**:1895–904.
20. Grice EA. The skin microbiome: potential for novel diagnostic and therapeutic approaches to cutaneous disease. *Semin Cutan Med Surg* 2014;**33**:98–103.
21. Percival SL, Emanuel C, Cutting KF, Williams DW. Microbiology of the skin and the role of biofilms in infection. *Int Wound J* 2012;**9**:14–32.
22. Brooks BD, Brooks AE. Therapeutic strategies to combat antibiotic resistance. *Adv Drug Deliv Rev* 2014;**78**:14–27.
23. Andrews SN, Jeong E, Prausnitz MR. Transdermal delivery of molecules is limited by full epidermis, not just stratum corneum. *Pharm Res* 2013;**30**:1099–109.
24. Alkilani AZ, Mccrudden MT, Donnelly RF. Transdermal drug delivery: innovative pharmaceutical developments based on disruption of the barrier properties of the stratum corneum. *Pharmaceutics* 2015;**7**:438–70.
25. Chantasart D, Li SK. Structure enhancement relationship of chemical penetration enhancers in drug transport across the stratum corneum. *Pharmaceutics* 2012;**4**:71–92.
26. Schoellhammer CM, Blankschtein D, Langer R. Skin permeabilization for transdermal drug delivery: recent advances and future prospects. *Expert Opin Drug Deliv* 2014;**11**:393–407.
27. Uchechi O, Ogbonna JD, Attama AA. Nanoparticles for dermal and transdermal drug delivery. In: Sezer AD, editor. *Application of nanotechnology in drug delivery*. Rijeka: InTech; 2014. p. 193–235.
28. Taylor E, Webster TJ. Reducing infections through nanotechnology and nanoparticles. *Int J Nanomedicine* 2011;**6**:1463.
29. Heffernan DS, Fox ED. Advancing technologies for the diagnosis and management of infections. *Surg Clin North Am* 2014;**94**:1163–74.
30. Zhu X, Radovic-Moreno AF, Wu J, Langer R, Shi J. Nanomedicine in the management of microbial infection—overview and perspectives. *Nano Today* 2014;**9**:478–98.
31. Tiwari G, Tiwari R, Sriwastawa B, Bhati L, Pandey S, Pandey P, et al. Drug delivery systems: an updated review. *Int J Pharm Investig* 2012;**2**:2–11.
32. Todo H, Oshizaka T, Kadhum WR, Sugibayashi K. Mathematical model to predict skin concentration after topical application of drugs. *Pharmaceutics* 2013;**5**:634–51.

33. Goyal R, Macri LK, Kaplan HM, Kohn J. Nanoparticles and nanofibers for topical drug delivery. *J Control Release* 2016;**240**:77–92.
34. Prausnitz MR, Elias PM, Franz TJ, Schmuth M, Tsai JC, Menon GK, et al. Skin barrier and transdermal drug delivery. *Dermatology* 2012;**3**:2065–73.
35. Paudel KS, Milewski M, Swadley CL, Brogden NK, Ghosh P, Stinchcomb AL. Challenges and opportunities in dermal/transdermal delivery. *Ther Deliv* 2010;**1**:109–31.
36. Som I, Bhatia K, Yasir M. Status of surfactants as penetration enhancers in transdermal drug delivery. *J Pharm Bioallied Sci* 2012;**4**:2–9.
37. Pathan IB, Setty CM. Chemical penetration enhancers for transdermal drug delivery systems. *Trop J Pharm Res* 2009;**8**:173–9.
38. Mitragotri S. Devices for overcoming biological barriers: the use of physical forces to disrupt the barriers. *Adv Drug Deliv Rev* 2013;**65**:100–3.
39. Aljuffali IA, Huang CH, Fang JY. Nanomedical strategies for targeting skin microbiomes. *Curr Drug Metab* 2015;**16**:255–71.
40. Kumar R, Philip A. Modified transdermal technologies: breaking the barriers of drug permeation via the skin. *Trop J Pharm Res* 2007;**6**:633–44.
41. Moghadam SH, Saliaj E, Wettig SD, Dong C, Ivanova MV, Huzil JT, et al. Effect of chemical permeation enhancers on stratum corneum barrier lipid organizational structure and interferon alpha permeability. *Mol Pharm* 2013;**10**:2248–60.
42. Pandey A, Mittal A, Chauhan N, Alam S. Role of surfactants as penetration enhancer in transdermal drug delivery system. *J Mol Pharm Org Process Res* 2014;**2**:1–10.
43. Chen Y, Quan P, Liu X, Wang M, Fang L. Novel chemical permeation enhancers for transdermal drug delivery. *Asian J Pharm Sci* 2014;**9**:51–64.
44. Prasanthi D, Lakshmi PK. Terpenes: effect of lipophilicity in enhancing transdermal delivery of alfuzosin hydrochloride. *J Adv Pharm Technol Res* 2012;**3**:216–23.
45. Chen J, Jiang QD, Wu YM, Liu P, Yao JH, Lu Q, et al. Potential of essential oils as penetration enhancers for transdermal administration of ibuprofen to treat dysmenorrhoea. *Molecules* 2015;**20**:18219–36.
46. Herman A, Herman AP. Essential oils and their constituents as skin penetration enhancer for transdermal drug delivery: a review. *J Pharm Pharmacol* 2015;**67**:473–85.
47. Gupta SK, Kumar S. Transdermal drug delivery enhancement by essential oils of *Eucalyptus globulus*. *Adv Biol Res* 2014;**8**:249–50.
48. Fang JY, Leu YL, Hwang TL, Cheng HC. Essential oils from sweet basil (*Ocimum basilicum*) as novel enhancers to accelerate transdermal drug delivery. *Biol Pharm Bull* 2004;**27**:1819–25.
49. Fox LT, Gerber M, Plessis JD, Hamman JH. Transdermal drug delivery enhancement by compounds of natural origin. *Molecules* 2011;**16**:10507–40.
50. Roustit M, Blaise S, Cracowski JL. Trials and tribulations of skin iontophoresis in therapeutics. *Br J Clin Pharmacol* 2014;**77**:63–71.
51. Giuliani F. Photodynamic therapy as a novel antimicrobial strategy against biofilm-based nosocomial infections: study protocols. *Methods Mol Biol* 2014;**1147**:287–98.
52. Yin R, Agrawal T, Khan U, Gupta GK, Rai V, Huang YY, et al. Antimicrobial photodynamic inactivation in nanomedicine: small light strides against bad bugs. *Nanomedicine* 2015;**10**:2379–404.
53. Liu Y, Qin R, Zaat SA, Breukink E, Heger M, Liu Y. Antibacterial photodynamic therapy: overview of a promising approach to fight antibiotic-resistant bacterial infections. *J Clin Transl Res* 2015;**1**:140–67.
54. Yoon I, Li JZ, Shim YK. Advance in photosensitizers and light delivery for photodynamic therapy. *Clin Endosc* 2013;**46**:7–23.
55. Debele TA, Peng S, Tsai HC. Drug carrier for photodynamic cancer therapy. *Int J Mol Sci* 2015;**16**:22094–136.
56. Yin R, Hamblin MR. Antimicrobial photosensitizers: drug discovery under the spotlight. *Curr Med Chem* 2015;**22**:2159–85.
57. Grinholc M, Nakonieczna J, Fila G, Taraszkiewicz A, Kawiak A, Szewczyk G, et al. Antimicrobial photodynamic therapy with fulleropyrrolidine: photoinactivation mechanism of *Staphylococcus aureus*, *in vitro* and *in vivo* studies. *Appl Microbiol Biotechnol* 2015;**99**:4031–43.
58. Dai T, Gupta A, Huang YY, Yin R, Murray CK, Vrahas MS, et al. Blue light rescues mice from potentially fatal *Pseudomonas aeruginosa* burn infection: efficacy, safety, and mechanism of action. *Antimicrob Agents Chemother* 2013;**57**:1238–45.

59. Wang J, Vermerris W. Antimicrobial nanomaterials derived from natural products—a review. *Materials* 2016;**9**:255.
60. Zhang Y, Zhu Y, Gupta A, Huang Y, Murray CK, Vrahas MS, et al. Antimicrobial blue light therapy for multidrug-resistant *Acinetobacter baumannii* infection in a mouse burn model: implications for prophylaxis and treatment of combat-related wound infections. *J Infect Dis* 2014;**209**:1963–71.
61. Akram FE, El-Tayeb T, Abou-Aisha K, El-Azizi M. A combination of silver nanoparticles and visible blue light enhances the antibacterial efficacy of ineffective antibiotics against methicillin-resistant *Staphylococcus aureus* (MRSA). *Ann Clin Microbiol Antimicrob* 2016;**15**:48.
62. Goldberg A, Yarmush ML, Sheridan RL, Austen WG, Broelsch GF, Rubinsky B, et al. The general hospital corporation, assignee. Systems and methods for delivering pulsed electric fields to skin tissue. United States patent application US 14/655,178 2013.
63. Golberg A, Broelsch GF, Vecchio D, Khan S, Hamblin MR, Austen Jr WG, et al. Pulsed electric fields for burn wound disinfection in a murine model. *J Burn Care Res* 2015;**36**:7–13.
64. Kvam E, Davis B, Mondello F, Garner AL. Nonthermal atmospheric plasma rapidly disinfects multidrug-resistant microbes by inducing cell surface damage. *Antimicrob Agents Chemother* 2012;**56**:2028–36.
65. Lloyd G, Friedman G, Jafri S, Schultz G, Fridman A, Harding K. Gas plasma: medical uses and developments in wound care. *Plasma Process Polym* 2010;**7**:194–211.
66. Heinlin J, Isbary G, Stolz W, Morfill G, Landthaler M, Shimizu T, et al. Plasma applications in medicine with a special focus on dermatology. *J Eur Acad Dermatol Venereol* 2011;**25**:1–11.
67. Ermolaeva SA, Varfolomeev AF, Chernukha MY, Yurov DS, Vasiliev MM, Kaminskaya AA, et al. Bactericidal effects of non-thermal argon plasma *in vitro*, in biofilms and in the animal model of infected wounds. *J Med Microbiol* 2011;**60**:75–83.
68. Tocco I, Zavan B, Bassetto F, Vindigni V. Nanotechnology-based therapies for skin wound regeneration. *J Nanomater* 2012;**2012**:4.
69. Abed N, Couvreur P. Nanocarriers for antibiotics: a promising solution to treat intracellular bacterial infections. *Int J Antimicrob Agents* 2014;**43**:485–96.
70. Bozzuto G, Molinari A. Liposomes as nanomedical devices. *Int J Nanomedicine* 2015;**10**:975–99.
71. Gupta M, Agrawal U, Vyas SP. Nanocarrier-based topical drug delivery for the treatment of skin diseases. *Expert Opin Drug Deliv* 2012;**9**:783–804.
72. Ainbinder D, Paolino D, Fresta M, Touitou E. Drug delivery applications with ethosomes. *J Biomed Nanotechnol* 2010;**6**:558–68.
73. Jose Morilla M, Lilia Romero E. Carrier deformability in drug delivery. *Curr Pharm Des* 2016;**22**:1118–34.
74. Godin B, Touitou E, Rubinstein E, Athamna A, Athamna M. A new approach for treatment of deep skin infections by an ethosomal antibiotic preparation: an *in vivo* study. *J Antimicrob Chemother* 2005;**55**:989–94.
75. Abdulbaqi IM. Ethosomal nanocarriers: the impact of constituents and formulation techniques on ethosomal properties, in vivo studies, and clinical trials. *Int J Nanomedicine* 2016;**11**:2279–304.
76. Zhao YZ, Lu CT, Zhang Y, Xiao J, Zhao YP, Tian JL, et al. Selection of high efficient transdermal lipid vesicle for curcumin skin delivery. *Int J Pharm* 2013;**454**:302–9.
77. Zorzi GK, Carvalho EL, Von Poser GL, Teixeira HF. On the use of nanotechnology-based strategies for association of complex matrices from plant extracts. *Rev Bras* 2015;**25**:426–36.
78. Zhang Y, Chan HF, Leong KW. Advanced materials and processing for drug delivery: the past and the future. *Adv Drug Deliv Rev* 2013;**65**:104–20.
79. El Maghraby GM, Ahmed AA, Osman MA. Penetration enhancers in proniosomes as a new strategy for enhanced transdermal drug delivery. *Saudi Pharm J* 2015;**23**:67–74.
80. Herkenne C, Alberti I, Naik A, Kalia YN, Mathy FX, Préat V, et al. *In vivo* methods for the assessment of topical drug bioavailability. *Pharm Res* 2008;**25**:87–103.
81. Mermel LA. Infection prevention and control during prolonged human space travel. *Clin Infect Dis* 2013;**56**:123–30.
82. Jie C, Changping H, Maoxin Z, Kaipin N. Ethosome gel filmcoating agent with multiple wound repair effects and preparation method of ethosome gel film-coating agent, CN103893394 (A), 2014.
83. Hurler J, Škalko-Basnet N. Advancements in burn therapy: promise of nanomedicine. In: McLaughlin ES, Paterson AO, editors. *Burns: prevention, causes and treatment.* New York: Nova Science Publishers; 2012. p. 39–63.

84. Page K, Wilson M, Parkin IP. Antimicrobial surfaces and their potential in reducing the role of the inanimate environment in the incidence of hospital-acquired infections. *J Mater Chem* 2009;**19**:3819–31.
85. Ghosh C, Haldar J. Membrane-active small molecules: designs inspired by antimicrobial peptides. *ChemMedChem* 2015;**10**:1606–24.
86. Seo MD, Won HS, Kim JH, Mishig-Ochir T, Lee BJ. Antimicrobial peptides for therapeutic applications: a review. *Molecules* 2012;**17**:12276–86.
87. O'Driscoll NH, Labovitiadi O, Cushnie TP, Matthews KH, Mercer DK, Lamb AJ. Production and evaluation of an antimicrobial peptide-containing wafer formulation for topical application. *Curr Microbiol* 2013;**66**:271–8.
88. Sood A, Granick MS, Tomaselli NL. Wound dressings and comparative effectiveness data. *Adv Wound Care* 2014;**3**:511–29.
89. Pereira RF, Barrias CC, Granja PL, Bartolo PJ. Advanced biofabrication strategies for skin regeneration and repair. *Nanomedicine* 2013;**8**(4):603–21.
90. Ramos DM, Subramanian A, Aravamudhan A, Harmon M, James R, Shelke NB, et al. Nanotissue engineering of skin cells. In: Baharvand H, Aghdami N, editors. *Stem-cell nanoengineering*. New Jersey: John Wiley & Sons; 2015. p. 315–26.
91. Jain S, Domb AJ, Kumar N. Drug delivery to wounds, burns, and diabetes-related ulcers. In: Domb AJ, Khan W, editors. *Focal controlled drug delivery*. New York: Springer; 2014. p. 585–605.
92. Pawar HV, Boateng JS, Ayensu I, Tetteh J. Multifunctional medicated lyophilised wafer dressing for effective chronic wound healing. *J Pharm Sci* 2014;**103**:1720–33.
93. Augustine R, Kalarikkal N, Thomas S. Advancement of wound care from grafts to bioengineered smart skin substitutes. *Prog Biomater* 2014;**3**:103–13.
94. Sevgi M, Toklu A, Vecchio D, Hamblin MR. Topical antimicrobials for burn infections—an update. *Recent Pat Antiinfect Drug Discov* 2013;**8**:161–97.
95. Heunis TD, Dicks LM. Nanofibers offer alternative ways to the treatment of skin infections. *J Biomed Biotechnol* 2010;**2010**. pii: 510682.
96. Hemmila MR, Mattar A, Taddonio MA, Arbabi S, Hamouda T, Ward PA, et al. Topical nanoemulsion therapy reduces bacterial wound infection and inflammation after burn injury. *Surgery* 2010;**148**:499–509.
97. Moghimi HR, Makhmalzadeh BS, Manafi A. Enhancement effect of terpenes on silver sulphadiazine permeation through third-degree burn eschar. *Burns* 2009;**35**:1165–70.
98. Kang L, Poh AL, Fan SK, Ho PC, Chan YW, Chan SY. Reversible effects of permeation enhancers on human skin. *Eur J Pharm Biopharm* 2007;**67**:149–55.
99. Karpanen TJ, Conway BR, Worthington T, Hilton AC, Elliott TS, Lambert PA. Enhanced chlorhexidine skin penetration with eucalyptus oil. *BMC Infect Dis* 2010;**10**:278.
100. Godin B, Touitou E. Transdermal skin delivery: predictions for humans from *in vivo*, *ex vivo* and animal models. *Adv Drug Deliv Rev* 2007;**59**:1152–61.
101. Thakur R, Jain N, Pathak R, Sandhu SS. Practices in wound healing studies of plants. *Evid Based Complement Alternat Med* 2011;**2011**:438056.
102. Lodhi S, Jain AP, Rai G, Yadav AK. Preliminary investigation for wound healing and anti-inflammatory effects of *Bambusa vulgaris* leaves in rats. *J Ayurveda Integr Med* 2016;**7**:14–22.
103. Rijo P, Matias D, Fernandes AS, Simões MF, Nicolai M, Reis CP. Antimicrobial plant extracts encapsulated into polymeric beads for potential application on the skin. *Polymers* 2014;**6**:479–90.
104. Montagu A, Saulnier P, Cassissa V, Rossines E, Eveillard M, Joly-Guillou ML. Aromatic and terpenic compounds loaded in lipidic nanocapsules: activity against multi-drug resistant *Acinetobacter baumannii* assessed *in vitro* and in a murine model of sepsis. *J Nanomed Nanotechnol* 2014;**5**:1–7.
105. Wang Y, Wu X, Chen J, Amin R, Lu M, Bhayana B, et al. Antimicrobial blue light inactivation of Gram-negative pathogens in biofilms: *in vitro* and *in vivo* studies. *J Infect Dis* 2016;**213**:1380–7.
106. Patra V, Byrne SN, Wolf P. The skin microbiome: is it affected by UV-induced immune suppression? *Front Microbiol* 2016;**7**:1235.

CHAPTER 17

Medical, Surgical Therapy, and Alternative Treatment of Infected Vascular Grafts

A.M. Settembrini*, P.G. Settembrini†
*Vascular Surgery, ASST Santi Paolo e Carlo, San Carlo Borromeo Hospital, Milan, Italy
†Professor of Vascular Surgery, University of Milan, Milan, Italy

1. INTRODUCTION

Since the first description by Szilagyi and coworkers[1] of infection of vascular synthetic grafts in the 1960s, the problem has grown. However, more knowledge of the problem led surgeons to prevention by accurate cleaning of the skin to be incised and by the introduction of continuously newer antibiotics. Furthermore, graft infections also decreased because the duration of the vascular interventions was significantly reduced and consequently the exposure of abdominal viscera and the whole operating field was limited, so the possibility of contamination was obviously less. During the ensuing decades, cleaning of the skin was best achieved through the application of liquid antiseptic substances and, most of all, by means of an accurate and deep washing of the skin in the most dangerous districts, such as the umbilicus, groin, genital organs in both genders and the perineal region. Furthermore, an accurate exploration of skin lesions of the legs (trophic ulcers of V–VI degree ischemia according to Rutherford's classification[2]) can be done, and these should be treated from an infective point of view.

On the other hand, a large diffusion of antibiotic therapy for viral infections (wrong!) caused a selection of new microbial strains and the appearance of unknown bacteria, or the transformation of some bacteria from a saprophytic to an aggressive status (e.g., *Staphylococcus epidermidis*) and activation of so-called "sleeping" bacteria (e.g., *Serratia* spp.). The causes of this microbial behavior should also be researched in the aggressive surgical treatment of aged and sometimes debilitated patients with a decreased immunological reaction.

The need for performing operations for limb salvage by using synthetic grafts in the lower limbs in the absence of venous substitutes favored contact between grafts and skin lesions, which is always a potential source of infections.

In the last few decades, we have witnessed the appearance of antibiotic resistant Gram-positive bacteria (e.g., *Staphylococcus aureus*), a cause of potentially devastating

The Microbiology of Skin, Soft Tissue, Bone and Joint Infections
ISSN 2451-9006, http://dx.doi.org/10.1016/B978-0-12-811079-9.00017-3

infections. But this obstacle has also been successfully resolved by the introduction of a new class of antibiotics. However, at present vascular graft infections have not disappeared, even though their incidence has been significantly lowered to <1% in the largest series.[3]

So, vascular graft infection is a rare but potentially devastating clinical problem. Nowadays the most-used standard technique for infected grafts is the complete graft excision, debridement of surrounding tissues, followed by an extra-anatomic bypass procedure through a noninfected field. The observations of Reilly et al.[4] improved the results of the extra-anatomic bypass, in terms of mortality and limb salvage, by first performing the bypass procedure followed by graft excision. Nevertheless, other authors have reported higher mortality rate and a lower percentage of limb salvage with recurrence of infection in 18% of patients.[5]

Patency rates of substituting grafts have been disappointing (primary patency rate of 43% at 3 years). In recent times, different techniques have emerged. More and more it seems advisable that, after graft excision, the reconstruction should be performed by means of an in situ treatment utilizing various conduits, including autogenous veins, like deep femoral veins in the aortic district,[6,7] cryopreserved arterial homografts[8] and synthetic fabric rifampicin-bonded or silver-impregnated.[9,10]

2. EPIDEMIOLOGY AND PATHOPHYSIOLOGY

The epidemiology of vascular graft infection[3] can be influenced by anatomical site, surgical indication, type of intervention and immunological status of the patient receiving the procedure. However, vascular graft infections can occur many years after implantation (15 years in one case we observed) with the incidence of infection at 5% during a 10-year period after prosthetic grafting of the aorta.[8]

Emergent surgical procedures (such as ruptured abdominal aortic aneurysm) can be easily complicated by a graft infection, which can also occur when the graft has been sutured to the femoral artery or runs in a subcutaneous tunnel (axillo-(bi)femoral or cross-femoral bypass). Development of infection can also appear after endoluminal stent deployment (lower, <1%).

The pathophysiology of graft infection is founded on the adhesion of bacteria to graft or stent surfaces and consequent inflammatory responses involving perigraft tissues and graft-artery anastomoses. It is important to remember that bacteria adhere more strongly to polyester grafts (Dacron) than to polytetrafluoroethylene (PTFE) grafts.

An insufficient graft embedding favors larger exposure of the implanted graft to bacteria themselves and a dissemination of the infections can run along the whole graft and into adjacent structures, causing formation of abscesses, anastomotic pseudoaneurysms, graft-cutaneous sinus tracts or aortoenteric or colonic fistula. The beginning of the infection is often marked by an insufficient graft healing and embedding into surrounding tissues and by collection of perigraft fluid.

Skin and lymph nodes can be a major reservoir of bacteria, especially at the groin. The groin, in fact, represents the crucial point of lymph drain of the lower limb. It is well known that lymph from peripheral territories can also be a vehicle for bacteria and/or toxins, which generally are stopped at the level of the groin lymph nodes. If the bacterial charge is high, a heavy inflammatory reaction can take place involving subcutaneous tissue and also the skin.

Therefore prosthetic grafts can be contaminated by bacteria through the surgical wound at the operation or by lymphatic sources flowing from remote sites of infection, such as ischemic and gangrenous foot or leg lesions.

Another mechanism for infection is erosion by the prosthetic graft of subcutaneous tissue, so reaching the surface of the skin. It is also possible that the graft can erode the wall of the stomach or intestine or ureters, causing a perigraft infection that can spread along the whole length of the graft, involving both anastomoses. Therefore, during the closure of the surgical wound close attention must be paid to avoid any direct contact between the graft and bowel and/or ureters by accurate interposition of retroperitoneal soft tissue. Nevertheless, bowel wall erosion can also occur due to the pulsatile movement of an aortic graft against adjacent bowel, involving the graft body or anastomotic sites. Reported incidence of bowel fistula is between 0.4% and 2%.

A basic principle of the microbiology and general pathology is that the conditions necessary for the development of an infection are three, namely: (1) amount of the aggressive infecting agents, (2) duration of the contact between infecting agents and infected organism, (3) immunological status of the infected host. The routes utilized by infecting agents (bacteria) are the direct contact or blood transport, so-called bacteremia.

Direct contact happens when there is infected skin where the surgical incisions have been performed or, rarely, when microlesions of intestinal wall occur and in this manner bacteria can flow outside of the bowel and contaminate the aortic prosthetic graft. Another condition occurring frequently originates from the so-called "focal" infections like tooth root infections, local skin infections, and infectious foci widespread in the body. A further possible source of infection is represented by the gallbladder or urine stagnation in the bladder, which can be contaminated by known urine bacteria, like *Escherichia coli*. In general, the infections from Gram-positive germs are usually more aggressive than those coming from Gram-negative germs. The most dangerous bacteria typically involved are *Staphylococcus, aureus* or *epidermidis*, especially if resistant to methicillin (MRSA).

Another chapter of this pathology is represented by in-hospital contracted infections, especially in the patients staying in the hospital after surgical intervention, and the more frequently if surgery was difficult, long lasting or complicated, or if many units of blood or frozen plasma were infused.

More recently, cases of infections from *Mycobacterium tubercolosis*[11] or HIV agents[12] were described. In these patients, the common element is the highly depressed immunological status. The bacterial contamination of the prosthetic graft kicks off the inflammatory process against foreign bodies, as when a splinter (metallic or wooden) casually penetrates the body by interrupting the skin integrity. The reaction against foreign bodies is well known if not extracted. *The existence of a graft infection along its course transforms it into a foreign body*. Moreover, it must be remembered that an implant of a vascular graft in an HIV patient can set off a violent inflammatory reaction which cannot be tolerated and can cause even the detachment of the graft.

Usually, the grafts are surrounded by fibrous tissue that represents the substance of the so-called "healing" process. This process needs many months to develop in a safe environment. Otherwise, the infection, through a massive migration of granulocytes "attacks" and destroys the developing fibrous tissue and excites the inflammatory reaction of the host tissues.

Skin and soft tissue infections are common, grading from superficial infections to life-threatening manifestations. Their classification is based on anatomical site, clinical severity, or microbial cause. Some authors simplify the difference into complicated and uncomplicated infections. Community-acquired infections are usually caused by staphylococci or streptococci, but remember that almost any organism can cause an inflammation.

How can this kind of reaction be avoided? First of all, the contamination can start from infected skin, which hosts usually many bacterial organisms and their exotoxins.

An accurate preparation of the patient undergoing a vascular prosthetic procedure is mandatory. The skin should be thoroughly washed in every district, particularly at groin, perineum, and periumbilical zones. Shaving the skin seems to be necessary only where surgical incisions will be carried out. This behavior is suggested by the risk of knife skin lesions, which could favor the penetration of bacteria and similar microbial agents and therefore cause an inflammatory response. Recently, recommendations have been published by the WHO. Consequently, every inflammatory site should be found and correctly treated before the operation, by means of surgical toilette, identification of the involved bacterium/a and antibiotic treatment.

The search must be addressed to find possible anatomical sites, like paranasal sinuses, throat (pharynx), tonsils, dental chronic granulomas, gallbladder stones, bladder chronic infections (particularly in the aged) and female genital organs, skin pimples and nail infections.

Particularly dangerous is the presence of preexistent skin lesions, as in diabetic patients or in peripheral arterial disease (PAD). It is of paramount importance to identify such possible sites of infection(s), which could kick off a bacteremia, especially in aged patients. Underhanded bacteremia can come from the bladder in an unsuspected way, carrying the most common urine bacterium, *E. coli*, to affect the foreign body, which means the prosthetic graft. Moreover, it must be remembered that the deposition of bacteria on the embedded graft can cause destruction of the bacterium itself, but at the same time the liberation both of endotoxins from the bacterial body and/or exotoxins. Every type of microbial toxin produces lytic enzymes that destroy the fibrous embedding of the graft soliciting the inflammatory reaction of the human body and the foreign body reaction, the aim of which is to eliminate the foreign body itself.

The infections can occur before, during, and after graft insertion. It is therefore mandatory that a correct evaluation of the patient be done from the point of view of his potential capacity of immunological reaction. For this aim the patient should be questioned about his preceding history of infectious diseases, even the so-called diseases of childhood, and in particular those involving teeth (risk of dental granulomas) and throat (beta-hemolytic *Streptococcus*). In the case of positive answers, an accurate investigation should be performed in order to set up an antibiotic prophylaxis. It is very important to remember that potential infective agents against prosthetic material can also be viruses, even the HIV agent. A perfectly healthy patient without any contact at all with infections is surely impossible to be found. In common medical practice, what makes the real difference among individuals is nutrition and immunological status. In general practice, it is well known at any level of medical knowledge that a young patient in healthy condition can react much better against infecting agents. On the other hand, an aged patient affected with many diseases (heart, metabolic, kidney, ischemic sequelae, liver insufficiency, psychiatric pathology) can deploy neither sufficient resistance nor sufficient fighting forces against infecting agents. In our opinion, we must not forget the words of the great Roman orator Cicero (first century BC) who said: "Senectus ipsa morbus" (age is a disease in itself). In fact, it has been observed that the vast majority of infections of prosthetic vascular grafts occur in aged patients, with a depressed immunological condition, lying for a long time in the hospital, and with poor capacity of communication inside their families and with doctors and nursing personnel. Moreover, it must be born in mind that the more superficial is the prosthetic graft (close to skin surface), the higher is the risk of contamination.

3. CLINIC

The clinical aspects of graft infection depend on the anatomic location and the virulence of the involved pathogen.

In some cases, a good outcome can be obtained through identification of the responsible microorganism and by ascertaining the extent of graft infection. It is further necessary to administer a specific antibiotic therapy according to the microbial culture, and then it is possible to plan the best surgical intervention. *Remember that each patient requires a tailored treatment*. An appropriate management may require: (a) *graft excision*, alone or in conjunction with extra-anatomic bypass grafting; (b) *graft preservation*; (c) *in situ graft replacement* following graft excision. *Each procedure must be adapted to the patient.*

Patients with late symptoms likely have a low-virulence graft infection. In these cases, in situ graft replacement with autogenous venous conduits, cryopreserved homografts or antibiotic or silver impregnated grafts should be preferred because the intervention is less heavy (anesthesia and surgical approach).

In clinical practice, it is mandatory to know that the contamination of deep or retroperitoneal prosthetic substitutes usually happens through a hematogenous route starting from focal infections or by filtration through a thin diseased wall of the bowel or the bladder.

General symptoms are septic chills and fever (up to 39–40°C, or 102–104°F). Very often in the site of a superficial or subcutaneous infection an abscess can develop rapidly, with a high probability of spraying into the neighboring tissues and in depth towards muscular fascia and the underlying graft, especially at the groin.

The presence of a seroma around the graft and for its whole length does not itself represent an infection, but, because it is constituted of proteins, the risk of bacterial contamination is easily understandable, especially close to the skin, and of a sure suppuration.

Any time when a patient with previous aortic surgery comes to be evaluated, it is important to always consider any clinical sign with a high degree of suspicion. Otherwise, if infection involves a prosthetic graft along the thigh or leg, the initial symptom can often be a local inflammation with or without a skin fistula. If an anomalous pulsation appears in the anastomotic site (groin), this is a reliable or even sure index of a pseudoaneurysm, which can explode. As said before, any patient with gastrointestinal bleeding and having an aortic graft should be presumed to have graft infection and bowel fistula. Bacteremia is a sign of an advanced graft infection.

An infected prosthetic graft must be suspected whenever an inflammatory process is detected in patients who received such a grafting, even many years before. In our experience, we observed a patient with an infected aortic graft after more than 10 years after implantation. Therefore, every patient having a vascular synthetic graft must remain in periodic medical observation for his whole life. The surgeon must make the patient aware of the heavy risk of any infectious process involving all organs of the body, i.e., throat, teeth, lungs, gallbladder, bladder, appendix, etc. Moreover, the doctor at home must clearly know the risks of the patient in every kind of infection.

Even though different treatments can be very useful for the healing of a graft infection, the main aim should be to avoid vascular infections. To minimize, if not completely avoid, graft infections, some conditions must be satisfied: (a) shorten preoperative hospital stay, i.e., prepare the patient outside the hospital; (b) patients should be accurately washed with an antibacterial soap the evening before the operation; (c) research out any possible remote infection before an elective operation; (d) shave operative site immediately before the operation, avoiding skin trauma; (e) protect vascular grafts from any external direct contact; (f) accurate closure of groin incisions, eliminating dead spaces between sheaths (do not tighten the knots of the suture on the skin, as correct tension avoids the development of dermal ischemia and wound edge necrosis); (g) antibiotic prophylaxis 30–60 min prior to implantation of a prosthetic graft. If infectious process appears: (h) at least a fortnight before surgery, a microbial culture can be very useful to choose the right antibiotic; (i) longer (>72 h) antibiotic prophylaxis in high-risk patients for surgical wound.

4. CLASSIFICATION

Classification of graft infections can be related to postoperative wound infection, according to Szilagyi,[1] and/or extent of graft involvement, according to Bunt[13,14] (Table 1).

Early infections are caused by hospital-acquired bacteria, with fever, leukocytosis, and bacteremia. Late infections are caused by graft colonization by low-virulence organisms, such as *S. epidermidis* or *Candida*.

The most-often involved bacteria are Gram-positive cocci, *S. epidermidis* in late infections, *S. aureus* in rapidly growing infections within a few months after the operation. In recent years (the last decade) a new clone has developed; *rectius* was selected by the natural selection of resistant bacteria, namely MRSA.

5. INSTRUMENTAL STUDIES

Before a vascular graft infection, it is fundamental to identify the correct site and nature of any infection. As said before, microbiological analysis is useful to identify the pathogen, but also the site of the infection allows the decision to be made as to which is the best treatment for the patient. Following these criteria, a precise imaging is recommended

Table 1 Szylagyi[1] classification of infections and its modification by Samson

Szylagyi grade	Definition	Samson grade	Definition
1	Only affection of the dermis	1	Extension no deeper than the dermis
2	Involvement of subcutaneous region but not the arterial implant	2	Involvement of subcutaneous but not observable direct contact with the graft
3	The arterial implant involved in the infection	3	Involvement of the body of the graft but not anastomosis
		4	Infections surround an exposed anastomosis, but nor bacteremia or anastomotic bleeding
		5	Involvement of graft to artery anastomosis and septicemia and/or bleeding

because often only one examination is not sufficient, rather a combination of different radiological exams is done.

In particular, the first exam is usually ultrasonography (US) followed by a CT scan and MRI for the definition of the anatomical site of the infection. By means of a gastrointestinal endoscopy, potential fistulas or bleeding can be seen. Functional radionuclide investigation by means of radiolabeled leucocyte imaging can confirm the diagnosis, if CT signs of perigraft infection are nonconclusive because of the glucidic metabolism of the bacteria.

Ultrasonography should be the initial step of utmost importance in an imaging study for even a simple suspicion of graft infection. In the hands of an experienced and skilled operator, US can show precise characteristics of fluid, detecting a purulent collection around the graft. In this investigation, the operator should not forget that a fluid collection around a graft is abnormal beyond 7–10 days after the operation. Persistence of fluid (rich in protein material) undoubtedly favors its colonization by bacteria.

During CT scan an infection can be revealed by the presence of inflammation and tissue subversion or fluid or gas around the graft. These are surely the first images of an infection, with an enlargement in the case of worsening of the infection itself. Other typical images of current infection are vertebral erosion, hydronephrosis and formation of pseudoaneurysm after a recent implantation (<6 months).

After a CT scan, if suspicion of infection persists, the most sensible exam to identify the grade and extension of the infection is the labeled leucocyte scanning. All radionuclide imaging techniques can demonstrate abnormal accumulation of leukocytes in perigraft tissue, but do not provide clear anatomic detail. These techniques should be correlated with CT to locate the anatomic extent of infection correctly.

If there is a suspicion of an aortic-enteric fistula, gastroduodenal endoscopy may be decisive in finding it. Often the fistula or even the presence of the graft inside the lumen may be found in the fourth tract of the duodenum. Therefore, an accurate exploration may be necessary. Due to the distance of the fourth duodenal tract, the usual instrument can be too short. In such a case, a typical trick is the use of a colonoscope.

6. MEDICAL THERAPY

One of the most important topics in graft infection research is the discovery and development of new antibiotic substances. About 10 years ago, pharmaceutical researchers developed two important antibiotics, daptomycin and tigecycline. The first one was discovered originally by researchers at Eli Lilly and Co. in the 1980s. It was promising in phases I and II of clinical tests for the treatment of infections caused by Gram-positive bacteria. But high doses of the substance were associated with adverse effects on skeletal muscles (myalgia and myositis). Further experiments were therefore stopped. In 1997 Cubist Pharmaceuticals bought the production rights and put it into commerce with the trade name Cubicin. Its mechanism of action is based on the inhibition of protein synthesis of RNA and DNA, causing the death of the bacterial cell. Daptomycin is effective only against Gram-positive cocci, including MRSA.

Tigecycline is a glycylglycinic antibiotic developed by Francis Tally. It was approved by FDA in 2005. It was developed following the resistance of some bacteria which developed output pumps, like *S. aureus* and *Acinetobacter baumannii*. The latter is increasingly found in patients having long hospital stays after heavy operations and difficult recoveries. Infections from *S. aureus* and *Acinetobacter* are difficult to identify and can be devastating on the affected organism. The core of the molecule is quite similar to the tetracycline ring and presents the same toxicity after a treatment period longer than 7–10 days. Explanted graft material usually shows positivity for *S. epidermidis* in more than 80% of grafts associated with anastomotic aneurysms, at the level of both the aorta and femoral arteries. The highest risk remains at the level of the aortic anastomosis because of contamination by enteric bacteria of the fibrous tissue embedding the anastomosis of the graft on the aorta itself. All bacteria produce eso- or endotoxins that attack and dissolve protein matrix.

A MRSA infection is associated with higher morbidity and mortality rates than other microbes.

Gram-negative bacteria (*E. coli*, *Pseudomonas*, *Klebsiella*, *Serratia* species) can produce challenging infections because of the presence of destructive endotoxins compromising the structural integrity of the vessel wall; they are always found in very debilitated patients who have lost most of their immunological capabilities.

7. SURGICAL AND ALTERNATIVE TREATMENT

As stated in the introduction, the ideal treatment of a vascular graft infection is represented by its removal (Fig. 1) and restoring of the vascular continuity, and also by means of an in situ reconstruction after a deep cleaning of the area and concomitant antibiotic therapy. This treatment can avoid recurrence and complications, but eradication of the bacterium is not certain.[15]

The treatment of an infected aortic graft is often challenging because it is mandatory to control the aorta in the supraceliac tract and the surgeon must be very careful regarding the involvement of the surrounding tissues and organs (ureters, duodenum) by evaluating the fibrotic reaction to the infection. After its free preparation, the surgeon removes the graft. At this point the choice can be made as to whether to carry out an extra-anatomical reconstruction or an in situ one. In the first case, the aortic stump must be sutured by applying single monofilament stitches, performing debridement of the involved area and covering it by sliding the omentum and performing an axillofemoral or bifemoral bypass with a Dacron or ePTFE graft.

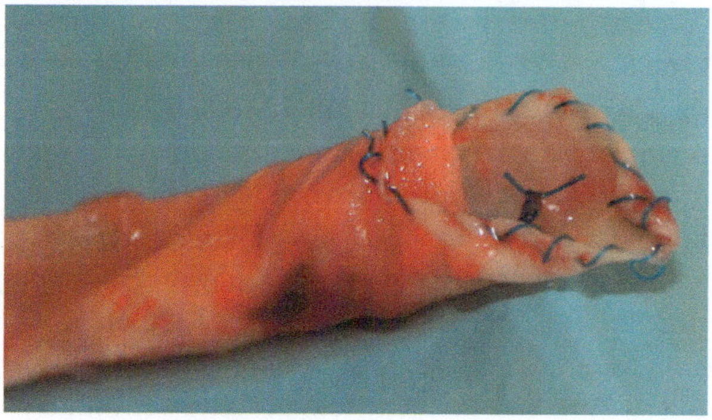

Fig. 1 Detached segment of the graft due to the infection. It is possible to see the intact running suture along the anastomosis with an intact knot. The arterial side was "digested" by the infection.

If the decision is made to perform an in situ reconstruction, different possibilities exist to achieve an autologous graft replacement,[16] Nevelsteen[6] introduced the use of both superficial femoral veins, which are sutured each to the other longitudinally, simulating an aortic bifurcation.

Another option is to replace the aorto-iliac tract with cryopreserved arterial homografts harvested from transplant donors and made nonantigenic by cryopreservation. The most important experience in this regard has been described by Kieffer et al.[8] who described homograft dilatation or stenosis and disruption in a significant number of patients during follow-up observation. The cryopreserved human homografts present a high risk of disruption if the suppuration has not been completely eradicated, because purulent fluids are able to "digest" the wall of the human homograft. Better results seem to come from the silver-coated grafts.

A modification of the Nevelsteen technique has been proposed by Clagett[7] who described a "pantaloon" technique, suturing together the two femoral veins to recon struct the aortic iliac bifurcation

A further option for the in situ replacement is the use of silver-impregnated Dacron grafts, which have shown good patency and resistance to infections over the long term.[16]

Also, antibiotic-bonded grafts are used for in situ replacement, but with poor results. In fact, this option is selected only in rare circumstances and the results are not conclusive; a relatively high percentage is associated with recurrent infection (10%–20% of cases, especially with Gram-negative bacteria and MRSA).[9] Recently a new method was proposed to face infections of vascular prosthetic conduits (Dacron, ePTFE).[16] The procedure is based on "washing" the graft with ultrasound (US) in a saline solution, with a major degree of energy if the graft involved is Dacron and more carefully in case of ePTFE, making a debridement of the tissues and the graft possible (Fig. 2). US debridement has proved to be a valuable aid in the treatment of infected grafts and poor medical conditions.[17] Used in conjunction with proper antibiotics, it allows more conservative

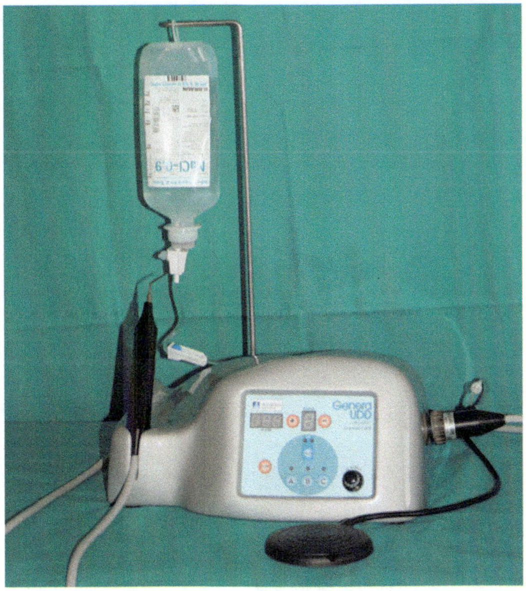

Fig. 2 The device utilized as the source of ultrasound (hanging is the saline solution). This device is produced by Genera, Milan.

treatment without compromising the chance of success, with reduced stress for the patient. In fact, extracavitary vascular graft infection is associated with a low rate of patient survival and poor limb salvage. Standard surgical treatment with graft removal and in situ reconstruction is burdened by high morbidity and mortality rates. US technology and its interaction with living tissues have been widely studied. For instance, it has been used in dentistry for deep cleaning of the teeth and other structures and in cardiac surgery to decalcify the valve, and it showed promising results in wound care and diabetic foot treatment. The debriding procedure is performed using a powerful ultrasonic generator. The graft and the surrounding tissues are treated by moving the probe over the surface until all necrotic parts and fibrin are removed and the color of the tissue returns to a macroscopically normal appearance. The probe operates directly onto the graft; for Dacron grafts, we noticed no harm, even when an intensive and prolonged debridement was done. Polytetrafluorethylene (PTFE) grafts need special care, because when the prosthesis was treated for too long, a slight loss of impermeability was noticed (Fig. 3); however, no major damage occurred to the grafts. This method can be used in cases involving older patients in poor clinical condition or having multiple comorbidities. These patients are often too fragile to consent to a radical surgical treatment. US debridement is also used if autogenous substitutes are not available. In our opinion, it is advisable to use the method in any case before performing the radical surgical removal of the graft, because vascular graft infections usually run more rapidly than other infections. Very often the procedure

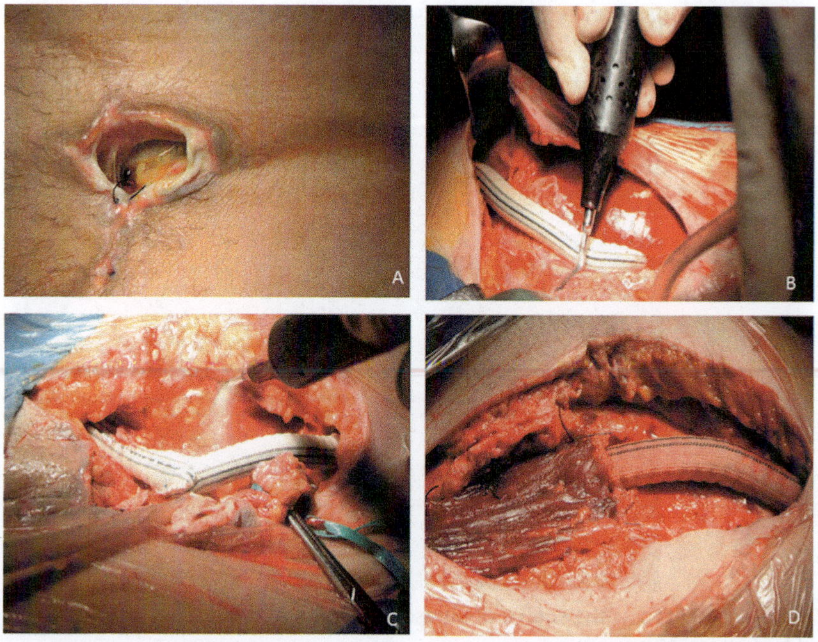

Fig. 3 (A) Open fistula. At the bottom of the wound, a Dacron graft is visible. (B) The US probe working directly on the Dacron graft. (C) The grafts (axillo-bifemoral bypass) at the end of treatment. The color of the cleaned grafts is clearly visible, *white* as before the implantation. (D) The sliding and suturing of the sartorius muscle to cover the anastomosis.

is not a single step but, especially in the groin, it can be completed with a *sartorius* muscle flap rotation or negative pressure wound treatment (NPWT) as adjunctive procedures. Different techniques have been used in such cases. Partial removal of the infected portion has claimed to be as effective as a radical excision of the graft; both Calligaro[18] and Hart[19] reported that, even in the case of bifurcated aortic graft infections, complete graft excision is not always necessary and, in prohibitive risk patients, partial or no removal of infected segments associated with best antimicrobial therapy led to comparable short- and long-term outcomes in terms of mortality and recurrence of the infection. In light of the various options of conservative treatment, surgical wound debridement (WD) and muscle transposition flap coverage have produced encouraging low amputation rates.

In our experience, NPWT was used only in one patient; in fact, most of the infections occurred before the NPWT devices became available. Moreover, this low use of the VAC device is likely due to the strategy we adopted, which is primarily to close the skin above the graft whenever a satisfactory cleaning of the wound is achieved and the skin tissues allow a tension free suture, therefore limiting the use of NPWT.[20]

Low-frequency US (LFUS) may play an important role in increasing the penetration of antibiotics in biofilms against certain species of bacteria, such as *E. coli*, but not

Pseudomonas aeruginosa as reported by Carmen et al.[21] In our experience, US debridement achieves the best tissue and/or graft cleaning and minimizes the invasiveness of surgery. Ancillary procedures such as a sartorius flap or NPWT promote a durable wound healing with a less invasive approach. In the case of MRSA infection, we usually perform the debridement associated with very strong antibiotic therapy with imipenem and daptomycin to cover all the spectrum of germs. According to the literature, we usually do not attempt a conservative approach when *P. aeruginosa* is the pathogen involved in graft infection.[21] In our opinion, US debridement can be a useful tool which can help greatly in facing this terrible complication.

8. ENDOVASCULAR TREATMENT

In recent years the widespread use of endovascular grafts has given the impression that graft infection could be eliminated by excluding any contact of the graft with the skin. Unfortunately, graft infections have appeared after some number of years, and also after endovascular abdominal aortic aneurysm repair (EVAR).

So, we should consider that the endovascular approach is not free from infectious complications, even though the incidence of infection in EVAR is lower than in open surgery (0.1%–1%).[22] The bacteria involved are more frequently Gram-negative with a low grade of virulence. This complication usually appears in old or very old patients who typically have depressed immunological power. When an endovascular graft is affected with infection, the solution is often not a simple one. In fact, from a technical point of view, the right solution should be the removal of the endograft, but it is well known that the extraction of an endograft is a very dangerous, life-threatening maneuver, because of the barbs or anchors attached to the aortic wall. Any attempt to remove the graft can cause a disruption of the aortic wall. Therefore, the possible treatment is founded on heavy antibiotic therapy, even though it cannot give the desired result. The adjunct risk in these patients is that the endograft becomes itself a source of spreading infectious emboli. For this reason, some authors have suggested performing a CT-assisted puncture to evacuate infected fluid from the aorta to reduce the bacteric charge and the risk of bacteremia, but this maneuver too is very dangerous. As in the treatment of a surgically implanted graft, the definitive treatment is the graft removal and extra-anatomic or in situ reconstruction.

An endograft for endovascular treatment of abdominal aortic aneurysm is today also considered a tool for bridge treatment in the case of aortic-enteric fistula.[23] In fact, endovascular treatment is less invasive than a laparotomy. In a bleeding emergency it could be no more difficult than elective surgery, so the deployment of an endograft to cover the perforated aorta, even in an infected bed for a short period, can be considered a useful option in case of hemodynamic instability. This experience began in the aortic area and today is sometimes used also in the arms or legs in the case of infected false aneurysm. Some episodes of definitive treatment through an endovascular treatment have been described in the literature, but in our opinion it would be better to move as soon as possible to a debridement and removal of the graft.

9. CONCLUSIONS

We are not yet at the end of the Graal query! Any apparently new solution includes the danger of infection because we always introduce foreign materials into the human body, and it is impossible to completely avoid the adverse reactions of the host organism. Therefore a targeted antibiotic therapy is still the milestone of treatment.

Alternative methods are represented by the use of an endograft as bridge, different surgical reconstructions and US debridement, which gave surprising results. In particular, US debridement can totally clean the graft from blood cells and components and, obviously, also from bacterial residuals without any sign of infection more detectable in the follow up also in case of high level of growth of bacteria. This very important result shows that a mechanical cleaning can also be decisive in eradicating infection from the graft, even in older and debilitated patients, without introducing foreign material into the body.

REFERENCES

1. Szilagyi DE, Smith RF, Elliott JP, Vrandecic MP. Infection in arterial reconstruction with synthetic grafts. *Ann Surg* 1972;**176**:321–33.
2. Rutherford RB, Baker JD, Ernst C, Johnston KW, Porter JM, Ahn S, et al. Recommended standards for reports dealing with lower extremity ischemia: revised version. *J Vasc Surg* 1997;**26**(3):517–38.
3. Young MH, Upchurch Jr. GR, Malani PN. Vascular graft infections. *Infect Dis Clin North Am* 2012; **26**(1):41–56.
4. Reilly LM, Stoney RJ, Goldstone J, Ehrenfeld WK. Improved management of aortic graft infection: the influence of operation sequence and staging. *J Vasc Surg* 1987;**5**:421.
5. Yeager RA, Taylor Jr. LM, Moneta GL, Edwards JM, Nicoloff AD, McConnell DB, et al. Improved results with conventional management of infrarenal aortic infection. *J Vasc Surg* 1999;**30**:76–83.
6. Nevelsteen A, Lacroix H, Suy R. Autogenous reconstruction with the lower extremity deep veins: an alternative treatment of prosthetic infection after reconstructive surgery for aortoiliac disease. *J Vasc Surg* 1995;**22**:129–34.
7. Clagett GP, Valentine RJ, Hagino RT. Autogenous aortoiliac/femoral reconstruction from superficial femoral-popliteal veins: feasibility and durability. *J Vasc Surg* 1997;**25**:25.
8. Kieffer E, Gomes D, Chiche L, Fléron MH, Koskas F, Bahnini A. Allograft replacement for infrarenal aortic graft infection: early and late results in 179 patients. *J Vasc Surg* 2004;**39**:1009.
9. Bandyk DF, Novotney ML, Johnson BL, Back MR, Roth SR. Use of rifampin-soaked gelatin-sealed polyester grafts for in situ treatment of primary aortic and vascular prosthetic infections. *J Surg Res* 2001;**95**:44–9.
10. Genovese EA, Avgerinos ED, Baril DT, Makaroun MS, Chaer RA. Bio-absorbable antibiotic impregnated beads for the treatment of prosthetic vascular graft infections. *Vascular* 2016;**24**(6):590–7.
11. Umer I, Mocherla S, Horvath J, Arora S, Ahmed Y. *Mycobacterium abscessus*: a rare cause of vascular graft infection. *Scand J Infect Dis* 2014;**46**(11):813–6.
12. Pillay B, Ramdial PK, Naidoo DP. HIV-associated large-vessel vasculopathy: a review of the current and emerging clinicopathological spectrum in vascular surgical practice. *Cardiovasc J Afr* 2015; **26**(2):70–81.

13. Bunt TJ. Synthetic vascular graft infections. *Surgery* 1983;**93**(6):733–46.
14. Samson RH, Veith FJ, Janko GS, Gupta SK, Scher LA. A modified classification and approach to the management of infections involving peripheral arterial prosthetic grafts. *J Vasc Surg* 1988;**8**(2):147–53.
15. Zetrenne E, McIntosh BC, McRae MH, Gusberg R, Evans GR, Narayan D. Prosthetic vascular graft infection: a multi-center review of surgical management. *Yale J Biol Med* 2007;**80**(3):113–21.
16. Batt M, Magne JL, Alric P. *In situ* revascularization with silver coated polyester grafts to treat aortic infection: early and midterm results. *J Vasc Surg* 2003;**38**:983–9.
17. Carmo M, Mazzaccaro D, Barbetta I, Settembrini AM, Roveri S, Fumagalli M, et al. Use of ultrasound debridement as an adjunctive tool for treating infected prosthetic vascular grafts in the lower extremities. *Ann Vasc Surg* 2015;**29**(3):607–15.
18. Calligaro KD, Veith FJ, Yuan JG, Gargiulo NJ, Dougherty MJ. Intra-abdominal aortic graft infection: complete or partial graft preservation in patients at very high risk. *J Vasc Surg* 2003;**38**(6):1199–205.
19. Hart JP, Eginton MT, Brown KR, Seabrook GR, Lewis BD, Edmiston Jr. CE, et al. Operative strategies in aortic graft infections: is complete graft excision always necessary? *Ann Vasc Surg* 2005;**19**(2):154–60.
20. Reiffel AJ, Henderson PW, Karwowski JK, Spector JA. An interdisciplinary approach to the prevention and treatment of groin wound complications after lower extremity revascularization. *Ann Vasc Surg* 2012;**26**(3):365–72.
21. Carmen JC, Roeder BL, Nelson JL, Ogilvie RL, Robison RA, Schaalje GB, et al. Treatment of biofilm infections on implants with low-frequency ultrasound and antibiotics. *Am J Infect Control* 2005; **33**(2):78–82.
22. Laser A, Baker N, Rectenwald J, Eliason JL, Criado-Pallares E, Upchurch Jr. GR. Graft infection after endovascular abdominal aortic aneurysm repair. *J Vasc Surg* 2011;**54**(1):58–63.
23. Setacci C, de Donato G, Setacci F. Endografts for the treatment of aortic infection. *Semin Vasc Surg* 2011;**24**(4):242–9.

CHAPTER 18

Dermoscopy Findings in Infectious and Parasitic Diseases

C. Scharf
Endomed Médicos Associados, Curitiba, Brazil
Hospital Universitário Evangélico de Curitiba, Curitiba, Brazil

1. INTRODUCTION

Dermoscopy is a diagnostic technique originally developed for evaluation of melanocytic and pigmented lesions. It allows histological and clinical correlation of the lesions, with analysis liable up to the reticular dermis. It is also known as surface microscopy or epiluminescent microscopy.[1–3]

The technique involves the use of a device composed of special lenses and a focus of light that is projected on the skin at an angle of 20 degrees. Using a fluid substance such as oil or gel, it eliminates the factor of light reflection, allowing specific characteristics to be displayed and the presence of melanin and hemoglobin in the different skin layers to be evaluated. Current typical dermoscopes rely on the presence of polarized light, which eliminates the need to use immersion fluids, facilitating the clinical exam even more. Common devices allow magnification of 10 times, while digital devices can reach up to 70 times without losing image definition.[1–8]

In the last 30 years, dermoscopy has been used in dermatology for the identification of different structures and colors not seen by the naked eye. Even though its primary application has been the diagnosis of pigmented lesions, dermoscopy and videodermoscopy have expanded their role in the diagnosis of skin diseases such as psoriasis, alopecia, parasites, nail disorders, collagen diseases, and others.[1] Since then, it has been possible to identify specific findings that may lead to a diagnosis.

Many dermoscopic patterns have been described for infectious diseases such as molluscum contagiosum, viral warts, and tinea nigra.[1] The term "entodermoscopy" was chosen to refer to the use of dermoscopy as an aid in the diagnosis of such infestations as scabies, pediculosis, tungiasis, cutaneous larva migrans, and tick infestations, as well as reactions to the spicules of spiders.[1,2]

The Microbiology of Skin, Soft Tissue, Bone and Joint Infections
ISSN 2451-9006, http://dx.doi.org/10.1016/B978-0-12-811079-9.00018-5

In this chapter, we will discuss dermoscopy findings in viral infections (molluscum contagiosum, warts), fungal infections (tinea capitis, tinea nigra), and infestations by parasites (larva migrans, myiasis, pediculosis, scabies, tungiasis).

2. VIRAL INFECTIONS

2.1 Molluscum contagiosum

Molluscum contagiosum is a viral infection of the epidermal keratinocytes caused by poxvirus. Lesions are generally numerous, characteristically 2–4 mm umbilicated skin-colored papules, with a translucent, glossy appearance, located at various sites on the skin surface. They are more likely to be seen in children and immunosuppressed patients, and the diagnosis is essentially clinical. In some cases, though, dermoscopy can work as an auxiliary instrument.[9–12]

The most usual pattern described in cases of molluscum contagiosum is the presence of a central yellowish-white structure and vessels around the lesion (vessels in a crown pattern). Also, grouped amorphous structures may be seen in the center of the lesion. (Fig. 1)

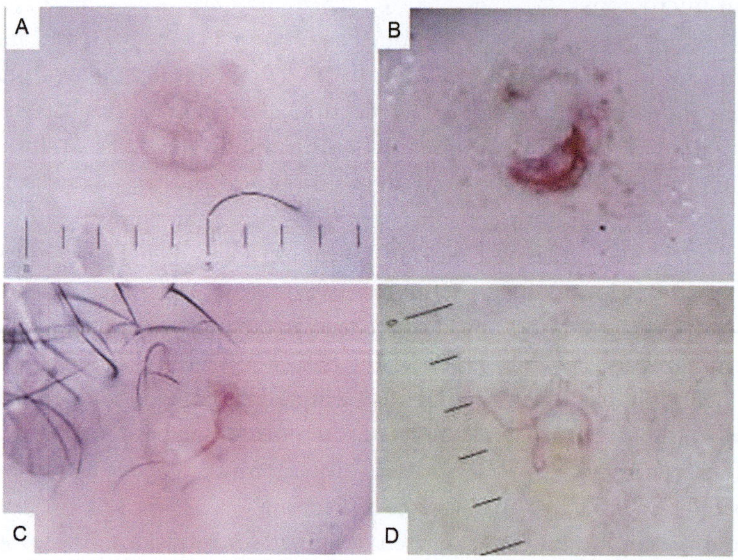

Fig. 1 Molluscum contagiosum: (A) Presence of vessel surrounding the lesions (crown vascular pattern) with an orifice in the center. (B) Presence of vessels perpendicular to the center that extend without passing through the orifice (radial vascular pattern). (C) Presence of vessels spread throughout the lesion in a punctiform pattern (punctiform vascular pattern). (D) Mixed vascular patterns (flower vascular pattern). *(Courtesy of Ianhez M, Cestari SCP, Enokihara MY, Seize MBPM. Padrões dermatoscópicos do molusco contagioso: estudo de 211 lesões confirmadas por exame histopatológico.* An Bras Dermatol *2011;**86**(1):74–9.)*

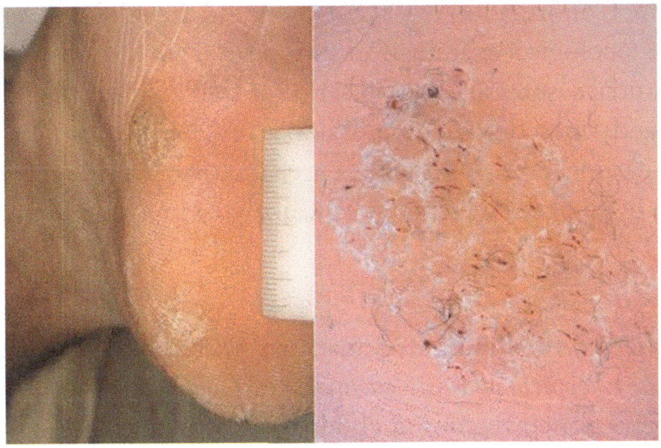

Fig. 2 Viral wart: Dermoscopy of a wart revealing tiny dotted (pinpoint) vessels on a light brownish background. Verruca palmo-plantaris typically shows dots and brown to red streaks that correspond to hemorrhages, which can help in selected cases to distinguish warts from callus. Original magnification 10×.

2.2 Warts

Warts are benign proliferations caused by human papillomavirus (HPV). The diagnosis is mostly clinical but sometimes dermoscopy is a very useful tool for differential diagnosis. The dermoscopic aspect of warts is characterized by the presence of dotted or homogeneously distributed staple shaped vessels, often associated with small hemorrhages (Fig. 2). Histologically, the red dots correspond to dilated capillaries in the dermal papilla and the hemorrhages correspond to thrombosed capillary loops in the same location.[1,2,13]

As they are usually keratotic lesions, *verruca vulgaris* presents papillary hemorrhages more frequently, while the flat wart has only dotted vessels of smaller caliber. The presence of this vascular pattern is especially important in the differential diagnosis of flat wart with other facial conditions such as acneiform lesions, comedones, or small cysts. Plantar warts also display thrombosed vessels, generally dotted or linear, allowing differentiation from callosities and acral amelanotic melanoma. Genital warts can also have dotted vessels, sometimes with an off-white halo around them, which allows the differentiation of the wart from penile papules.

Dermatoscopy can also be a great aid in the monitoring and treatment of warts.

3. FUNGAL INFECTIONS

3.1 Tinea Capitis

Tinea capitis is an infection of the scalp caused by dermatophytes. It is more common in children, and presents clinically as an alopecia plaque with tonsured hair, erythema and scaling on the scalp. Confirmation of the diagnosis and the etiological agent of *tinea capitis* is made by direct mycological examination and culture, but dermoscopy could be of great assistance in the initial suspicion. The presence of the "comma" hair is characteristic of this infection. The dermoscopic finding of corkscrew hair seems to be a variation of the comma hair, manifesting in black patients. This particular pattern is probably due to infestation of the hair by hyphae, which makes it likely to break and bend (Fig. 3).[14–17]

3.2 Tinea Nigra

Tinea nigra is caused by the fungus dematiaceous *Phaeoannellomyces werneckii*, also called *Exophiala werneckii*, *Exophiala phaeoannellomyces*, or *Cladosporum werneckii*. Lesions mainly affect the palmar and plantar regions of children and young adults. The dermoscopy can be a fast and accurate method of diagnosis of Tinea nigra, as the lesion can be easily recognized as nonmelanocytic.[18–23] The presence of multiple light brown filaments that cross each other and do not respect the dermatoglyphic lines forming a maze is characteristic of this infection (Fig. 4).

A B

Fig. 3 Tinea capitis. (A) Videodermoscopy 20 × and (B) videodermoscopy 40 ×, comma and corkscrew hair. *(Courtesy of Pinheiro AMC, Lobato LA, Varella TCN. Dermoscopy findings in tinea capitis. Case report and literature review.* An Bras Dermatol *2012;**87**(2):313–4.)*

A B

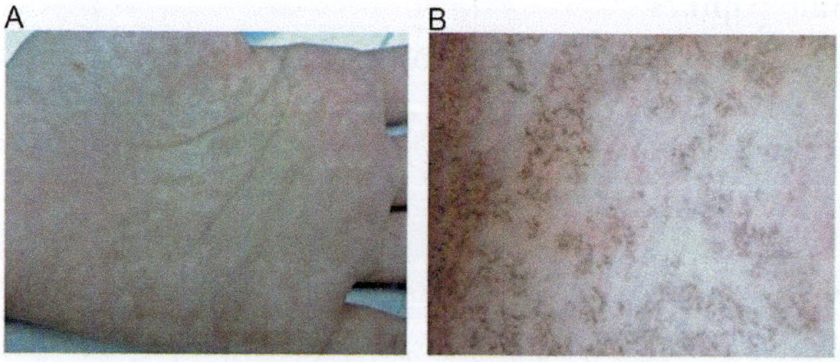

Fig. 4 (A) Tinea nigra: a well-demarcated, nondesquamative, irregularly pigmented macula on the left hand and (B) dermoscopy 10 ×: superficial fine, wispy pigmented spicules. *(Courtesy of Negroni R. Historical aspects of dermatomycoses.* Clin Dermatol *2010;**28**:125–32. Rossetto AL, Cruz RC. Tinea nigra nas formas geográficas em "coração" e "bico do papagaio."* An Bras Dermatol *2011;**86**:389–90.)*

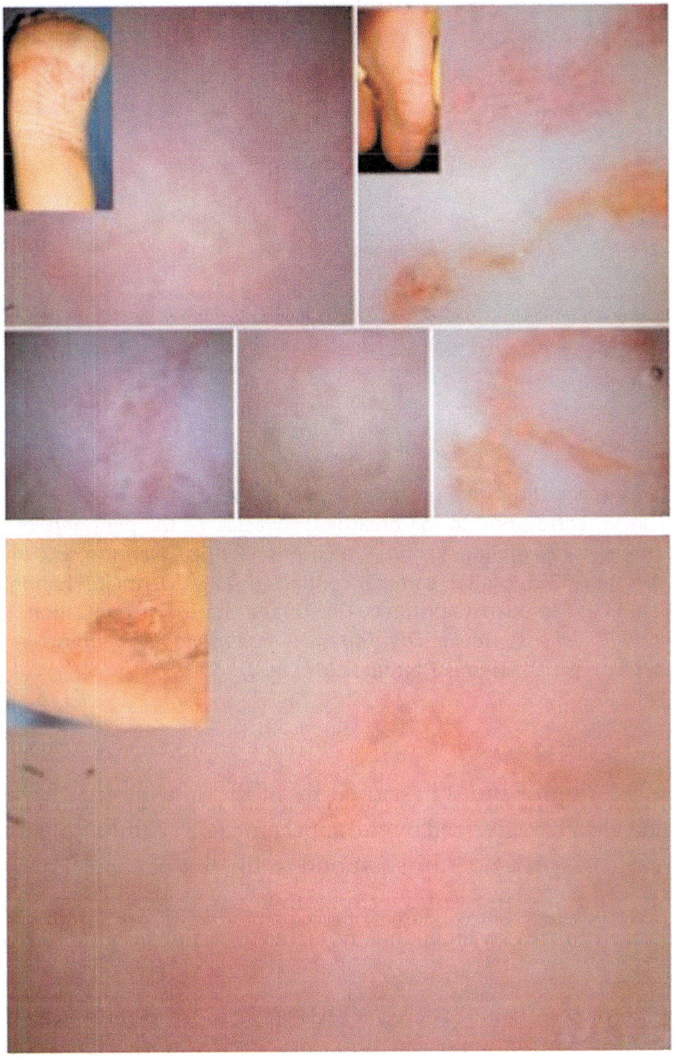

Fig. 5 Larva migrans: *reddish spots* that relate to empty paths.

4. INFESTATIONS BY PARASITES

4.1 Larva Migrans

Larva migrans is caused by infestation by the helminth *Ancylostoma braziliense*. The contact is through the soil, contaminated by the feces of cats and dogs. The clinical aspect is very characteristic, consisting of extremely pruritic erythematous linear lesions that occur mainly in the buttocks and feet. A dermoscopic pattern was described for *Larva migrans* with areas with translucent brownish structures distributed linearly. At any rate, clinical manifestations generally are sufficient for the diagnosis, leaving dermoscopy as an additional asset[1,24] (Figs. 5).

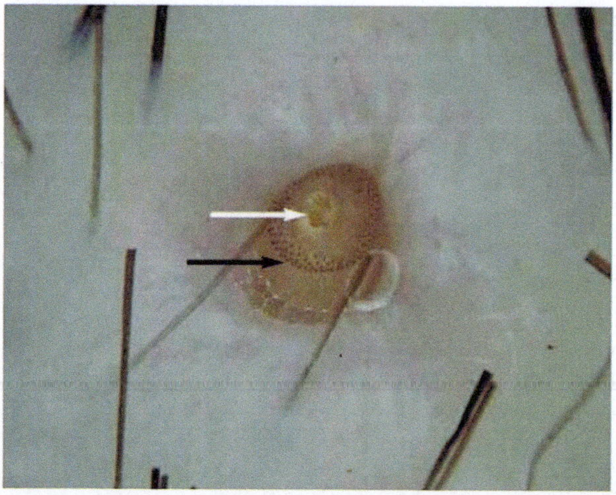

Fig. 6 Myiasis: Dermoscopy shows the posterior aspect of the larva. In the center, two bird's feet-like structures *(white arrow)* correspond to the breathing spiracles. At the periphery of the *creamy-white larva, black dots* are seen resembling a thorn crown *(black arrow)*. Dermoscopic images were taken with a Sony Cybershot W55 digital camera connected to a DermLite handheld II Hybrid m dermoscope, 3Gen, LLC, San Juan Capistrano, CA, United States; magnification 10×. *(Courtesy of Abraham LS, Azulay-Abulafia L, Aguiar DP, Torres F, Argenziano G. Dermoscopy features for the diagnosis of furuncular myiasis.* An Bras Dermatol *2011;**86**(1):160–2.)*

4.2 Myiasis

Myiasis is an infestation of the skin caused by of the skin by the fly larva *Dermatobia hominis*. It is clinically characterized by the appearance of a painful inflammatory nodule with a central orifice which does not respond to the use of topical or systemic antibiotics. The injury is unique, most of the time; however, the infestation by several larvae can cause multiple injuries, which does not always make myiasis the most suspected diagnosis.[25,26]

Dermoscopy can be extremely useful for diagnosis. Using a nonpolarized light dermoscopy in need of contact with the skin, you can view the larva through the hole formed in the nodule. The rear portion of the larva, which is visible, has a yellowish color and small black spicules. The frequent movements that it performs are important to monitor viability, and thus perform their removal. After being occluded for 24 hours, if the larvae does not display any movements after a few minutes, this means that it is dead and can be easily removed (Fig. 6).[25,27–29]

4.3 Pediculosis

Infestation of the scalp by *Pediculus humanis capitis* is more common in children and is easily contracted by close contact. It is characterized by itching scalp and is diagnosed

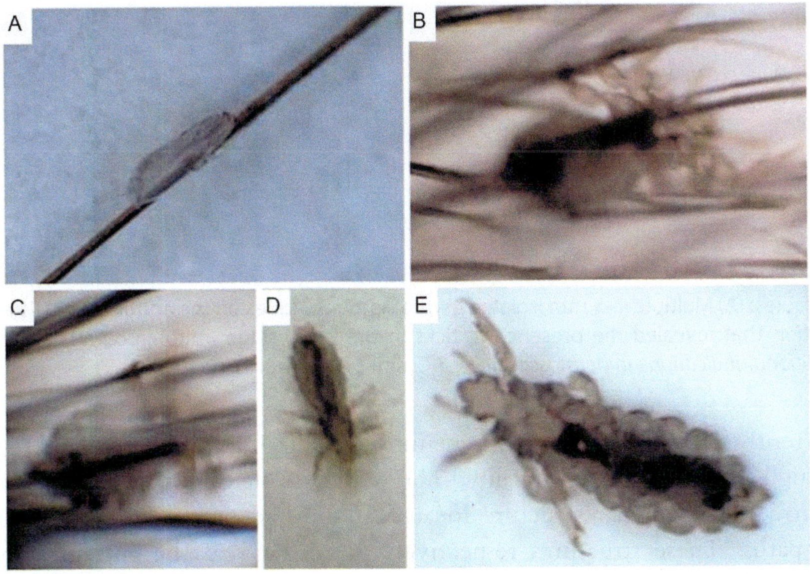

Fig. 7 Pediculosis: (A) Empty nit affixed to a strand of hair, seen at noncontact dermatoscopy. (B) Observation of *Pediculus humanus capitis* on the patient's hair. (C) Detail using the zoom function of the Sony DS9 digital camera, showing the parasite affixed to strands of hair. (D) Parasite removed with tweezers and placed on a smooth surface. (E) Detail of *Pediculus humanus capitis*, with the *brownish-burgundy color* indicating the digestive tract filled with the patient's blood. *(Courtesy of Criado PR. Entodermoscopy: dermoscopy for the diagnosis of pediculosis.* An Bras Dermatol *2011;**86**(2):370–1.)*

by the visualization of the nits adhered to the hair and by the lice circulating in the scalp. Dermoscopy with polarized light can be useful in visualizing the lice and nits on the surface of the scalp. In this way, not only can the diagnosis be made using dermoscopy, but also the effectiveness of the treatment can be monitored by checking whether the nits adhered to the hair are completed by a louse or are empty. The phthiriasis, pediculosis of the pubic region, is caused by the louse *Phthirus pubis*. Recently, dermoscopy was also described as an aid to diagnosis of this parasite (Fig. 7).[26,30]

4.4 Scabies

Scabies is caused by infestation by the mite *Sarcoptes scabiei hominis* variant. It is a fairly frequent infestation that can often occur in epidemic form. Its clinical presentation is quite varied, but most often presents as excoriated erythematous papules in body areas such as armpits, wrists, interdigital, periumbilical region, pubic region and scrotum. From the clinical suspicion of scabies, the diagnosis is confirmed by scarification of the skin and identification of the mites in the optical microscope. Dermoscopy is

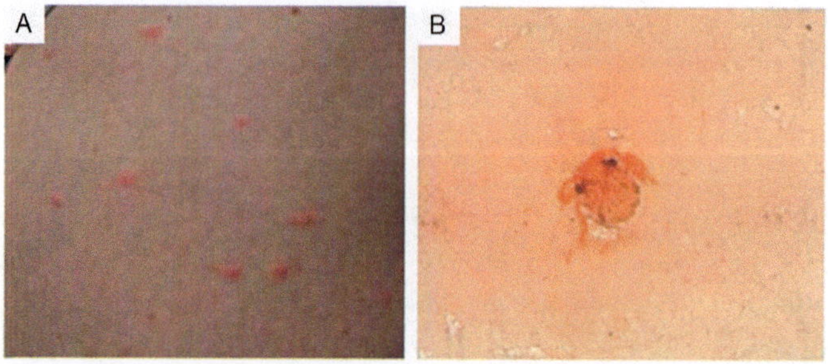

Fig. 8 Scabies: (A) Multiple 4–5 mm isolated erythematous plaques on the abdomen. (B) Dermoscopic examination that revealed the presence of ticks in the larval stage (mite). *(Courtesy of Bastos CAS. Non-traditional indications in dermoscopy.* Surg Cosmet Dermatol *2012;***4***(2)203–5.)*

also currently considered one of the main diagnostic methods for scabies, especially for its high accuracy and practicality. The dermoscopic aspects of scabies include a small brownish triangular structure located at the end portion of a linear zone of curved paths. These structures respectively correspond to the anterior portion of the mite and the scabiotic tunnel. In the tunneliform path, sometimes it is possible to identify small black dots that match the feces of the mite. When using an episcope for achieving a greater magnitude (50–70 ×), mites are shown in greater detail (Fig. 8).[1,26,31]

4.5 Tungiasis

Tungiasis is a skin infestation by the flea *Tunga penetrans*. It is a benign self-limited dermatosis; however, in some cases, late diagnosis might lead to patient discomfort, secondary infection and other complications. Toes, plantar regions and nail folds are the most affected areas. The dermoscopic pattern is quite characteristic. It is possible to identify a pigmented ring in the central portion of the lesion, corresponding to chitin that covers the rear opening of the flea exoskeleton. Laterally, violet and white globular chains are often observed, that correspond to the flea eggs that will later be expelled and can be seen on the skin surface at later stages (Fig. 9).[32–37]

5. CONCLUSIONS AND FUTURE PROSPECTS

With the advent of polarized dermatoscopes, dermoscopy has increased in popularity among dermatologists worldwide and dermoscopic features of a huge range of dermatoses have been described. Although there is not much challenge in diagnosing many conditions, such as warts or scabies, I believe that dermoscopy would be a time-saving and helpful tool in doubtful cases.

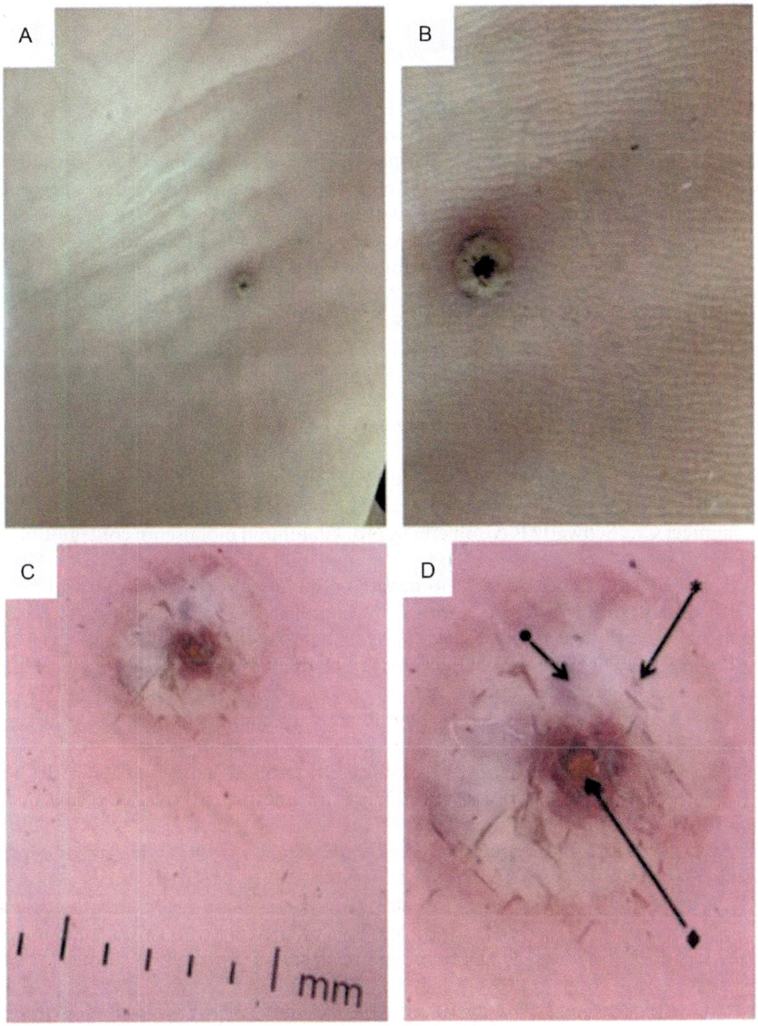

Fig. 9 Tungiasis: (A) *Yellowish-white plantar papule* with a *central black dot*. (B) Panel (A) seen in greater detail. (C) in vivo dermoscopy performed without skin contact showing a white halo with a central orifice containing a *light-brown structure* (original augmentation 10 ×). (D) Panel (C) seen in greater detail, showing *bluish-gray areas* (likely corresponding to the parasite's exoskeleton or his gut) (●), *white halo* (corresponding to the parasite's dilated abdomen containing the eggs) (*), *light-brown structure* appearing in the center of the orifice (posterior segment of the parasite) (◆). *(Courtesy of Criado PR, Landman G, Reis VMS, Belda W Jr. Tungiasis under dermoscopy: in vivo and ex vivo examination of the cutaneous infestation due to Tunga penetrans.* An Bras Dermatol *2013;**88**(4):649–51.)*

The more time we spend examining skin lesions, the more detailed descriptions of specific conditions we will have, which might help others who are beginning their practice or health providers around the globe with less experience in a particular infectious or parasitic disease not that usual in the area.

REFERENCES

1. Rezze GG, Paschoal FM, Hirata SH. *Atlas de Dermatoscopia Aplicada*. 2nd ed. São Paulo: Lemar Livraria E Editora Marina; 2014.
2. Ferreira CMM, Barcaui C, Pineiro-Maceira J. *Atlas de Dermatoscopia—Aplicação Clínica e Correlação Histopatológica*. 1st ed. Rio de Janeiro: DI Livros; 2010.
3. Grin CM, Friedman KP, Grant-Kels JM. Dermoscopy: a review. *Dermatol Clin* 2002;**20**(4):641–6, viii.
4. Soyer HP, Smolle J, Hodl S, Pachernegg H, Kerl H. Surface microscopy: a new approach to the diagnosis of cutaneous pigmented tumors. *Am J Dermatopathol* 1989;**11**:1–10.
5. Yadav S, Vossaert KA, Kopf AW, Silverman M, Grin-Jorgensen C. Histopathologic correlates of structures seen on dermoscopy (epiluminescence microscopy). *Am J Dermatopathol* 1993;**15**:297–305.
6. Nachbar F, Stolz W, Merkle T, Cognetta AB, Vogt T, Landthaler M, et al. The ABCD rule of dermatoscopy: high prospective value in the diagnosis of doubtful melanocytic skin lesions. *J Am Acad Dermatol* 1994;**30**:551–9.
7. Stolz W, Braun-Falco O, Bilek P, Lanthaler M, Cognetta AB. *Colour atlas of dermatoscopy*. Oxford: Blackwell Scientific; 1994. p. 3.
8. Menzies SW, Crotty KA, Ingvar C, McCarthy WH. *An atlas or surface microscopy of pigmented skin lesions*. New York, NY: McGraw-Hill; 1996. p. 1.
9. Ianhez M, Cestari SCP, Enokihara MY, Seize MBPM. Padrões dermatoscópicos do molusco contagioso: estudo de 211 lesões confirmadas por exame histopatológico. *An Bras Dermatol* 2011;**86**(1):74–9.
10. Brown J, Janniger CK, Schwartz RA, Silverberg NB. Childhood molluscum contagiosum. *Int J Dermatol* 2006;**45**:93–9.
11. Lewis EJ, Lam M, Crutchfield 3rd CE. An update on molluscum contagiosum. *Cutis* 1997;**60**:29–34.
12. Zaballos P, Ara M, Puig S, Malvehy J. Dermoscopy of molluscum contagiosum: a useful tool for clinical diagnosis in adulthood. *J Eur Acad Dermatol Venereol* 2006;**20**:482–3.
13. Tschandl P, Argenziano G, Bakos R, Gourhant JY, Hofmann-Wellenhof R, Kittler H, et al. Dermoscopy and entomology (entomodermoscopy). *J Dtsch Dermatol Ges* 2009;**7**(7):589–96.
14. Pinheiro AMC, Lobato LA, Varella TCN. Dermoscopy findings in tinea capitis. Case report and literature review. *An Bras Dermatol* 2012;**87**(2):313–4.
15. Slowinska M, Rudnicka L, Schwartz RA, Kowalska-Oledzka E, Rakowska A, Sicinska J, et al. Comma hairs: a dermatoscopic marker for tinea capitis. A rapid diagnostic method. *J Am Acad Dermatol* 2008;**59** (5 Suppl.):S77–9.
16. Sandoval ABC, Ortiz JAS, Rodrígues JM, Vargas AG, Quintero DGV. Hallazgos dermoscópicos en tinea capitis. *Rev Iberoam Micol* 2010;**27**:151–2.
17. Hughes R, Chiaverini C, Bahadoran P, Lacour JP. Corkscrew hair: a new dermoscopic sign for diagnosis of tinea capitis in black children. *Arch Dermatol* 2011;**147**:355–6.
18. Delgado L, Criado PR, Alonso G. Dermoscopy revealing a case of Tinea Nigra. *An Bras Dermatol* 2013;**88**(1):128–9.
19. Cerqueira AG. *Keratomycosis nigricans palmaris [tese]*. Salvador (BA): Faculdade de Medicina da Bahia; 1916.
20. (a) Negroni R. Historical aspects of dermatomycoses. *Clin Dermatol* 2010;**28**:125–32. (b) Rossetto AL, Cruz RC. Tinea nigra nas formas geográficas em "coração" e "bico do papagaio". *An Bras Dermatol* 2011;**86**:389–90.
21. Pilliouras P, Allison S, Rosendahl C, Buettner PG, Weedon D. Dermoscopy improves diagnosis of Tinea nigra: a study of 50 cases. *Australas J Dermatol* 2011;**52**:191–4.
22. Rezusta A, Gilaberte Y, Betran A, Gene J, Querol I, Arias M, et al. Tinea nigra: a rare imported infection. *J Eur Acad Dermatol Venereol* 2010;**24**:89–91.
23. Xavier MH, Ribeiro LH, Duarte H, Saraça G, Souza AC. Dermatoscopy in the diagnosis of Tinea nigra. *Dermatol Online J* 2008;**14**:15.
24. Crocker-Sandoval AB, et al. *Hallazgos dermatoscópicos en larva migrans. Dermatol Rev Mex* 2015;**59** (2):98–101. http://www.medigraphic.com/pdfs/derrevmex/rmd-2015/rmd152d.pdf.
25. Abraham LS, Azulay-Abulafia L, Aguiar DP, Torres F, Argenziano G. Dermoscopy features for the diagnosis of furuncular myiasis. *An Bras Dermatol* 2011;**86**(1):160–2.

26. Zalaudek I, Giacomel J, Cabo H, Di Stefani A, Ferrara G, Hofmann-Wellenhof R, et al. Entodermoscopy: a new tool for diagnosing skin infections and infestations. *Dermatology* 2008;**216**:14–23.
27. Bakos RM, Bakos L. Dermoscopic diagnosis of furuncular myiasis. *Arch Dermatol* 2007;**143**:123–4.
28. Llamas-Velasco M, Navarro R, Santiago Sánchez-Mateos D, De Argila D. Dermoscopy in furuncular myiasis. *Actas Dermosifiliogr* 2010;**101**:894–6.
29. Maier H, Hönigsmann H. Furuncular myiasis caused by *Dermatobia hominis*, the human botfly. *J Am Acad Dermatol* 2004;**50**(2 Suppl.):S26–30.
30. Criado PR. Entodermoscopy: dermoscopy for the diagnosis of pediculosis. *An Bras Dermatol* 2011;**86**(2):370–1.
31. Bastos CAS. Non-traditional indications in dermoscopy. *Surg Cosmet Dermatol* 2012;**4**(2):203–5.
32. Criado PR, Landman G, Reis VMS, Belda Jr. W. Tungiasis under dermoscopy: in vivo and ex vivo examination of the cutaneous infestation due to *Tunga penetrans*. *An Bras Dermatol* 2013;**88**(4):649–51.
33. Dunn R, Asher R, Bowling J. Dermoscopy: ex vivo visualization of fleas head and bag of eggs confirm the diagnosis of Tungiasis. *Australas J Dermatol* 2012;**53**:120–2.
34. Bauer J, Forschner A, Garbe C, Röcken M. Dermoscopy of tungiasis. *Arch Dermatol* 2004;**140**:761–3.
35. Di Stefani A, Rudolph CM, Hofmann-Wellenhof R, Müllegger RR. An additional dermoscopic feature of tungiasis. *Arch Dermatol* 2005;**141**:1045–6.
36. Bakos RM, Bakos L. 'Whitish chains': a remarkable in vivo dermoscopic finding of tungiasis. *Br J Dermatol* 2008;**159**:991–2.
37. Cabrera R, Daza F. Dermoscopy in the diagnosis of tungiasis. *Br J Dermatol* 2009;**160**:1136–7.

INDEX

Note: Page numbers followed by "*f*" indicate figures, "*t*" indicate tables, and "*b*" indicate boxes.

R

9780128110799